Market Development for Genetically Modified Foods

Market Development for Genetically Modified Foods

Edited by

Vittorio Santaniello

Dipartimento di Economia e Istituzioni
Universita degli Studi Roma 'Tor Vergata'
Rome, Italy

Robert E. Evenson

Economic Growth Center
Department of Economics
Yale University
Connecticut, USA

and

David Zilberman

Department of Agricultural and Resource Economics
University of California
Berkeley, USA

CABI *Publishing*

CABI *Publishing* is a division of CAB *International*

CABI Publishing
CAB International
Wallingford
Oxon OX10 8DE
UK

Tel: +44 (0)1491 832111
Fax: +44 (0)1491 833508
Email: cabi@cabi.org
Web site: www.cabi-publishing.org

CABI Publishing
10 E 40th Street
Suite 3203
New York, NY 10016
USA

Tel: +1 212 481 7018
Fax: +1 212 686 7993
Email: cabi-nao@cabi.org

A catalogue record for this book is available from the British Library, London, UK

Library of Congress Cataloguing-in-Publication Data
Market development for genetically modified foods / edited by Vittorio Santaniello, Robert E. Evenson, and David Zilberman.
p. cm.
Includes bibliographical references and index.
Papers presented at a conference.
ISBN 0–85199–573-X (alk. paper)
1. Genetic engineering industry--Europe--Congresses. 2. Genetically modified foods--Europe--Congresses. 3. Food industry and trade--Europe--Congresses. 4. Genetic engineering industry--North America--Congresses. 5. Genetically modified foods--North America--Congresses. 6. Food industry and trade--North America--Congresses. 7. Market surveys--Europe--Congresses. 8. Market surveys--North America--Congresses. I. Santaniello, V. II. Evenson, Robert E. (Robert Eugene), 1934-III. Zilberman, David.

HD9999.G453 E8516 2001
381'.45664--dc21 2001043187

ISBN 0 85199 573 X

Typeset in Adobe Garamond by Wyvern 21 Ltd, Bristol.
Printed and bound in the UK by Cromwell Press, Trowbridge.

Contents

Contributors

Corinne Alexander, *Department of Agricultural and Resource Economics, University of California, One Shields Avenue, Davis, CA 95616, USA*

Kevin Allison, *Economics and Management of Agrobiotechnology Center, Department of Agricultural Economics, University of Missouri-Columbia, 125D Mumford Hall, Columbia, MO 65211, USA.* Present address: *Harvard University, John F. Kennedy School of Government, USA*

Anthony Artuso, *Department of Agricultural Food and Resource Economics, 55 Dudley Road, Cook College, Rutgers University, New Brunswick, NJ 08901, USA*

Andreas Böcker, *Department of Agricultural Economics and Market Research, Justus-Liebig-University, Senckenbergstrasse 3, 35390 Giessen, Germany*

Michael Boland, *Department of Agricultural Economics, 342 Waters Hall, Kansas State University, Manhattan, KS 66506, USA*

Michael Burton, *Department of Agricultural and Resource Economics, Faculty of Agriculture, University of Western Australia, 35 Stirling Highway, Crawley (Perth), WA 6009, Australia*

Leslie J. Butler, *Department of Agricultural and Resource Economics, Room 2111, Social Science and Humanities, University of California, One Shields Avenue, Davis, CA 95616, USA*

Janet E. Carpenter, *National Center for Food and Agricultural Policy, 1616 P Street NW, First Floor, Washington, DC 20036, USA*

Wen S. Chern, *Department of Agricultural, Environmental and Development Economics, The Ohio State University, 2120 Fyffe Road, Columbus, OH 43210, USA*

David A. Darr, *Department of Agricultural, Environmental and Development Economics, The Ohio State University, 2120 Fyffe Road, Columbus, OH 43201, USA*

Christine Domegan, *Department of Marketing, National University of Ireland, Newcastle Road, Galway, Ireland*

Christopher E. Duffield, *Department of Agricultural and Food Economics, PO Box 237, The University of Reading, Reading, RG6 6AR, UK*

Robert E. Evenson, *Economic Growth Center, Department of Economics, Yale University, New Haven, CT 06520, USA*

Jorge Fernandez-Cornejo, *Economic Research Service, USDA, 1800 M. Street, NW Room 4052, Washington, DC 20036-5831, USA*

Xavier Gellynck, *Faculty of Agriculture and Applied Biological Sciences, Department of Agricultural Economics, Division of Agro-Marketing, Ghent University, Coupure Links 653-9000, Ghent, Belgium*

Leonard P. Gianessi, *National Center for Food and Agricultural Policy, 1616 P Street NW, First Floor, Washington, DC 20036, USA*

Rachael E. Goodhue, *Department of Agricultural and Resource Economics, University of California, One Shields Avenue, Davis, CA 95616, USA*

Maria Teresa Gorgitano, *Department of Agricultural Economics, University of Naples – Federico II, Via Universita 96, 80055 Portici, Italy*

Richard S. Gray, *Department of Agricultural Economics, 3D34 Agriculture Building, University of Saskatchewan, 51 Campus Drive, Saskatoon, Saskatchewan, S7N 5A8, Canada*

Claus-Hennig Hanf, *Department of Agricultural Economics, Chair of Agribusiness Management, Christian-Albrechts-University, Olshausenstrasse 40, 24118 Kiel, Germany*

Irene Henriques, *Schulich School of Business, York University, 4700 Keele Street, Toronto, Ontario, M3J 1P3, Canada*

Wallace E. Huffman, *Department of Economics, 260 Heady Hall, Iowa State University, Ames, IA 50011, USA*

Sallie James, *Department of Agricultural and Resource Economics, Faculty of Agriculture, University of Western Australia, 35 Stirling Highway, Crawley (Perth), WA 6009, Australia*

Lovell S. Jarvis, *College of Agricultural and Environmental Sciences, University of California, 150 Mrak Hall, Davis, CA 95616, USA*

Nicholas Kalaitzandonakes, *Economics and Management of Agrobiotechnology Center, Department of Agricultural Economics, University of Missouri-Columbia, 125D Mumford Hall, Columbia, MO 65211, USA*

Taeho Kim, *207D Armsby Building, Pennsylvania State University, University Park, PA 16802, USA*

Brian F. Lavoie, *Department of Agricultural, Environmental and Development Economics, The Ohio State University, 2120 Fyffe Road, Columbus, OH 43210, USA*

William W. Lin, *ERS, USDA, 1800 M Street NW, Suite S5206, Washington, DC 20036-5831, USA*

Bob Linder, *Faculty of Agriculture, University of Western Australia, 35 Stirling Highway, Crawley (Perth), WA 6009, Australia*

Marianne McGarry Wolf, *Agribusiness Department, California Polytechnic State University, San Luis Obispo, CA 93407, USA*

Heather McNeill, *Department of Agricultural Economics, 3D34 Agriculture Building, University of Saskatchewan, 51 Campus Drive, Saskatoon, Saskatchewan, S7N 5A8, Canada.* Present address: *903-101 Cumberland Avenue, Saskatoon, Saskatchewan, S7N 1L5, Canada*

Stavroula Malla, *Department of Agricultural Economics, 3D34 Agriculture Building, University of Saskatchewan, 51 Campus Drive, Saskatoon, Saskatchewan, S7N 5A8, Canada*

Leonie A. Marks, *Economics and Management of Agrobiotechnology Center, Department of Agricultural Economics, University of Missouri-Columbia, 125D Mumford Hall, Columbia, MO 65211, USA*

Catherine A. Mendenhall, *Environmental Health Sciences, Department of Epidemiology and Public Health, Yale University, 60 College Street, PO Box 208034, New Haven, CT 06520, USA*

Siân Mooney, *Department of Agricultural Economics, Montana State University, Bozeman, MT 59717, USA*

Jyrki Niemi, *MTT Agrifood Research Finland, Economic Research (MTTL), PO Box 3 (Luutnantinnie 13), FIN-00411, Helsinki, Finland*

Peter W.B. Phillips, *Department of Agricultural Economics, 3D34 Agriculture Building, University of Saskatchewan, 51 Campus Drive, Saskatoon, Saskatchewan, S7N 5A8, Canada*

Jo Pluske, *Department of Agricultural and Resource Economics, Faculty of Agriculture, University of Western Australia, 35 Stirling Highway, Crawley (Perth), WA 6009, Australia*

Vittorio Santaniello, *Dipartimento di Economia e Istituzioni, Universita degli Studi Roma 'Tor Vergata', Rome, Italy*

Günter Schamel, *Faculty of Agriculture and Horticulture, Institute of Economic and Social Sciences, Humboldt University at Berlin, Luisenstrasse 56, 10099, Berlin, Germany*

Ian M. Sheldon, *Department of Agricultural, Environmental and Development Economics, The Ohio State University, 2120 Fyffe Road, Columbus, OH 43210, USA*

Stuart Smyth, *Department of Agricultural Economics, 3D34 Agriculture Building, University of Saskatchewan, 51 Campus Drive, Saskatoon, Saskatchewan, S7N 5A8, Canada*
Valeria Sodano, *Department of Agricultural Economics, University of Naples – Federico II, Via Universita 96, 80055 Portici, Italy*
Nickolas M. Spetsidis, *Faculty of Agriculture and Horticulture, Institute of Economic and Social Sciences, Humboldt University of Berlin, Luisenstrasse 56, 10099 Berlin, Germany.* Present address: *55 A. Papandreou Avenue, GR-71305, Iraklion – Crete, Greece*
Loren W. Tauer, *Cornell University, 451 Warren Hall, Ithaca, NY, 148503, USA*
Abebayehu Tegene, *Resource Economic Division, ERS, USDA, 1800 M Street NW, Room 54194, Washington DC, 20036-5837, USA*
W. Bruce Traill, *Department of Agricultural and Food Economics, PO Box 237, The University of Reading, Reading, RG6 6AR, UK*
Wim Verbeke, *Faculty of Agriculture and Applied Biological Sciences, Department of Agricultural Economics, Division of Agro-Marketing, Ghent University, Coupure Links 653-9000, Ghent, Belgium*
Annelies Verdurme, *Faculty of Agriculture and Applied Biological Sciences, Department of Agricultural Economics, Division of Agro-Marketing, Ghent University, Coupure Links 653-9000, Ghent, Belgium*
Jacques Viaene, *Faculty of Agriculture and Applied Biological Sciences, Department of Agricultural Economics, Division of Agro-Marketing, Ghent University, Coupure Links 653-9000, Ghent, Belgium*
Meri Virolainen, *MTT Agrifood Research Finland, Economic Research (MTTL), PO Box 3 (Luutnantinnie 13), FIN-00411, Helsinki, Finland*
Robert D. Weaver, *207D Armsby Building, Pennsylvania State University, University Park, PA 16802, USA*
Ludmila Zakharova, *Economics and Management of Agrobiotechnology Center, Department of Agricultural Economics, University of Missouri-Columbia, 129D Mumford Hall, Columbia, MO 65211, USA*
David Zilberman, *Department of Agricultural and Resource Economics, University of California, Berkeley, CA 94720, USA*

Acknowledgements

This book has arisen from the conference entitled *Biotechnology, Science and Modern Agriculture: a New Industry at the Dawn of the Century*, convened by the International Consortium on Agricultural Biotechnology Research (ICABR), held in Ravello, Italy on 15–18 June 2001.

The editors would like to acknowledge the sponsorship provided by the following institutions: CEIS–University of Rome 'Tor Vergata'; Center for Sustainable Resource Development, University of California at Berkeley; and Economic Growth Center, Yale University. The editors would also like to thank the Italian National Research Council (CNR) for its financial support of the conference.

Introduction

Robert E. Evenson, Vittorio Santaniello and David Zilberman

The investment climate for firms producing genetically modified (GM) agricultural products has changed considerably in recent months. In the US, transgenic crop production has reached very high levels. The global area planted to transgenic crops reached approximately 40 Mha in 1999 (12 countries planted transgenic crops and 84% of transgenic area was in industrialized countries). This was a very promising beginning for GM products.

However, consumer resistance has turned out to be much stronger than anticipated. Consumer attitudes regarding GM foods are being expressed in a setting where premiums and discounts for GM food products and GM-free food products have generally not yet emerged. At least part of the problem can be seen as a coordination problem. The suppliers of ag-biotech products sell to farmers who produce GM foods. These foods in turn must be processed and marketed by the food industry. Consumer attitudes (after all, 'the customer is always right') must be reflected in farm product markets and in farm input (ag biotech) markets.

Consumer movements have already promoted government actions in the form of import restrictions and prohibitions of the sale of GM products. These actions have the support of groups with allied interests. Farm producer groups in importing countries and in the European Union generally have an interest in maintaining both tariff and non-tariff barriers to trade. They thus have common interests with consumer groups opposing GM food sales, even though they may not share consumer attitudes.

The opposition to GM foods has also been incorporated into broader political movements opposing 'globalization' and related issues. The growth in the political opposition to the expansion in world trade and to the international organizations supporting this expansion, notably the World Trade Organization (WTO), but including the World Bank and the International Monetary Fund, has also been surprising. This movement has many conflicting cross currents and reflects a number of interests. Because biotechnology interests and particularly GM concerns have been incorporated into this broader movement, policy issues associated with GM and GM-free food markets are more urgent.

This urgency is further heightened by the current (early 2001) rise in concern over a related food issue, the 'Mad Cow' disease issue, in a number of European countries. As there are more reports of the incidence of this disease and of related incidence of Creutzfeldt-Jakob disease in human populations, confidence in governments' capacity to monitor and prevent food-related illnesses has fallen. This loss of confidence in food safety and regulatory agencies spills over to biotechnology issues associated with the environment. It is perhaps not an exaggeration to suggest that public confidence in food safety and regulatory agencies was, particularly in Europe, lower at the end of the 20th century than at almost any other time during the century.

All parties engaged in agricultural biotechnology activities – the firms developing biotech products, the farmers producing food and other products, the food and related agribusiness industrial firms and the consumers of food – will be affected by changes in regulatory, trade and food safety regimes. The present system is in disequilibrium. Changes are rapidly being made. Economists have an opportunity and a responsibility to inform and influence these changes to produce a new equilibrium that is in the public interest in all countries. This new equilibrium may well take different forms in different countries, but international issues are involved.

It is difficult to predict, at this point, how the move towards a new equilibrium will unveil itself. We almost certainly will see strong labelling requirements and segregated markets. The current GM products are being sold as 'cost-saving' products to farmers. Future products will very likely have food quality elements for consumers. As price differentials and quality differentials become available to consumers, new equilibria will emerge. The food industry has a record of success in introducing many new products into food markets each year.

The acceptance of biotech products in the health field suggests that consumers are likely to accept GM products that they believe contribute to their welfare. As new GM foods incorporate desirable health-related features, it is quite likely that they will gain consumer acceptance. Recent developments in Europe indicate that the scientific establishment is attempting to develop better information for the public regarding safety issues. With the recent publication in *Science* and *Nature* of the Human Genome, a landmark in biological science has been achieved. There is little doubt that a scientific revolution is underway and that it will affect the way we do science and the way we produce technology in the future.

The International Consortium on Agricultural Biotechnology Research (ICABR) hosted the fourth in a series of conferences on Economics of Agricultural Biotechnology at Ravello, Italy, on 24–28 August 2000. Papers presented at that conference (and at the preceding conferences as well) addressed many of the relevant economic policy issues inherent in the move to a new market equilibrium. A subset of those papers is included in this volume, which addresses market development issues in developed countries, primarily in Europe and North America. This volume focuses on consumer reactions to GM food information and regulatory issues, farmer acceptance of biotech products, and changes in industrial organization in the life science and food sector.

The volume is organized in four parts. Part I includes chapters evaluating consumer attitudes to GM foods. Part II addresses the acceptance by farmers of biotech products. Part III addresses the role of information systems and of associated regulatory developments. Part IV addresses industry structure issues.

The studies reported in this volume do not constitute a comprehensive treatment of all policy issues associated with GM product market development. The industries and economic entities are dealing with changing conditions and problems continuously. In this context, these studies should inform and improve this complex process.

The initial chapter by Burton *et al.* provides a general framework for assessing the nature of the new equilibrium under conditions of market development as opposed to an equilibrium produced by a high degree of regulation and prohibition of production and sale of GM food products. The alternatives of a partially regulated market outcome with labelling, segregation and certification and a highly regulated market outcome with trade prohibitions and selective food prohibitions are real. The failure to move towards a partially regulated market outcome will lead to the alternative highly regulated outcome.

Chapter 1 sets forth the fundamentals for market development, with labelling and segregation. A model is developed and applied to a GM crop (canola). While the model itself is technical, it is instructive in that it illustrates the types of premiums and discounts in the food markets that emerge from consumer preferences, which are then reflected in farm price premiums and discounts for canola with and without GM content. The model incorporates costs of identity preservation. Alternative simulations of the model illustrate the complexities of moving to a new market equilibrium. The simulations show the importance of identity preservation costs and of the incorporation of a technology fee by the suppliers of the biotech seeds. Net welfare measures are also calculated.

Part I of this volume includes five chapters addressing consumer attitudes and preferences. As the first chapter demonstrates, if sufficient consumer preference exists, premiums for non-GM goods will emerge (alternatively, discounts for GM foods will emerge). The magnitude of these premiums (discounts) will ultimately depend not just on the current attitudes of consumers, but on attitudes and on a 'willingness to pay' non-GM premiums over a long period of time. One of the difficulties of assessing consumer attitudes under conditions where the premiums are not actually paid is that there is uncertainty about how consumers will actually behave when faced with premiums.

Chapter 2 (Wolf and Domegan) reports a comparison of consumer attitudes in the USA and Ireland. The authors use market survey research techniques to develop this comparison. Their surveys sought to determine differences in European and US attitudes. Interestingly, the chapter reports a similar level of familiarity with GM goods between consumers in the USA and Ireland. A minority of consumers in both countries indicated a willingness to pay a non-GM food premium. Consumers in Ireland also attributed more negative attributes to GM foods.

Chapter 3 (Verdurme *et al.*) reports findings from a literature survey, group discussions and in-depth interviews of European consumers where a distinction is made between premium branded and non-branded food products. This study finds that consumer attitudes towards GM technology were more negative than attitudes towards specific GM food products. The chapter also reports that, among GM products, premium branded products are better accepted than generic products.

Chapter 4 (Hanf and Böcker) focuses on the non-GM premium and the perception of health and other dangers associated with GM foods. The authors conclude that consumers are likely to continue to hold attitudes regarding dangers that are difficult to overcome with cost reductions. They also conclude that high quality GM foods are unlikely to overcome consumer hesitation regarding GM foods.

Chapter 5 (Mendenhall and Evenson) reports a small survey of consumers' 'willingness to pay' a premium for non-GM foods. This survey illustrates an important dimension of market development. Consumers will ultimately have to pay a premium sufficient to cover the identity preservation (segregation and certification) costs associated with non-GM foods if these markets are to develop. The authors find that consumers want labelling but that not all are willing to pay a premium for non-GM foods. The chapter concludes, however, that for food products where the ratio of consumer value to farm value is high, even small non-GM premiums can support non-GM markets.

Chapter 6 (Spetsidis and Schamel) provides a review of studies of new product development and discusses consumer acceptance of GM food products in the context of this literature. The chapter provides a useful European perspective to GM food issues.

The studies in Part I of this volume are generally based on attitude surveys and related market survey techniques. To date, few true market experiments where non-GM foods are actually priced at a premium have been conducted. As non-GM premiums (GM discounts) emerge, more studies of consumer attitudes, and especially of 'willingness to pay', will be required to establish guidelines for non-GM product market developments. These new studies will be informed by the studies reported here.

Part II of the volume includes eight chapters addressing farmer acceptance and farm profitability of ag biotech products. Three chapters (7, 8 and 9) deal with farmer acceptance of bovine somatotropin (bST), one of the earliest biotech products introduced to farmers. Other chapters deal with crop biotech products. These chapters find that farmers have adopted several biotech products rapidly. In the case of some crop products, diffusion rates have been comparable to hybrid maize diffusion rates in the 1930s and green-revolution wheat and rice diffusion rates in the 1960s.

Farmer acceptance of biotech products has been driven by profitability. However, acceptance of these products to date has been under conditions where few GM discounts have been in place. The chapters are important in indicating the extent of profitability and hence of the possible effects of GM discounts on this market.

Chapters 7 and 8 report bST adoption in New York (Chapter 7, Tauer) and California (Chapter 8, Henriques and Butler). The New York study reports repeated surveys of 138 dairy farms for the years 1994 through 1997. Statistical procedures for selectivity bias are utilized in the study. This study finds that bST use results in increased milk production per cow, but that higher than average profits for bST are generally not being realized. The study finds, under one statistical procedure, that only the well-managed, high-profit farms make profits from bST. The California study of bST adoption considered the complementary use of feed management practices. This study concluded that feed management techniques are important to bST adoption and use. The study has implications for extension services.

Chapter 9 (Jarvis) provides a comprehensive review of bST adoption in the USA and in 16 other countries where the technology has been adopted. This study also concludes that bST rewards management skills and higher quality feed. Attitudinal as well as economic factors are considered in the study.

Chapters 10–13 report evaluations of crop biotech adoption. Canola is the subject of Chapter 10 (Gray *et al.*). Canola is exceptional in that it emerged from minor crop status in the 1950s to that of a major crop in Canada today. Plant breeders' rights were important in encouraging private sector investment in canola research. Chapter 10 reports estimates of rates of return to canola research. It concludes that biotechnology has yet to produce measurable high social rates of return in the canola sector.

Chapter 11 (Alexander *et al.*) reports a study of adoption of GM maize and soybeans by Iowa maize–soybean farmers. The study concludes that early adopters of GM maize and soybean fit the stylized facts of early adopters for other agricultural innovation.

Chapter 12 (Darr and Chern) reports a survey of Ohio farmers designed to measure attitudes, beliefs and behaviours. Adoption rates are estimated for GMO maize and soybeans.

Chapters 13 and 14 report assessments of adoption of GM crops in Europe. Chapter 13 (Niemi and Virolainen) reports adoption rates of GM crops in Finland and concludes that GM technology will lower farm costs, increase yields and provide improved insurance against pests. The chapter also considers the implications for consumers and for farm income.

Chapter 14 (Gorgitano and Sodano) evaluates the potential effects of biotechnology on the processed tomato sector in southern Italy. This study considers local agricultural development issues.

Part III of the volume includes eight chapters addressing the information, regulatory and institutional issues associated with market development.

Chapter 15 (Huffman and Tegene) provides an analytical review of models of information and communication of information. Verifiable information is critical to efficient market development. It can reduce the scope for disputes and uncertainty. One of the responsibilities of social scientists is to objectively evaluate data and provide policy makers with analytical insight. The development of more verifiable information will have considerable social value.

Chapter 16 (Smyth and Phillips) also addresses the role of information in shaping the regulatory environment for biotech products. A contrast between responses to controversial scientific studies in the USA (where regulatory agencies have high credibility) and the UK (where they do not) is drawn. The chapter concludes that the UK regulatory responses to consumer concerns were likely to be more costly and less efficient than the US regulatory response to similar concerns.

Chapter 17 (Marks *et al.*) focuses on how the risks of ag biotech have been communicated via the media in the USA and UK. The media can be seen as one of many influences in the overall acceptance of ag biotech by consumers. Acceptance has differed between the two countries and media reporting has likewise differed significantly.

Chapter 18 (Marks *et al.*) documents a potential gap in perceived food safety risks and benefits between food scientists and reporters. The authors find, from the journals investigated, that no safety tests of GM foods have been published in them and, hence, are not in the public domain. A tentative conclusion is that the media (and the public) do not have easy access to studies of the safety (or lack thereof) of GM foods from peer-reviewed sources, although this conclusion cannot be generalized to all journals.

Chapter 19 (Carpenter and Gianessi) reports a case study of risk assessment practice in the case of a major agricultural biotech product, Roundup Ready soybeans. The study concludes that 'no indication of greater health or environmental risks were found compared to those associated with conventional varieties'. The study also reports estimates of economic benefits from reduction in weed control costs.

Chapters 20 and 21 report studies addressing two important components of identity preservation: labelling, and segregation and certification costs.

Chapter 20 (Phillips and McNeill) reviews the current state of labelling requirements in a number of countries. The study reports that 18 countries plus the EU, 29 manufacturers, 21 retailers and six restaurant chains have signalled intentions to adopt voluntary or mandatory labels for GM foods. The study found little convergence in labelling standards.

Chapter 21 (Lin) provides information on the costs of segregation and certification that are essential to the development of non-GM food markets. There are very few data on segregation costs, and this study provides badly needed data. It concludes that costs depend on tolerance levels for biotech content. The chapter also discusses implications for the grain handling industry.

In Chapter 22 (Artuso), a formal model of optimal product regulation is presented. The link between consumer risk perceptions and regulatory policy is emphasized. Under certain conditions it is shown that regulatory requirements may be called for other than those based on scientific risk assessment.

Part IV of the volume includes four chapters addressing industrial development to date. It should be noted that the ag-biotech industry is in a state of disequilibrium at present.

Chapter 23 (Boland) examines agricultural or 'Life Science' firm performance. The technique applied is to determine value of Tobin's q (the ratio of a firm's market role to its asset value) for firms in the ag-biotech industries. All firms studied had q values greater than one with pharmaceutical firms having the highest values.

Chapter 24 (Traill and Duffield) examines the European agro-food biotechnology industry. The authors note that the industry has developed 'alliances' with universities and specialized or dedicated biotech firms. The chapter relies on the 'dynamic capabilities' business strategy model to suggest that these alliances are likely to be maintained over long periods.

Chapter 25 (Lavoie and Sheldon) also examines the dynamics of the biotechnology industry. The role of foreign-based multinational firms in the USA is considered. The US comparative advantage in these industries is analysed in a real options framework. This framework explains the earlier advantage of US firms. It is also used to explain why foreign multinationals find it attractive to be in the new consolidating industrial structure in the USA.

The final chapter in the volume, Chapter 26 (Weaver and Kim), addresses industry structure in an integrative supply chain assessment. As noted in the early part of this review, ag-biotechnology decisions made in one part of the supply chain (the biotech-supplying firms) have implications for other parts of the supply chain (farmers and the food industry). At present, important tensions and incompatibilities must be resolved between different parts of the supply chain. Chapter 26 offers insights for managing technologies for different parts of the supply chain.

1 A Way Forward for Frankenstein Foods

Michael Burton, Sallie James, Bob Lindner and Jo Pluske
Faculty of Agriculture, University of Western Australia, Perth, Australia

Introduction

Up until the end of the 1990s the future for genetically modified organisms (GMOs) looked bright. Plantings of transgenic soybean, maize, cotton and canola by American, Argentine and Canadian farmers set new benchmarks for the rate of adoption of a new agricultural technology. Virtually the only cloud on the horizon was widespread consumer resistance to GMOs in Europe. Industry assumed that this was a *temporary* problem that could be overcome by an 'educational programme' that provided more information about the benefits of GMOs (Marshall, 1998).

Now, though, the outlook has changed. Consumer resistance to GMOs has intensified rather than waned in Europe. Furthermore, it has now spread to many other countries as well. Even in Canada and the USA, there are press reports of supermarket chains stocking items labelled as genetically-modified (GM)-free.[1] Moreover, the effect of education programmes may be questionable given findings by Zechendorf (1998) that suggest consumer acceptance depends on people's socio-cultural attitudes as well as their knowledge about the benefits of biotechnology.

Despite the uncertainty surrounding consumer acceptance of GMOs, there are lessons to be learnt from studies of innovation adoption. First and foremost, for an innovation to be adopted enduringly, it must not only create value but also must deliver meaningful net benefits to all potential adopters. That is, the benefits of adoption must be distributed all along the supply chain, including to consumers.

Genetically engineered crops that are already being grown commercially include tobacco, cotton, soybean, corn/maize, canola/rapeseed, tomato and potato. In a review, James (1999) noted that seven transgenic crops were grown commercially by 1996 on approximately 2.8 Mha, mostly in the US and Canada. Between 1996 and 1998, there was a further increase in the global area of transgenic crops to 27.8 Mha (James, 1999). As James (1999) points out adoption rates have been some of the highest ever for new agricultural technologies, and reflect grower satisfaction with significant benefits ranging from more flexible crop management, higher productivity and a safer environment through decreased use of conventional pesticides and herbicides.

To date, the overwhelming majority of GM foods are the products of first generation GM crops. The principal transgenic traits in 1999 were herbicide tolerance, insect and viral resistance, and hybrid technology (James, 1999). As explained by Fulton and Keyowski (1999a), these 'input traits' lower average costs of production through some combination of reduced costs of control of, and/or smaller losses from weed, pest and disease infestation, and

[1] As reported by Thomas Walkom, *The Toronto Star*, Editorial, 'GMO folks have a little surprise from Alberta', 2 May 2000 (www.plant.uoguelph.ca/safefood).

through increased yields. Because these beneficial traits can be introduced into a plant without disturbing the rest of the plant's genetic code, the resulting varieties are potentially much more profitable for growers. Realized profitability will fall short of potential profitability to the extent that a product price discount applies to the GM crop, and/or to the extent that growers have to pay a premium to grow the GM crop relative to comparable 'conventional' crop varieties. Even though these crops may deliver lower costs of production to farmers, they typically deliver no or least negligible benefits to consumers[2] unless some of the lower production costs are passed on as lower retail prices for GM food relative to non-GM food. This has not happened to date, and will not happen until the necessary preconditions of retail labelling underpinned by a credible and verifiable system of identity preserved production and marketing are implemented.

Second generation or quality enhanced GM crops, most of which are still under development, incorporate crop attributes that provide direct benefits to the consumer, or in some cases to intermediate producers. Delayed ripening tomatoes, oilseed rape with modified fatty acid, high oleic acid soybean, and carnations with extended shelf-life and modified colour, are examples of second generation crops that are already in commercial production. Some companies are predicting that the third generation of GM crops will be nutraceuticals. Nutraceuticals are foods that prevent or treat diseases or otherwise provide medical or health benefits. GMOs that include genes coding for pharmaceutical drugs are touted as GM crops of the future. In contrast to first generation crops, quality enhanced crops have not been widely adopted so far (James, 1999).

Alerted by activists, consumers are increasingly aware of public health concerns about GMOs and this appears to be the most important consideration regarding their development and use. Within the scientific community, there also are worries about the long-term effects on human health (for example, through the use of antibiotic resistant marker genes and the risk of allergen transfer) and the environment from widespread use of genetically engineered crops. Other concerns about GM crops include the influence of multinational seed companies on countries' economies; and the possible demise of the small-scale farmer. While various special interest groups share these concerns, it is a growing reluctance to eat GM food by the general public that is limiting the size of the market for GMOs, threatening the realization of substantial value creation from genetic manipulation technologies and perhaps even threatening the financial viability of some life science companies.

Some advocates of GM foods point out that consumer reactions to real price differentiated choices (as opposed to hypothetical choices) between conventional (i.e. non-GM) and GM foods have yet to be observed (see, e.g. Caulder, 1998) for a significant number of foodstuffs. They anticipate that when consumers are more regularly exposed to GM foods, and compare them favourably (or at least neutrally) with conventional foods, the anticipated price differential between GM and non-GM foods will create a viable market for GM foods (Caulder, 1998). If correct, such a response will belie the survey results that suggest such a market is likely to be small at best.

Work completed by Gamble *et al.* (2000) indicates that when second generation foods become available, the market for these foods may be larger than for first generation foods as consumers appreciate the extra direct benefits they offer (such as longer shelf-life, or enhanced nutritional value). It is possible, depending on the extent of extra benefits and the willingness of consumers to accept any perceived risk associated with the technology, that these products might command a price *premium* over conventional foods.

That is, for now, a hypothetical scenario. The immediate challenge facing producers and advocates of GM foods is to convince consumer and environmental groups that regardless of its generation, any food produced using recombinant gene technology is safe to consume and to produce. General acceptance of second and third generation crops will not be realized if the market potential of first generation crops is thwarted by health and environmental fears, regardless of their legitimacy.

For the biotechnology industry, there is a clear lesson about how to solve the consumer 'problem' and current lack of demand for GMOs. Trying to allay consumer concerns about the health risk from eating GMOs by relying on scientific argument has

[2] There may be indirect benefits to consumers by knowing farmers benefit from lower on-farm costs or by valuing reduced environmental costs.

not, and will not succeed. Consumers want to be assured about the origins of their food, and ways must be found to allow them to knowingly choose between GM and non-GM foods. Moreover, products from first generation GM crops will have to sell at a discount (relative to the GM-free equivalents) to induce significant numbers of consumers to buy them.

The technology component of GM foods is, for our purposes here, a credence attribute. This means that the technology used to produce a (first generation) food is indistinguishable to the consumer both before and after purchase (Caswell and Mojduszka, 1996; Caswell, 2000). Thus the only way a non-GM food producer is able to elicit a price premium for his or her product is by indicating the status of the product by the use of a label supported by a credible testing and assurance programme.

So long as identity preservation remains haphazard and labelling regulations ambiguous, price premiums for any type of good – GM or non-GM, first, second or third generation – are likely to be small.[3] That is, so long as there is no way of differentiating between types of good, consumers have no certain or reliable way of knowing whether the food they eat is GM-free. Producers of GM-free products are unable to advertise (with any integrity) the status of their product and, as such, rational consumers will be unwilling to pay more than the 'non-segmented' market price. This is especially true for first generation GM foods since there are no enhanced attributes from which consumers could derive extra (direct) benefit. Consumer rejection of GM foods is rational if they are not offered it at a lower price. After all, what rational consumer would accept a 'bad' characteristic in the absence of no offsetting benefits such as a lower price?[4]

To deliver a price premium for non-transgenic food, industry must provide verifiable labelling and maintain credible identity preserved production and marketing (IPPM) systems, thus facilitating choice by consumers of food products that align with their preferences. However, work completed by KPMG (1999) indicated that the introduction of an IPPM system could prove to be prohibitively expensive. This conclusion, however, is difficult to sustain given the current widespread practice of segregating different grades of non-GM crops to separate higher added value products from other commodities in order to exploit niche markets. In the case of GM crops, it is the absence of genetic engineering in food that is the key 'attribute' being demanded, so only non-transgenic food would need to be segregated in the marketing chain, labelled and subject to some form of verification. A particularly apposite case is marketing systems for organic food.

Buckwell *et al.* (1999) estimated that the increased cost of segregating GM products could range between 5 and 15% of the usual farm gate price. Despite this cost, the same authors explain that there could be benefits both to consumers and to farmers as long as consumers are willing to pay the added cost of separating GM from non-GM crops. Labelling is likely to be the most efficient alternative because market forces would determine the acceptance of the new technology. So long as most people demand food that is GM-free, the advantage of labelling may be minimal, and arguably even unnecessarily expensive if IPPM costs exceed cost savings from growing first generation GM crops. If and when demand for non-transgenic food declines in the longer run to the point where it becomes a speciality product, then requiring compulsory labelling of GMOs is likely to prove unduly costly. Ultimately, the magnitude and cost structure of an IPPM system will determine, *inter alia*, the market determined equilibrium level of the price differential between GM and non-GM foods at farm gate, and at retail level.

The rest of this chapter reports the findings from some preliminary analyses of the impact on prices at farm gate and at retail of introducing a system of retail labelling of non-GM food. Estimated price differentials obviously depend on the likely costs of introducing and maintaining a credible IPPM system. Of interest here is the nature of market segmentation and price differentiation after a first generation GM crop (canola, in this example) is produced and marketed. The simple model that follows explores some possible scenarios for future prices for GM and non-GM canola under a few key assumptions. Firstly, the GM crop is of the first

3 Price premiums for non-GM food will, of course, be market determined, and depend on the proportion of supply that is non-GM. Recent market reports for maize and soybean have, however, reported premiums of 10–15% and 5–35%, respectively (Miranowski *et al.*, 1999).

4 This clearly ignores the possibility of consumers buying GM food because they derive utility from knowing it has been produced using techniques beneficial to the environment.

generation such that consumers will not purchase food produced using GM canola unless it is sold at a lower price than conventional canola. Clearly the model would need to be adjusted to allow for any positive attributes associated with GM foods, such as those in second or third generation GM foods. Secondly, the production function for both types of foods is assumed to be constant returns to scale with a constant elasticity of substitution. The market is characterized by perfect competition. Identity preservation costs are presented as simple fixed costs in each market.[5]

Model Specification

Formal modelling of the market with segregated production/consumption will be presented in two parts. In the first, a simplified model is developed which allows for an analytical solution, but is restrictive. In the second, a more general model is presented which can be solved numerically, but not analytically. These models are similar in structure to those developed by Fulton and Keyowski (1999b) and Falck-Zepeda *et al.* (1999) but provide a number of extensions. There is a more formal representation of the technology than used in either. Our models also remove the restriction of a fixed output level used by Fulton and Keyowski, and do not assume that all consumers accept the product, as assumed by Falck-Zepeda *et al.*

The simple model

Representative demand functions for the two types of good, non-GM (subscript, *n*) and GM (subscript, *g*) are assumed to be simple linear functions of (normalized) own price. It is assumed that at the individual level, consumers make a decision to purchase either one or the other, but not both. Hence, the price of the alternative form of the product is not an argument to the representative demand functions:

$$d_n = a_0 + a_1 P_n \qquad (1.1)$$

$$d_g = a_0 + a_1 P_g \qquad (1.2)$$

However, the relative price of the two goods does determine which form of the product is selected. As the good under consideration is a first generation GM crop, there are no intrinsic benefits in consumption. Hence we assume there will be no demand for the GM version if the prices of the two forms are equal. This is consistent with the argument above that even the slightest residual perception of risk from consuming the product, or concern about potential non-consumptive issues (on farm ecological effects for example) will lead consumers to reject GMOs unless there is price differential.

Assume that there is some latent index of concern, c, and some underlying discrete choice process which means consumer i will consume the GM product if:

$$f(P_g / P_n) < c_i \qquad (1.3)$$

That is, if the price differential is large enough they will be induced to change. Assuming the function $f(\bullet)$ is linear in the price ratio, and that c is distributed across the population as a uniform variable from 0 to 1, the share of the population that consumes non-GM is determined by:

$$S_n = P_g / P_n \qquad (1.4)$$

With no loss of generality, normalizing the consumer population size to unity gives aggregate demands of:

$$D_n = (a_0 + a_1 P_n)\, P_g / P_n \qquad (1.5)$$

$$D_g = (a_0 + a_1 P_g)\, (1 - P_g / P_n) \qquad (1.6)$$

The supply side of the model is represented by the marginal cost of production:

$$MC_n = b_0 + b_1 (Q_n + Q_g) \qquad (1.7)$$

$$MC_g = b_0 - b_t + b_1 (Q_n + Q_g) \qquad (1.8)$$

where b_t is the cost advantage enjoyed by the GM crop. Assuming b_1 is positive implies a rising marginal cost, determined by the aggregate production of both crops. Given the similarity in the two goods, it would be expected that there will be interactions between the two goods in production, leading either to joint decreasing marginal productivity of resources, or common impacts on costs through the input markets.

Assuming profit maximization and perfect competition allows one to equate the marginal cost

[5] Market forces (specifically the relative elasticities of supply and demand) will determine the incidence of IPPM costs. The likely scenario is that farmers will have to bear part of the cost of segregating, testing and marketing of non-GM crops and consumers will bear the rest through higher prices.

with the product price, and the market clearing condition of $D_j = Q_j$ allows a solution to be identified.

Defining Q as aggregate quantity (i.e. $D_n + D_g$) of both crops leads to:

$$Q = (a_0 + a_1 P_n) P_g / P_n + (a_0 + a_1 P_g)(1 - P_g / P_n) \quad (1.9)$$

$$P_n = b_0 + b_1 Q + IP_n \quad (1.10)$$

$$P_g = b_0 - b_t + b_1 Q + IP_g \quad (1.11)$$

where P_n and P_g are prices at retail, and IP_j is the marginal cost of identity preservation of crop j when the GM crop is introduced.

Substituting Equations 1.10 and 1.11 into Equation 1.9 leads to a single equation which can be solved for Q (see Appendix I), which can then be used to identify prices and quantities of the individual commodities.

The model leads to a number of intuitive conclusions. So long as there are no identity preservation (IP) costs, the extent of market penetration of the GM crop will be directly related to the degree of cost advantage it enjoys. Furthermore, there will be an increase in the aggregate market for the commodity as the average cost of producing GMOs falls because the increase in the GM segment of the market will be larger than the non-GM segment it displaces. Since marginal cost is specified to rise with increasing aggregate output, the marginal cost and hence price of the non-GM commodity must rise following the introduction of the GM crop. In turn, this will cause a movement along the representative non-GM demand curve, which will compound the reduction in non-GM demand due to the segmentation of the market. The greater the cost advantage enjoyed by the GM crop, the greater the size of these effects. Introduction of an IP cost on the non-GM product widens the gap between GM and non-GM prices, although the rise in non-GM prices will depend on the elasticities of non-GM demand and supply curves: the standard incidence argument. However, that widening will cause further restructuring of the market between the two crops. The introduction of IP costs on the GM product alone can simply be seen as an offset for the technological cost reduction. If the IP cost is sufficiently large, the GM product may not be able to penetrate the market.

If the IP costs fall on both sectors, and they are large enough, it is possible for both consumer prices to rise relative to the pre-GM situation, and for aggregate consumption/production to be less. This would lead to the interesting situation that aggregate welfare would be reduced, and yet all markets would be in equilibrium, and there would be no competitive pressure for production of GM to cease.

The model structure used here is rather restrictive. In particular, the segmentation of the market is a linear function of the ratio of prices, and one might expect that the expansion of the GM sector would accelerate as the price differential expanded. Secondly, the production side of the market is very simplified, with no differentiation of technical change and input market effects on the marginal cost of production.

The extended model

Attempts to derive analytical solutions from more elaborate models were not successful. Consequently, it was decided to resort to numerical methods to obtain solutions from a somewhat more realistic model outlined below. Dropping the requirement for the model to be solved analytically allowed a number of changes to be made, including explicit introduction of production functions utilizing two inputs, characterized respectively as a seed and herbicide complex on the one hand, and on the other a composite factor for all other inputs, including land. Nevertheless, it should be noted that the model is still very simplified. In particular, the model does not explicitly include trade; processing and marketing activities are subsumed into supply/demand functions; and it is assumed that there is a single consumer good generated from the farm product (i.e. joint or by-products are not considered).

The representative demand functions are expressed as a constant elasticity form:

$$d_n = a_0 P_n^{a1} \quad (1.12)$$

$$d_g = a_0 P_g^{a1} \quad (1.13)$$

The function determining the share of the market allocated to non-GM is extended, as the proportion of the people consuming the GM product may rise non-linearly as the price differential increases:

$$S_n = (P_g / P_n)^{\lambda} \qquad \lambda > 1 \quad (1.14)$$

This gives aggregate demands of:

$$D_n = a_0 P_n^{a1}(P_g/P_n)^{\lambda} \tag{1.15}$$

$$D_g = a_0 P_g^{a1}(1 - (P_g/P_n)^{\lambda}) \tag{1.16}$$

The production functions are given by a two-input constant returns to scale (CRTS), constant elasticity of substitution (CES) production function:

$$Q_n = \gamma\left[\delta k_{1n}^{-\theta} + (1-\delta)k_{2n}^{-\theta}\right]^{1/-\theta} \tag{1.17}$$

$$Q_g = \gamma t_\gamma\left[\delta\left(t_1 k_{1g}\right)^{-\theta} + (1-\delta)\left(t_2 k_{2g}\right)^{-\theta}\right]^{1/-\theta} \tag{1.18}$$

where k_{1n} is the amount of input 1 used by the non-GM sector, k_{2n}, the amount of input 2 and so on.

In the GM sector t_1 and t_2 are the input augmenting technical change associated with the new innovation. Setting $t_1 > 1$ implies that k_{1g} is becoming more effective. Hicks neutral technical change can be represented by setting $t_\gamma > 1$.

The other parameters in the function can be interpreted as follows: γ is a general scaling factor; δ determines (in part) the slope of the isoquant and must lie between 0 and 1 for it to be downward sloping; while θ determines the elasticity of substitution (σ) between the two inputs ($\sigma = 1/(1 + \theta)$).

For this CRTS, CES the marginal costs of production are given by:

$$MC_n = \frac{P_1}{\gamma}\delta^{1/\theta}\left(\left(\frac{1-\delta}{\delta}\right)^{\sigma}\left(\frac{P_2}{P_1}\right)^{\theta\sigma} + 1\right)^{1/\theta} + \frac{P_2}{\gamma}(1-\delta)^{1/\theta}\left(\left(\frac{\delta}{1-\delta}\right)^{\sigma}\left(\frac{P_1}{P_2}\right)^{\theta\sigma} + 1\right)^{1/\theta} \tag{1.19}$$

$$MC_g = \frac{P_1}{\gamma t_\gamma t_1}\delta^{1/\theta}\left(\left(\frac{1-\delta}{\delta}\right)^{\sigma}\left(\frac{P_2}{P_1}\right)^{\theta\sigma} + 1\right)^{1/\theta} + \frac{P_2 + tf}{\gamma t_\gamma t_2}(1-\delta)^{1/\theta}\left(\left(\frac{\delta}{1-\delta}\right)^{\sigma}\left(\frac{P_1}{P_2}\right)^{\theta\sigma} + 1\right)^{1/\theta} \tag{1.20}$$

where P_1 and P_2 are the prices of k_1 and k_2 respectively, and t_f is the technology fee charged by the provider of the improved inputs. The form of this fee will be described later, when the model is parametrized.

Assuming profit maximizing, perfectly competitive behaviour, one can directly infer that at equilibrium the product price and marginal cost will be equal, allowing for any identity preservation costs that may arise:

$$P_n = MC_n + IP_n \tag{1.21}$$

$$P_g = MC_g + IP_g \tag{1.22}$$

With fixed input prices (P_1, P_2), marginal costs are not dependent on the scale of production, and Equations 1.19 to 1.22 will define the product price for the two commodities, and hence the resulting demands. However, things are more interesting if one makes the input markets endogenous.

The optimal input demand, for a given level of output, is given by:

$$K_{1n} = \frac{Q_n\delta^{1/\theta}}{\gamma}\left(\left(\frac{1-\delta}{\delta}\right)^{\sigma}\left(\frac{P_2}{P_1}\right)^{\theta\sigma} + 1\right)^{1/\theta} \tag{1.23}$$

$$K_{2n} = \frac{Q_n(1-\delta)^{1/\theta}}{\gamma}\left(\left(\frac{\delta}{1-\delta}\right)^{\sigma}\left(\frac{P_1}{P_2}\right)^{\theta\sigma} + 1\right)^{1/\theta} \tag{1.24}$$

$$K_{1g} = \frac{Q_g\delta^{1/\theta}}{\gamma t_\gamma t_1}\left(\left(\frac{1-\delta}{\delta}\right)^{\sigma}\left(\frac{P_2}{P_1}\right)^{\theta\sigma} + 1\right)^{1/\theta} \tag{1.25}$$

$$K_{2g} = \frac{Q_g(1-\delta)^{1/\theta}}{\gamma t_\gamma t_2}\left(\left(\frac{\delta}{1-\delta}\right)^{\sigma}\left(\frac{P_1}{P_2}\right)^{\theta\sigma} + 1\right)^{1/\theta} \tag{1.26}$$

The (inverse) supply curve associated with each input is assumed to be linear and a function of aggregate input demand:

$$P_1 = b_0 + b_1(K_{1n} + K_{1g}) \tag{1.27}$$

$$P_2 = b_2 + b_3(K_{2n} + K_{2g}) \tag{1.28}$$

So, although there are no direct interactions between the two types of goods on the cost of production as a result of changing output levels, there are indirect effects through the input markets.

Parametrization of the Model

Numerical solutions to models require some foundation in specific data, or the results are little more than curiosities. The simplified nature of the model, and the unknowable aspects of the consumer response to GM products when they enter the market, means that the linkage between parameters and data is not exact. In the following sections, the method of parameter identification and the relevant baseline data are outlined. A summary of parameter values is reported in Tables 1.1 and 1.2.

Production and price data

As this study is based on a world market scenario, it is not country specific and therefore there are no ramifications associated with trade. The basic model is parametrized to give a stylized representation of a

Table 1.1. Baseline and scenario parameter values: k_1 input share = 16.7%.

Baseline parameters and values

Parameter	Canola price (US\$ t^{-1})	Canola Prod. (million t)	θ	σ	δ	γ	b_0	b_1	b_2	b_3[a]
Value	296	35.87	2	1/3	7.936×10^{-3}	4.42×10^{-3}	1	0	0.2	9.04×10^{-5}

With elasticity of demand set at −0.75 and used for Table 1.3	a_1	−0.75	a_0	2560
With elasticity of demand set at −0.5 and used for Table A1	a_1	−0.50	a_0	617

Technology, yield, identity preservation and technology fees used in Scenarios 1 to 7

	Baseline	S1	S2	S3	S4	S5	S6	S7
t_1	1	1.87[b]	1	1.87	1.87	1.87	1.87	1.87
t_γ	1	1	1.0845	1.0845	1.0845	1.0845	1.0845	1.0845
IP_n	0	0	0	0	0	44	44	44
IP_g	0	0	0	0	44	44	0	0
t_f	0	0	0	0	0	0	0	0.085

[a]Implied elasticity of supply of k_2 = 1.25, implied elasticity of supply of output = 1.5, share of k_1 in total costs = 16.66%.
[b]The assumption is that the technology increases the effect of k_1 by a factor of 87%, i.e. generates a 47% saving in cost of that factor. With this input ratio that is equivalent to a 'neutral' 8.45% increase in yield.

Table 1.2. Baseline and scenario parameter values: k_1 input share = 40%.

Baseline parameters and values

Parameter	World canola price (US\$ t^{-1})	Total world Prod. (million t)	θ	σ	δ	γ	b_0	b_1	b_2	b_3[a]
Value	296	35.87	2	1/3	0.2286	6.384×10^{-3}	1	0	−0.111	1.74×10^{-4}

With elasticity of demand set at −0.75 and used for Table A2	a_1	−0.75	a_0	2560
With elasticity of demand set at −0.5 and used for Table A3	a_1	−0.50	a_0	617

Technology, yield, identity preservation and technology fees used in Scenarios 1 to 7

	Baseline	S1	S2	S3	S4	S5	S6	S7
t_1	1	1.87[b]	1	1.87	1.87	1.87	1.87	1.87
t_γ	1	1	1.0845	1.0845	1.0845	1.0845	1.0845	1.0845
IP_n	0	0	0	0	0	44	44	44
IP_g	0	0	0	0	44	44	0	0
t_f	0	0	0	0	0	0	0	0.118

[a]Implied elasticity of supply of k_2 = 0.9, implied elasticity of supply of output = 1.5, share of k_1 in total costs = 40%.
[b]The assumption is that the technology increases the effect of k_1 by a factor of 87%, i.e. generates a 47% saving in cost of that factor. With this input ratio that is equivalent to a 'neutral' 22% increase in yield.

canola market. In the model the aggregate quantity of canola grain is set at 35.87 Mt and is based on FAO data for world production of canola grain in 1998 (FAOSTAT, 2000). The price for canola grain is assumed to be US$296 t^{-1} which, according to the USDA's Foreign Agricultural Service, was the canola grain price for 1998 (FAS Online, 2000).

Farm input costs and benefits

Without GM inputs

The two inputs in the production function are characterized as k_1, the canola complex (seed and herbicide) and k_2, all other inputs. Fulton and Keyowski (1999a) suggest that the share of the canola complex is some 16% in total cost, and the parameters of the production function are selected to generate this result at the pre-GM equilibrium. Given an elasticity of substitution (σ) set exogenously at 1/3, and normalizing the input prices to unity allows the input ratio to be determined (the ratio of Equations 1.23 and 1.24) as a function of a single parameter, δ.

The scale parameter γ is then identified by equating marginal cost (Equation 1.19) with canola price.

As an alternative, the original share of 16% is increased to 40%, which may be closer to the value for Australia. This leads to alternative values for δ and γ (see Table 1.2).

Having identified all parameters of the production function, the equilibrium input quantities are identified (Equations 1.23, 1.24). The units in which these are measured cannot be interpreted, as the input prices have been normalized to unity, and they will change with the differing assumption about the input share (or σ, if it were altered). However, once the input quantities have been established at equilibrium, the parameters of their (inverse) supply functions can be obtained.

With GM inputs

Fulton and Keyowski (1999b) suggest that farmers who have adopted some form of reduced tillage system are more likely to profit from using HR (herbicide resistant) seed. Production of GM canola requires a one-pass chemical operation (as opposed to two passes required by non-GM canola) so eliminates the cost of additional machine operations over the field; enables control of the entire spectrum of weeds so giving farmers much more flexibility in terms of the timing and type of weed control; and has the potential to improve the crop yield by removing competition for moisture and nutrients (Fulton and Keyowski, 1999a). Even so, the benefits of the new technology will only be recognized if returns increase through reduced weed control costs and/or increased yields (CCGA, 2000). Therefore, where weed control is not a major concern farmers are unlikely to benefit and may be better off using conventional varieties (CCGA, 2000). Ballenger *et al.* (2000) state that producers in different countries will consider the relative prices for biotech and non-biotech crops in relation to their local farming conditions when deciding what to plant. For the purpose of this study, this statement is extended further to assume that producers will use GMO technology only if it is beneficial to their production method thereby capturing benefits associated with herbicide reduction and yield increase. Therefore, while acknowledging the argument by Fulton and Keyowski (1999a) that total benefits derived from GMO technology will depend on agronomic, management and technology factors facing individual farmers, we assume here that farmers included in this study are alike.

The GM innovation is assumed to have two potential modes of action in the production function that may occur separately or together. The first is a change in the effectiveness of the canola complex inputs. This is represented by assuming that the effectiveness of this input rises by some 87% (t_1 = 1.87), a figure derived using results found by Fulton and Keyowski (1999a). The implied reduction in marginal costs (before allowing for substitution and input price effects) is approximately 8.5% if the input share is 16%.

The second mode of action is a Hicks neutral shift in the production function. Fulton and Keyowski (1999a) found a yield decrease of around 7%[6] with the introduction of Roundup Ready™ canola; James (1998) found that the average canola yield in Canada increased by 7.5% between 1996 and 1997.[7] Here we set this value at 8.5%, so that

[6] As Fulton and Keyowski (1999a) note, farmers were not differentiated in the study and those who have not adopted conservation practices are unlikely to receive the same benefits as those who have.

[7] From Fulton and Keyowski (1999a) almost 4% of canola grown in Canada was GM canola in 1996 compared to 33% in 1997 and therefore it could be assumed that part of the overall yield increase could be contributed to production of GM canola.

the change in marginal cost due to this change is equivalent to that induced by the input-specific shift.

It should be noted that when the input share of k_1 is raised to 40%, the economic impact of the innovation is significantly increased for the same increase in effectiveness: equivalent to a 22% reduction in marginal cost. In the simulations with this higher share the yield effect is retained at 8.5%.

Elasticities of supply and demand

The elasticity of supply of the canola complex (k_1) is assumed to be infinite. Given constant returns to scale, imposing an elasticity of supply on the other inputs effectively determines the long-run equilibrium response of output to changes in the canola price. While Johnson *et al.* (1996) found elasticity of supply to be up to 0.85, for the purpose of this project it was deemed that the long-run supply elasticity would be set at 1.5. If the input share of k_2 is set at 84% (because the share of k_1 has been set at 16% as described above), this implies an elasticity of supply of k_2 of 1.25. If the input share is 60%, then the elasticity of supply of k_2 is set at 0.9. In both cases the parameters of the linear marginal supply function for the input can subsequently be identified.

Johnson *et al.* (1996) indicate the elasticity of demand of canola oil to be –0.6 for Canada, –0.69 for the USA and for the EU, –0.56. Goddard and Glance (1989) quote elasticity of demand for canola oil ranging from –1.17 to –0.31 with a mean and median of –0.78. Here, we simply specify a single derived demand for oilseed, with no differentiation by end-use, and an elasticity of demand of either –0.75 or –0.5. This, combined with the base quantity/price data allows the parameters of the representative demand functions to be identified.

Market share

Phillips (1999) and Buckwell *et al.* (1999) argue that for non-GM products to enter the market, the market would have to segment and the cost that this segmentation could bear would depend on the willingness of consumers to pay extra for non-GM products. In the long run consumers around the world will decide on the premiums they will pay for non-biotech products (Ballenger *et al.*, 2000). Miranowski *et al.* (1999) add that the price premium for a non-GM crop will depend on the supply of that crop and costs of identity preservation. The size of these premiums is unknown and any market intelligence concerning GM food is scarce. Differences in regional attitudes towards GM foodstuffs complicate the picture, as does the fact that canola generates two products, oil and meal, with human and animal feed end-markets. If public concerns about GM products do not extend to products from animals raised on GM feeds, then the derived demand for meal will not segment.

In the face of uncertainty about possible responses, λ is set at an arbitrary value of 3, a value which leads to significant segmentation at relatively low price differentials (e.g. for a GM product priced at 10% less than non-GM product, the market share for the GM product would be 27% of individuals).

Identity preservation costs

Smyth and Phillips (1999) found the cost of an identity preservation and marketing system for canola in Canada to be between 12 and 15% of the farm gate price. This cost is assumed to impose a wedge between farm and retail price by raising the effective marginal cost of supply at retail level. In the simulations that follow, the cost is set at US\$44 t^{-1}, or approximately 15% of the pre-GM farm gate price.

Farmers must pay for GM canola seed as it is assumed that they are not able to retain any seed for planting from the previous year. From Fulton and Keyowski (1999a) the increase in GM canola seed price over non-GM was found to be 2% of the total return. To acquire Roundup Ready™ canola seed, farmers must attend a sign-up meeting and agree to a Technology Use Agreement, pay US\$37 ha^{-1} technology fee and buy a package of seed and Roundup™ herbicide (Phillips, 1999). In this study, these costs are referred to as 'the technology fee that is associated with the GM technology' and is applied to the 'other' input (k_2), leading to an effective increase in its price. This is emerging as a common practice in the industry, with the technology fee applied to land area planted, rather than output levels or seed. In an initial simulation the technology fee is set at 7% of total revenue (at pre-GM equilibrium input quantities and output price).

Summary of data

Tables 1.1 and 1.2 summarize the data used in the model. Table 1.1 generates the production parameters on the assumption that the input share for k_1 is 16.7%, and reports alternative values for the demand elasticity. The lower half reports the values for the technological change parameters, identity preservation costs and technology fee for seven scenarios. Table 1.2 is generated on the basis of an input share of 40%.

The scenarios explore the response of the model to:

- different forms of technical change;
- different incidence of identity preservation costs;
- the impact of a technology fee.

Scenarios 1–3 introduce a factor saving innovation, a Hicks neutral innovation and a combination of both, but no identity preservation costs. The most likely outcome from adoption of herbicide resistant GM canola is assumed to be significant savings in the cost of the canola complex input, as well as an additional factor neutral yield increase.

Scenarios 4–6 build on Scenario 3 by introducing identity preservation costs that respectively affect the cost of retail supply of the GM commodity alone, on both commodities and finally on the non-GM commodity alone. Again for reasons outlined above, it is thought that the latter is the most likely outcome in the market. Finally, a technology fee is added to the system and is fixed exogenously at the rate suggested by Phillips (1999) (Scenario 7).

Results

The full set of simulation results is reported in Appendix II. For simplicity, the discussion will focus on one set of results presented in Table 1.3 based on an input share for the canola complex of 16% and an elasticity of demand of –0.75 (i.e. based on parameter values drawn from Table 1.1).

The first column of Table 1.3 reports the baseline simulation, with the equilibrium price and quantity as initially set. In Scenario 1, the GM technology is depicted as a factor saving technical change with an impact on k_1 alone. The market segments, with the GM crop taking some of the market, with non-GM price rising slightly and a substantial fall in the GM price compared with the initial equilibrium. The expansion in demand by

Table 1.3. Simulated values for canola price and quantity, input quantities and welfare impacts.

		Baseline	S1	S2	S3	S4	S5	S6	S7
P_n		296	298	296	298	296	328	338	333
P_g		n/a	275	273	254	296	285	250	265
Q_n		35.8	28.0	28.1	22.0	35.8	21.8	13.1	16.5
Q_g		n/a	8.1	8.2	15.5	0.1	12.7	24.2	19.4
	$\mathbf{Q}$	**35.8**	**36.1**	**36.3**	**37.5**	**35.9**	**34.5**	**37.3**	**35.9**
P_2		1	1.01	0.99	1.01	1	0.95	0.99	0.97
k_{1n}		1770	1388	1389	1087	1765	1061	647	809
k_{1g}		0	215	375	377	2.3	305	588	468
	$\mathbf{k_1}$	**1770**	**1603**	**1764**	**1464**	**1366**	**1366**	**1235**	**1277**
k_{2n}		8850	6921	6949	5423	8828	5395	3244	4088
k_{2g}		0	2005	1876	3517	21	2900	5517	4423
	$\mathbf{k_2}$	**8850**	**8926**	**8825**	**8940**	**8849**	**8295**	**8761**	**8511**
Δcs_n			–48	16	–44	0	–715	–579	–629
(share)			*(0.78)*	*(0.78)*	*(0.85)*	*(1.0)*	*(0.65)*	*(0.40)*	*(0.50)*
Δcs_g			81	100	319	0	–94	253	62
Δps_1			0	0	0	0	0	0	376
Δps_2			61	–20	72	0	–430	–71	–266
	$\mathbf{\Delta W}$		**94**	**96**	**347**	**0**	**–1239**	**–397**	**–457**

Note: The bottom six rows of the table contain estimates of changes in consumer and producer welfare relative to the baseline scenario representing no production of GM food. The estimate of share is the proportion of the consumers consuming the non-GM food.
Δcs_j is the change in consumer surplus for those consuming good j, Δps_1 the change in producer surplus for supplier of input 1(k_1), and ΔW the aggregate effect. See Appendix III for further details on the method of calculation.

those who switch to GM product leads to an expansion in aggregate output. While this increased output is produced using less of k_1 (due to the technical change), an expansion in k_2 is necessary due to the scale effect. The latter causes the price of k_2 to be bid up, which is the cause of the increased cost (and hence price) of non-GM output. Note that the reduction in demand for k_1 gives no compensating relief, as its price does not vary with output.

In the reported estimates of welfare effects, it should be noted that the changes in consumer surplus are reported for the sub-populations of consumers. The proportion of the market that remains with non-GM food is reported. Per capita estimates of welfare changes could be obtained by multiplying the aggregate change in welfare by the share. Those who remain with non-GM product are worse off due to the increased price, while those who switch to the GM alternative are better off due to the lower price. Net, there is an increase in welfare, which is increased when the increase in producer surplus of those supplying k_2 is included.[8] Parenthetically, comparison of changes in consumer welfare between scenarios should be conducted with care, because the size of the sub-populations involved varies.

Under Scenario 2, the input specific technical change is replaced with a Hicks neutral effect, which, at initial quantities and prices, leads to the same reduction in marginal cost. However, the distributional effects on the input side differ. Demand for both inputs falls, despite a slight increase in output and hence input price P_2 falls. This reduces the marginal cost of producing the non-GM product, leading to welfare gains for both sets of consumers (the price P_n declines by less than the rounding factor used in Table 1.3).

Scenario 3 combines both forms of technical change. The increased cost advantage allows the GM market to expand significantly, leading to gains in consumer surplus of that group. However, due to the scale effect, adoption of the GM crop again results in increased demand for k_2 and consequential increased cost (and hence price) of non-GM food. Thus consumers who continue to purchase the non-GM product despite the price differential are worse off than they were prior to the introduction of the GM crop, even though there are no identity preservation costs under this scenario.

Under Scenario 4, the incidence of the cost of identity preservation is assumed to fall exclusively on the GM crop. In this case, these costs almost outweigh the benefits of the technical change and, although GM enters the market, it does so only marginally.

In Scenario 5, the incidence of IP costs falls on both commodities. As a result, there is a net reduction in welfare even for those who switch to GM crop, despite the fact that the price of GM lies below the baseline level of 296 (see Appendix III for an explanation of this). Combined with the losses felt by both the consumers of non-GM product and the producers of k_2, who suffer from the reduction in aggregate supply, aggregate welfare also declines as a result of the innovation.

Scenario 6 also maintains both forms of technical change, while introducing an identity preservation cost for the non-GM product only. This increases the price of the non-GM product, but not by the full US$44 (the incidence is about 90%). The GM market expands further, but aggregate output falls. This, combined with a greater concentration in the lower input GM sector leads to reduced input demands and hence lower input prices. The marginal cost and hence price of GM product falls. Consumers of non-GM product and suppliers of k_2 lose, while GM consumers gain.

As noted above, Scenario 7 presents the 'best guess' for possible configurations of a technology innovation fee and identity preservation costs. In fact, the level of the technology fee will depend on the behaviour of the technology provider who may be expected to set this fee so as to maximize rents. This is an area of research that will be explored at a later date.

In Scenario 7, a technology fee is introduced which allows the supplier of the new seed/herbicide complex to extract part of the rent associated with the innovation. IP costs are only placed on the non-GM crop. Compared with Scenario 6 (which has the equivalent IP cost, but no technology fee), the non-GM market recovers some ground, as its lower cost competitor is now facing the technology fee. The rent earned by the fee is reported as the change in producer surplus of input 1 (Δps_1).

Increasing the share of k_1 in production increases the significance of the technological innovation: it is clearly more valuable. The results for this simulation are reported in Table A2 in Appendix II. As the marginal cost of producing the GM product drops further, the price differential widens and it

[8] Details of how the change in consumer surplus is calculated are reported in Appendix III.

absorbs more of the market. As a result, the IP and technology fees used here are no longer sufficient to outweigh the benefits of the innovation, and net welfare increases in all scenarios when comparing Tables 1.3 and A2.

Comparing Table 1.3 with Table A1 and Table A2 with Table A3 reveals that changing the elasticity of demand from –0.75 to –0.5 does little to either the qualitative or quantitative results from the model. In general, both sets of consumers are better off when the elasticity of demand is set at –0.5 as compared to –0.75, and producers of input 2 are worse off except when input share is 16.7%.

Conclusions

The simulation results presented here are predicated on a consumer market that can differentiate between the production technology used to produce the good, with a heterogeneity of preferences within the population which mean that the consumer cares about the technology used to produce the good.[9] This assumption requires segregation and credible labelling of the product.

When this occurs, the market segments. The aggregate and distributional effects of that segmentation will depend on the economic and technical parameters of the underlying system and, in particular, the degree to which the market segments. The latter is the great unknown in the GM debate: if presented with genuine choices, how resilient would the consumer concerns that are expressed in surveys be to price discounts? This chapter does not attempt to answer that question, but derives some implications conditional on an imposed response.

In general the price of conventional, non-GM product is increased, owing to the requirement that it bears identity preservation costs. In the case where there is either zero cost, or the cost is borne solely by the GM crop, it may be the case that the price of non-GM product falls, due to changes in the input markets induced by the innovation, but these are small.

The key distributional impacts are between consumers. Those for whom the new technology holds no qualms in general benefit from access to cheaper commodity, while those who remain with the non-GM product can suffer significant losses: typically greater than the gains to the industry supplying the innovation. There are also losses to the suppliers of other inputs to the industry, as long as the innovation is input enhancing rather than Hicks neutral. These losses are extended where a technology fee is associated with these other inputs: the quantity demanded is being adversely affected by the increase in effective price, without its benefits.

Net welfare effects can be positive or negative. If the identity preservation costs are sufficiently high relative to the cost savings, and in particular if they fall on both commodities, aggregate welfare may fall, but with no competitive incentives for the innovation to be dis-adopted. Furthermore, the possibility of monopoly rent seeking on the part of the supplier of innovation increases that likelihood.

The range of conditions under which the latter holds true seems a particularly fruitful area of further development in the model.

References

Ballenger, N., Bohman, M. and Gehlhar, M. (2000) Biotechnology: implications for US corn and soybean trade. *Agricultural Outlook* April 2000, 24–28. Economic Research Service/USDA. www.ERS.USDA.gov/epubs/pdf/agout/apr2000/ao270g.pdf

Buckwell, A., Brookes, G. and Bradley, D. (1999) Economics of identity preservation for genetically modified crops. Final report of a study for Food Biotechnology Communications Initiative (FBCI).

Caswell, J.A. (2000) Labeling policy for GMOs: to each his own? *AgBioForum* 3(1), 53–57. www.agbioforum.org

Caswell, J. and Mojduszka, E. (1996) Using informational labelling to influence the market for quality in food products. *American Journal of Agricultural Economics* 78(5), 1248-1253.

Caulder, J. (1998) Agricultural biotechnology and public perceptions. *AgBioForum* 1(1), 38-39. www.agbioforum.missouri.edu

CCGA (2000) Canadian Canola Growers Association – The canola producers' perspective on biotechnology, December 1999. www.canola-council.org/biotechweb.htm

Falck-Zepeda, J.B., Traxler, G. and Nelson, R.G. (1999) *Rent Creation and Distribution from the First Three*

[9] This heterogeneity is represented by the preferences that lead to the choice between the two. An alternative specification that extends that heterogeneity to the individual's demand function is reported in Appendix IV, but the additional complexity does little to change the basic story told here.

Years of Planting Bt Cotton, ISAAA Briefs No.13, ISAAA, Ithaca, New York.

FAOSTAT (2000) Agriculture data. www.apps.fao.org/cgi-bin/nph-db?subset=agriculture

FAS Online (2000) Oilseed prices. www.fas.usda.gov/oilseeds/circular/2000/00-06/Prices.pdf

Fulton, M. and Keyowski, L. (1999a) The producer benefits of herbicide-resistant canola. *AgBioForum* 2(2), 85–93. www.agbioforum.missouri.edu

Fulton, M. and Keyowski, L. (1999b) The impact of technological innovation on producer returns: the case of genetically modified canola. Paper presented at the Transitions in Agbiotech: Economics of Strategy and Policy Conference, 24–25 June 1999, Washington, DC.

Gamble, J., Muggleston, S., Hedderley, D., Parminter, T. and Richardson-Harman, N. (2000) Genetic engineering: the public's point of view. Report to Stakeholders. The Horticulture and Food Research Institute of New Zealand Ltd. www.hort.cri.nz/media/gm_report.pdf

Goddard, E.W. and Glance, S. (1989) Demand for fats and oils in Canada, United States and Japan. *Canadian Journal of Agricultural Economics* 37(3), 421–443.

James, C. (1998) *Global Review of Commercialised Transgenic Crops: 1998*, ISAAA Briefs No. 8. ISAAA, Ithaca, New York.

James, C. (1999) *Global Review of Commercialised Transgenic Crops: 1999*, ISAAA Briefs No. 12. ISAAA, Ithaca, New York.

Johnson, D.D., Satyanarayana, V., Dahl, B.L. and Dooley, F.J. (1996) *Trade in Minor Oilseeds: a Spatial Equilibrium Analysis of Sunflower and Canola*, Agricultural Economics Report No. 353. Department of Agricultural Economics, North Dakota State University, Fargo.

KPMG (1999) Report on the compliance costs facing industry and government regulators in relation to labeling genetically modified foods. KPMG, Canberra.

Marshall, K. (1998) What's in a label? *AgBioForum* 1(1), 35–37. www.agbioforum.org

Miranowski, J.A., Moschini, G., Babcock, B., Duffy, M., Wisner, R., Beghin, J., Hayes, D., Lence, S., Baumel, C.P. and Harl, N.E. (1999) Economic perspectives on GMO market segregation. Staff Paper No. 298, October.

Phillips, P.W.B. (2000) Intellectual property rights, canola and public research in Canada. In: Santaniello, V., Evenson, R.E., Zilberman, D. and Carlson, G.A. (eds) *Agriculture and Intellectual Property Rights: Economic, Institutional and Implementation Issues in Biotechnology*. CAB International, Wallingford, pp. 223–248.

Smith, N. (2000) *Seeds of Opportunity: an Assessment of the Benefits, Safety and Oversight of Plant Genomics and Agricultural Biotechnology*. Report transmitted to the Committee on Science for the 106th Congress Second Session, 13 April. www.house.gov/science

Smyth, S. and Phillips, P. (1999) Competitors co-operating: Establishing a supply change to manage genetically modified canola. Paper presented at the The Economics of Quality Control in Agriculture meeting, Saskatoon, 2–4 December 1999.

Zechendorf, B. (1998), Agricultural biotechnology: why do Europeans have difficulty accepting it? *AgBioForum* 1(1), 8–13. www.agbioforum.org

Appendix I: Solution for Aggregate Output for the Simple Model

$$Q = \frac{\begin{array}{l}[a_0^2\, b_1^2 + 2a_0\, b_1\, (b_0 + IP_n) + 4\, a_1^2\, b_1^2\, (b_t^2 + 2\, b_t (IP_n - IP_g) + (IP_g - IP_n)^2\,) \\ -\, 4\, a_1\, b_1\, (b_t^2 + 2\, b_t (IP_n - IP_g) + (IP_g - IP_n)^2\,) + (b_0 + IP_n)^2\,]^{0.5} \\ +\, a_0\, b_1 + (b_0 + IP_n)\, (2\, a_1\, b_1 - 1)\end{array}}{2b_1(1 - a_1\, b_1)}$$

Appendix II: Alternate Simulation Results

Table A1. Simulated values for canola price and quantity, input quantities and welfare impacts, for parameters described in Table 1.1, with elasticity of demand = –0.5.

		Baseline	S1	S2	S3	S4	S5	S6	S7
P_n		296	297	295	296	296	330	336	333
P_g		n/a	274	272	252	296	286	248	265
Q_n		35.8	28.1	28.2	22.1	35.8	22.3	13.5	17.0
Q_g		n/a	8	8.1	14.9	0.1	12.5	23.4	18.8
	Q	**35.8**	**36.1**	**36.3**	**37**	**35.9**	**34.8**	**36.9**	**35.8**
P_2		1	1.01	1	0.99	1	0.96	0.98	0.97
k_{1n}		1770	1389	1388	1089	1765	1088	667	833
k_{1g}		0	211	367	364	2	301	566	455
	k_1	**1770**	**1600**	**1755**	**1453**	**1767**	**1389**	**1233**	**1288**
k_{2n}		8849	6935	6954	5446	8827	5518	3349	4209
k_{2g}		0	1971	1840	3403	18	2859	5322	4299
	k_2	**8849**	**8906**	**8794**	**8849**	**8845**	**8377**	**8671**	**8508**
Δcs_n			–35	35	0	1	–766	–561	–636
(share)			*(0.78)*	*(0.78)*	*(0.62)*	*(1.0)*	*(0.65)*	*(0.40)*	*(0.50)*
Δcs_g			83	103	338	0	–113	287	64
Δps_1			0	0	0	0	0	0	365
Δps_2			43	–46	–2	–2	–369	–140	–270
	ΔW		**91**	**92**	**336**	**–1**	**–1268**	**–414**	**–477**

Note: Δcs_j is the change in consumer surplus for those consuming good *j*, Δps_k the change in producer surplus for supplier of input *k*, and ΔW the aggregate effect.

Table A2. Simulated values for canola price and quantity, input quantities and welfare impacts, for parameters described in Table 1.2, with elasticity of demand = –0.75.

	Baseline	S1	S2	S3	S4	S5	S6	S7
P_n	296	305	295	305	296	334	348	341
P_g	296	249	272	230	266	261	229	243
Q_n	35.8	19.2	28.2	15	26.1	15.6	9.1	11.6
Q_g	0	18.5	8.2	24.7	10.6	20.6	31	26.6
Q	**35.8**	**37.7**	**36.4**	**39.7**	**36.7**	**36.2**	**40.1**	**38.2**
P_2	1	1.05	0.99	1.06	1	0.97	1.05	1.01
k_{1n}	4246	2292	3333	1797	3088	1838	1083	1373
k_{1g}	0	1185	900	1460	619	1196	1828	1560
$\mathbf{k_1}$	**4246**	**3477**	**4233**	**3257**	**3707**	**3034**	**2911**	**2933**
k_{2n}	6370	3384	5004	2650	4633	2788	1600	2054
k_{2g}	0	3270	1351	4026	1738	3393	5049	4363
$\mathbf{k_2}$	**6370**	**6654**	**6355**	**6676**	**6371**	**6181**	**6649**	**6417**
Δcs_n		–169	13	–143	0	–623	–506	–555
(share)		*(0.54)*	*(0.78)*	*(0.43)*	*(0.72)*	*(0.48)*	*(0.28)*	*(0.36)*
Δcs_g		383	99	755	159	97	644	375
Δps_1		0	0	0	0	0	0	515
Δps_2		317	–21	343	–4	–212	310	49
$\mathbf{\Delta W}$		**531**	**91**	**955**	**155**	**–738**	**448**	**384**

See note to Table A1.

Table A3. Simulated values for canola price and quantity, input quantities and welfare impacts, for parameters described in Table 1.2, with elasticity of demand = –0.5.

	Baseline	S1	S2	S3	S4	S5	S6	S7
P_n	296	302	295	301	295	334	343	339
P_g	296	247	272	226	265	261	225	240
Q_n	35.9	19.3	28.2	15.1	26.1	16.1	9.4	11.9
Q_g	0	17.9	8.1	23.6	10.3	19.9	29.5	25.6
Q	**35.9**	**37.2**	**36.3**	**38.7**	**36.4**	**36**	**38.9**	**37.5**
P_2	1	1.04	0.99	1.03	0.99	0.96	1.02	0.99
k_{1n}	4246	2303	3331	1803	3090	1893	1115	1417
k_{1g}	0	1141	881	1383	603	1159	1731	1491
$\mathbf{k_1}$	**4246**	**3444**	**4212**	**3186**	**3693**	**3052**	**2846**	**2908**
k_{2n}	6371	3415	5009	2681	4644	2875	1661	2131
k_{2g}	0	3164	1325	3845	1694	3292	4825	4193
$\mathbf{k_2}$	**6371**	**6579**	**6334**	**6526**	**6338**	**6167**	**6486**	**6324**
Δcs_n		–125	32	–73	26	–625	–463	–527
(share)		*(0.55)*	*(0.78)*	*(0.42)*	*(0.72)*	*(0.48)*	*(0.28)*	*(0.35)*
Δcs_g		411	102	818	165	105	745	426
Δps_1		0	0	0	0	0	0	495
Δps_2		230	–45	170	–40	–227	125	–54
$\mathbf{\Delta W}$		**516**	**89**	**915**	**151**	**–747**	**407**	**340**

See note to Table A1.

Appendix III: Calculation of the Change in Consumer Surplus, Following Introduction of the GM Alternative

Estimation of the aggregate change in consumer surplus is based initially on changes for the individual consumers, which are then aggregated according to whether the consumer has switched from non-GM product to GM product. For those that do not switch, the conventional approach can be applied: for a demand function of the form:

$$d_{ni} = a_0 P_n^{a1} \tag{A1}$$

leads to the standard measure for the change in consumer surplus of:

$$\Delta cs_i = \int_{P_{n2}}^{P_{n1}} d_{ni}\, \mathrm{d}p \tag{A2}$$

where P_{n1} is the price of non-GM canola before the introduction of GM, and P_{n2} the new market price, *ex post.* (Note the introduction of a further subscript: 1 denotes the initial period, 2 the post-GM period.) This gives:

$$\Delta cs_i = \frac{a_0\left(P_{n1}^{\ a_1+1} - P_{n2}^{\ a_1+1}\right)}{a_1+1} \tag{A3}$$

As all individuals who remain consuming non-GM product are identical, the aggregate change in welfare for that group is obtained by multiplying Equation A3 by the proportion who remain with non-GM:

$$\Delta cs = \frac{a_0\left(P_{n1}^{\ a_1+1} - P_{n2}^{\ a_1+1}\right)}{a_1+1}\left(\frac{P_{g2}}{P_{n2}}\right)^{\lambda} \tag{A4}$$

The groups who switch to GM product present more of a problem, because by definition, they do not have a 'base line' GM price, from which the change in consumer surplus can be identified. However, such a price can be inferred.

Recall that c_i is defined as an index of concern, and is the basis on which the decision to switch consumption is made that is, i will consume the GM product if:

$$c_i > \left(\frac{P_g}{P_n}\right)^{\lambda} \tag{A5}$$

We now introduce the notion of an *equivalent price* for GM. This is the price at which consumer i is indifferent between consuming GM and non-GM product:

$$EP_g = c_i^{1/\lambda} P_{n1} \tag{A6}$$

That is, if the consumer switches to GM canola, and can purchase it at EP_g there will be no change in their welfare, as compared with their pre-GM consumption of non-GM product. If they switch to GM product, and can pay a price less than EP_g then their welfare will be increased, by the conventional amount, defined as the wedge below the GM demand curve and between the effective price and the market price of GM:

$$\Delta cs_i = \frac{a_0\left(\left(c_i^{1/\lambda} P_{n1}\right)^{a_1+1} - P_{g2}^{\ a_1+1}\right)}{a_1+1} \tag{A7}$$

Given the distribution of c there will be a range of welfare impacts, ranging from large (for those individuals who are effectively indifferent between the two products, and hence need very little price differential to switch) to negligible for those who are more concerned, and whose equivalent price of GM coincides with the market price.

Note that it is the post-GM price of non-GM that governs the decision to switch, while it is the pre-GM price of non-GM canola which determines the welfare impact. As a result it is possible for consumer surplus for the individual to fall.

Assume that the initial price of non-GM canola is 296, and the equivalent price for individual i is 266; that is, they require a 10% discount before they will switch. Let the post-GM price of non-GM canola rise to 320 (due to IP costs), and the GM price be 266. This individual will certainly switch (there is now a 15% price differential), but the change in welfare will be zero, by definition. Furthermore, if the GM price were 270, they would still switch (the price differential still exceeds 10%) but they would be at a lower level of welfare *compared with the pre-GM position.* Indeed, they would be prepared to switch at prices of GM up to 288. The decision to switch is still rational, in that it minimizes the losses associated with the new price regime, after the introduction of the GM crop.

The aggregate change in welfare for all those who switch is given by integrating across the population who switch:

$$\int_{\left(\frac{P_{g2}}{P_{n2}}\right)^{\lambda}}^{1} \Delta cs_i = \frac{a_0\left(\left(c_i^{1/\lambda}P_{n1}\right)^{a_1+1} - P_{g2}{}^{a_1+1}\right)}{a_1+1} \mathrm{d}c_i \quad \text{(A8)}$$

Appendix IV: Heterogeneity of Demand

The demand decision has been treated as a two-stage process: consumers first decide which type of commodity they will consume (depending on relative prices of GM and non-GM), and then the quantity of commodity. So far the representative demand curve for the two commodities has been identical: there is heterogeneity in preferences between the two commodities, but homogeneity with respect to the actual demand function. This restriction will now be relaxed.

We assume that those who are willing to convert at a relatively low price differential are consumers who are more price responsive in general. Thus we model the elasticity of demand of individual i within the population as a function of their latent 'concern' variable c.

$$d_{ni} = a_0 P_n^{a1+a2c_i} \quad \text{(A9)}$$

$$d_{gi} = a_0 P_g^{a1+a2c_i} \quad \text{(A10)}$$

If $a_2 < 0$ then those most likely to switch from non-GM to GM product (see (A5) above) are those in the population that have the highest price elasticity of demand.

Identifying aggregate market demand in the two segments will now require aggregation over the individual demands within the two. Defining c^* as the level of the c held by the marginal consumer (who is indifferent between non-GM and GM) then aggregate demands are given by:

$$D_n = \int_0^{c^*} a_0 P_n^{a1+a2c} \mathrm{d}c \quad \text{(A11)}$$

$$D_g = \int_{c^*}^{1} a_0 P_g^{a1+a2c} \mathrm{d}c \quad \text{(A12)}$$

Given the assumption that c is distributed as a uniform variable, and maintaining $a_2 \uparrow 0$ then

$$D_n = P_n^{a_1}\left(\frac{a_0 P_n^{a_2c^*}}{a_2 \ln(P_n)} - \frac{a_0}{a_2 \ln(P_n)}\right) \quad \text{(A13)}$$

$$D_g = P_g^{a_1}\left(\frac{a_0 P_g^{a_2}}{a_2 \ln(P_g)} - \frac{a_0 P_g^{a_2c^*}}{a_2 \ln(P_g)}\right) \quad \text{(A14)}$$

and noting that

$$c^* = (P_g/P_n)^{\lambda}$$

allows the identification of aggregate demand as a function of the two product prices.

2 A Comparison of Consumer Attitudes towards GM Food in Ireland and the United States: a Case Study Over Time[1]

Marianne McGarry Wolf[1] and Christine Domegan[2]
[1]*Agribusiness Department, California Polytechnic State University, San Luis Obispo, California, USA;* [2]*Department of Marketing, National University of Ireland, Galway, Ireland*

Introduction

New products and technologies are becoming available for the production of food. In particular, biotechnology is being used to produce genetically modified organisms (GMOs) that are used in food production. Food producers have used the new biotechnologies. Recent negative consumer response to these products in Europe, Japan and Australia has caused farmers to question whether or not to adopt the new technologies. There is concern that American consumer attitudes may follow those of the Europeans. However, there is a void of knowledge about consumer acceptance, regulation, trade implications and public opinion.

The objective of this research is to use a case study to compare consumer attitudes towards genetically modified (GM) food in the USA and Europe using two communities during two time periods. The communities examined are in California and Ireland. This research examines differences in the following between the US and Irish respondent: attitudes towards science and food purchasing behaviour, familiarity with GM food and consumer attitudes towards GM food. In addition, attitudes of respondents in each country were examined based on two subgroups: those familiar with GM food compared with those not familiar with GM food; and those that are likely to purchase GM food compared with those that are not are likely to purchase GM food. There are demographic and attitudinal differences between the subgroups in each country.

This research shows that there is a similar level of familiarity with GM food in Ireland and the USA. Approximately 43% of respondents in both countries indicated that they were familiar with GM food. However, familiar Irish consumers are aware from more sources than the US consumer. The Irish consumers appear to be more interested in GM food because they were more likely to indicate that they had discussions with family, friends and colleagues concerning GM food. Higher educated respondents in both countries are more likely to be familiar with GM food.

Most consumers in both countries indicated that government imposition of mandatory labelling is important, 95% in Ireland and 81% in the USA. The more familiar US and Irish consumers are with GM food, the more likely they are to indicate that mandatory labelling is very important. Further, almost three-quarters of US respondents that are likely to purchase a GM food product indicated that mandatory labelling is important.

A minority of consumers in each country said

[1] Research assistance provided by: Jacinta M. Ryan and Corey Hartung, Research Assistants, California Polytechnic State University.

Table 2.1. Demographics.

	Ireland (%)	US (%)	Chi square
Sex	$N = 197$	$N = 680$	
Female	56.3	51.6	
Male	43.7	48.4	1.369
Age	$N = 196$	$N = 682$	
Under 20 years	2.6	2.9	
20 to 24 years	17.9	13.6	
25 to 44 years	42.3	38.6	
45 to 54 years	25.5	21.4	
55 to 59 years	6.1	8.9	
60 + years	5.6	14.5	14.731**
Marital status	$N = 197$	$N = 683$	
Married	51.3	46.4	
Living with a partner	12.7	8.6	
Single	29.9	31.3	
Separated/divorced	3.0	9.1	
Widowed	3.0	4.5	11.493**
Education	$N = 196$	$N = 683$	
Grade school or less	9.7	0.6	
Some high school	34.2	1.3	
High school graduate	9.2	11.0	
Some college	15.8	35.1	
College graduate	21.9	38.8	
Postgraduate work	9.2	13.2	271.790**
Employment	$N = 197$	$N = 682$	
Employed, full time	58.9	63.0	
Employed, part time	20.8	16.6	
Not employed	20.3	20.4	1.998
Dual income household	$N = 197$	$N = 683$	
Yes	47.7	49.3	
No	52.3	50.7	0.162
Income	$N = 177$	$N = 670$	
Under US$20,000	21.5	12.4	
US$20,000 to US$29,999	25.4	14.9	
US$30,000 to US$39,999	25.4	17.5	
US$40,000 to US$54,999	11.3	16.4	
US$55,000 to US$69,999	8.5	14.5	
US$70,000 or more	7.9	24.3	46.003**

** Significant difference at 0.05 level.

that they were likely to purchase a GM food product. However, the more familiar respondents in Ireland and the US are with GM food, the more likely they are to purchase a GM food product. Familiarity with GM food seems to increase positive attitudes towards it. Therefore, educational programmes concerning the process of producing GM food are likely to generate more positive attitudes.

Familiar respondents in both countries described GM food as: 'grown in the US, improves the production of food', and 'is made by splicing genes from one plant or organism to another'. The three characteristics describing GM foods the least were: 'is the same as a hybrid, has superior taste', and 'is harmful to consumers'. There is a difference in attitudes between the Irish consumer and consumers in the US towards GM food. The familiar US respondents perceive GM food to have neutral or positive attributes. The Irish consumer attributed more negative attributes to GM food. Further, they are more likely to indicate that mandatory labelling is important and less likely to purchase a GM food product.

Attitudes were examined for differences over time in both Ireland and the USA for two time periods, 3 months apart. Attitudes did not differ between the two time periods in either country. Additional research will be conducted over time to determine if attitudes change over longer time frames.

Research Methodology

A simulated before–after experimental design was used to conduct this research. The first phase of this research was conducted in October 1999 in the USA and in November 1999 in Ireland. The second phase of the research was conducted during January and February 2000.

The research used a survey instrument that was administered through the use of a personal interview. The first phase of this research examined 423 randomly selected food purchasers in San Luis Obispo, California, and Galway, Ireland. The second phase of this research examined 459 randomly selected food purchasers in San Luis Obispo and Galway. San Luis Obispo has a population of about 42,000 and Galway has a population of approximately 57,000.

Demographics of Sample

The demographics for the sample from Galway were compared with national statistics provided by the Central Statistics Office, Cork, Ireland. The age of the sample respondents was slightly younger than the 1996 Central Statistics Office national figures with a smaller proportion of respondents in the 60 years and older age group. Gender and income were similar to those reported by the Central Statistics Office (Central Statistics Office, 1999).

The demographics of the San Luis Obispo sample were older and had a higher education and income than the 10-year-old US Census Statistics (Census, 1991). However, they were similar to statistics generated by other recent local research projects that may better reflect the current demographics of the area (Wolf *et al.*, 2000).

Table 2.1 shows that the Irish respondent was more likely to be younger and married or living with a partner than the US respondent. Further, the Irish respondent was more likely to have completed some high school and earned US$39,000 or less compared with the US respondent. The US respondent was more likely to be older, married, a college graduate and have a higher income than the Irish respon-

Table 2.2. Scientific research is an important factor in the improvement of the quality of life.

	Ireland (%) (*N* = 197)	US (%) (*N* = 682)	Chi square
Strongly agree	39.6	48.5	
Agree	54.8	43.3	
Disagree	4.6	6.0	
Strongly disagree	1.0	2.2	8.754**

** Significant difference at 0.05 level.

Table 2.3. Organic food consumption within past year.

	Ireland (%) (*N* = 196)	US (%) (*N* = 678)	Chi square
Yes	52.6	62.9	
No	47.4	30.8	18.579**

** Significant difference at 0.05 level.

Table 2.4. Nutritional label readership and purchase decision.

	Ireland (%) ($N = 197$)	US (%) ($N = 683$)	Chi square
Very often	27.9	47.7	
Somewhat often	28.4	30.9	
Not very often	28.9	14.8	
Not at all	14.7	6.9	41.755**

** Significant difference at 0.05 level.

Table 2.5. Ingredient label readership and purchase decision.

	Ireland (%) ($N = 195$)	US (%) ($N = 681$)	Chi square
Very often	22.6	38.3	
Somewhat often	31.8	30.8	
Not very often	30.3	21.3	
Not at all	15.4	9.5	20.860**

** Significant difference at 0.05 level.

dent. Both groups of consumers had similar gender, employment status and dual income household status.

Attitudes towards science and food purchasing behaviour

US respondents appear to have more positive attitudes towards science (Table 2.2) and are more likely to have purchased organic food in the past year (Table 2.3) than Irish respondents. Food labelling appears to be more important to US respondents than Irish respondents when purchasing food since nutritional labels and ingredient information are read more often by US respondents (Tables 2.4 and 2.5).

Familiarity with GM food

A similar proportion of US and Irish respondents, approximately 43%, indicated that they were somewhat or very familiar with GM food (Table 2.6). However, an examination of the sources of information about GM food among respondents who were at least somewhat familiar indicates that the Irish respondents had learned about GM food from a wide variety of sources. Almost all of the familiar Irish respondents had heard about GM food from the newspaper or television news, while only two-thirds of familiar US respondents had heard about GM food from these sources. Slightly over one-quarter of familiar US respondents had heard about GM food from the radio, while over four-fifths of the familiar Irish respondents had heard about GM food from radio news. It is clear that GM food is an important issue to the familiar Irish respondent because almost three-fourths indicated that they engaged in discussions with family, friends and colleagues. However, only one-third of familiar US respondents indicated they had engaged in discussions with family, friends and colleagues about GM food (Table 2.7).

Table 2.6. Familiarity with GM food.

	Ireland (%) ($N = 196$)	US (%) ($N = 671$)	Chi square
Very familiar	5.1	7.0	
Somewhat familiar	38.3	36.6	
Not very familiar	39.3	39.4	
Not at all familiar	16.8	17.0	1.076

Table 2.7. Sources of GM food awareness among very or somewhat familiar.

	Ireland (%) (*N* = 86)	US (%) (*N* = 297)	Chi square
Newspaper	97.6	68.3	28.93**
Television news	97.6	64.2	34.88**
Radio news	86.3	27.9	90.82**
Discussion w/ family, friends, etc.	71.8	36.7	30.43**
News magazines	52.2	40.4	3.63**
Employment, work in farming or food process	27.6	16	4.35**
Consumer reports magazine	25.4	20.5	10.40**
Internet	22.8	15.1	2.04
Other	0	11.3	6.01**

** Significant difference at 0.05 level.

Table 2.8. Crops believed to be grown domestically/internationally using GM methods among very or somewhat familiar.

	Ireland (%) (*N* = 86)	US (%) (*N* = 297)	Chi square
Tomatoes	79.5	69.4	2.89*
Maize	70.7	65.3	4.87*
Wheat	63.6	40.8	11.23**
Barley	62.3	34.5	17.78**
Potatoes	62.2	33.1	25.2**
Soybean	56.2	36.1	14.03**
Sugarbeet	55.4	25.5	21.95**
Cotton	29.5	39.3	2.28
None of the above	0	1.8	0.89
Do not know	10.2	15.7	6.49**

* Significant difference at 0.10 level; ** significant difference at 0.05 level.

Most respondents familiar with GM food were able to identify tomatoes and maize as crops that are grown domestically or internationally using GM methods. The familiar Irish respondents identified more crops as being grown using GM methods (Table 2.8).

Attitudes towards GM Food

Most consumers in both countries indicated that government imposition of mandatory labelling is important, 95% in Ireland and 81% in the USA. Although the Irish respondents were less likely than the US respondents to read nutritional labels or ingredient labels when making a food purchase decision, they were more likely than the US respondents to indicate that government imposition of mandatory labelling of GM food is very important. Almost three-quarters of the Irish respondents indicated that mandatory labelling of GM food is very important compared with approximately half of US respondents (Table 2.9).

The Irish consumer is less likely to purchase food that has been genetically modified than the US consumer. Over half of the Irish respondents indicated that they were not likely to purchase food that has been genetically modified compared with only one-quarter of the US consumers (Table 2.10).

Perceptions of GM food

In order to examine respondents' perceptions about GM food products, they were asked to rate 15 statements on a five-point scale concerning how each statement describes GM foods. The following question was used to determine respondents' knowledge about GM food products:

Table 2.9. Government imposition of mandatory labelling for GM food.

	Ireland (%) ($N = 197$)	US (%) ($N = 681$)	Chi square
Very important	70.6	52.3	
Somewhat important	25.4	28.9	
Not very important	3.6	12.2	
Not at all important	0.5	6.6	31.709**

** Significant difference at 0.05 level.

Table 2.10. Likelihood of GM food purchase.

	Ireland (%) ($N = 197$)	US (%) ($N = 679$)	Chi square
Definitely	3.6	6.9	
Probably	19.3	28.3	
Maybe	24.9	37.6	
Probably not	28.4	18.9	
Definitely not	23.9	8.4	52.005**

** Significant difference at 0.05 level.

To the best of your knowledge, how well do each of the following statements describe GM food? As I read each statement, please think about how well it describes GM food. Even if you are not too familiar with certain aspects of GM food, please rate it based on your impressions. To answer, please tell me the number from the scale I read which best describes GM food on that statement.

5 = Describes GM food completely
4 = Describes GM food very well
3 = Describes GM food somewhat
2 = Describes GM food slightly
1 = Does not describe GM food at all
(0 = Not at all familiar with GM food)

A mean rating was calculated for each descriptive statement among respondents that used the 1–5 rating scale. Respondents that answered 0, not at all familiar with GM food, were excluded from this analysis of each statement. The results of the analysis of means indicated that the familiar respondents considered the descriptive statements to be broken into two distinct groups: describes GM food somewhat to very well and describes GM food slightly to somewhat (Table 2.11). The top three characteristics from the list of 15 statements describing GM foods were: 'grown in the US, improves the production of food', and 'is made by splicing genes from one plant or organism to another'. The three characteristics describing GM foods the least were: 'is the same as a hybrid, has superior taste', and 'is harmful to consumers'.

There are differences between the attitudes of the Irish respondents and the respondents from the US (Table 2.12). The top three characteristics remain the same for the two groups. However, the Irish gave higher ratings to the descriptive statements: 'is made by splicing genes from one plant or organism into another plant, has seeds from Monsanto, is modified to kill pests, may be harmful to the environment, is grown in Ireland', and 'is harmful to consumers'. The respondents from the US rated the following statements higher: 'is grown in the US, is modified to help plants withstand weed killers, is beneficial to the environment', and 'has superior taste'. The familiar US respondents appear to perceive GM food to have neutral or positive attributes. Since the Irish respondents gave a lower rating to 'is beneficial to the environment', and a higher rating to 'is harmful to consumers and environment', it appears that the Irish respondent has a less positive attitude towards GM foods than the respondent from the US. This attitude difference is reflected in a higher importance of mandatory labelling and a lower likelihood of purchasing GM food for the Irish consumer.

Table 2.11. Total sample descriptive ratings of GM food characteristics among familiar.

	Mean rating	Standard error	*N*
Describes somewhat to very well			
Is grown in the US	3.7790	0.05	751
Improves the production of food	3.5481	0.07	771
Is made by splicing genes from one plant or organism into another plant	3.5311	0.05	659
Is the same as bioengineered food	3.2103	0.05	623
Is modified to help plants withstand weed killers	3.0567	0.05	688
Describes slightly to somewhat			
Is altered to improve nutrition	2.9577	0.06	734
Has seeds from Monsanto	2.9386	0.08	342
Is modified to kill pests	2.8025	0.05	729
May be harmful to the environment	2.7371	0.05	715
Is grown in Ireland	2.6320	0.07	413
Has seeds from Novartis	2.4435	0.09	230
Is beneficial to the environment	2.4342	0.05	714
Is the same as a hybrid	2.4227	0.05	608
Has superior taste	2.3110	0.05	582
Is harmful to consumers	2.2098	0.05	696

Table 2.12. Comparison of descriptive ratings of GM food characteristics.

	Ireland	US
Describes somewhat to very well		
Is grown in the US** ($t = -2.49$)	3.5460	3.8435
Improves the production of food ($t = -1.46$)	3.4556	3.5740
Is made by splicing genes from one plant or organism into another plant* ($t = 1.45$)	3.6713	3.4922
Is the same as bioengineered food ($t = 0.70$)	3.1377	3.2309
Is modified to help plants withstand weed killers** ($t = -0.97$)	2.9673	3.0822
Describes slightly to somewhat		
Is altered to improve nutrition ($t = -1.310$)	2.8280	2.9931
Has seeds from Monsanto** ($t = 1.781$)	3.2143	2.8676
Is modified to kill pests** ($t = 0.147$)	2.8160	2.7986
May be harmful to the environment** ($t = 4.090$)	3.1218	2.6297
Is grown in Ireland** ($t = 3.792$)	2.9931	2.4366
Has seeds from Novartis ($t = 1.486$)	2.7895	2.3750
Is beneficial to the environment** ($t = -5.092$)	1.9740	2.5607
Is the same as a hybrid* ($t = 1.802$)	2.6330	2.3768
Has superior taste** ($t = -2.413$)	2.0752	2.3808
Is harmful to consumers** ($t = 4.677$)	2.6341	2.0789

*Significant difference at 0.10 level using an independent sample *t*-test; **significant difference at 0.05 level using an independent sample *t*-test.

Familiarity with GM Food

Respondents in each country were examined based on two subgroups: those familiar with GM food compared with those not familiar with GM food. Respondents that are identified as *familiar* indicated that they were very or somewhat familiar with GM food. Respondents that are identified as *not familiar* indicated that they were not very or not at all familiar with GM food.

In Ireland and the US familiarity varies with demographics (Table 2.13). In Ireland, respondents

Table 2.13. Demographics.

	Ireland			US		
	Familiar (%)	Not familiar (%)	Chi square	Familiar (%)	Not familiar (%)	Chi square
Sex	$N = 86$	$N = 110$	0.617	$N = 297$	$N = 384$	0.20
Female	53.5	59.1		50.5	52.2	
Male	46.5	40.9		49.5	47.8	
Age	$N = 85$	$N = 110$	12.468**	$N = 297$	$N = 383$	4.26
Under 20 years	1.2	3.6		2.2	3.1	
20 to 24 years	18.8	16.4		10.8	15.9	
25 to 44 years	54.1	33.6		39.4	32.1	
45 to 54 years	20.0	30.0		21.9	20.6	
55 to 59 years	3.5	8.2		9.4	8.6	
60+ years	2.4	8.2		15.8	13.6	
Marital status	$N = 86$	$N = 110$	3.115	$N = 297$	$N = 384$	4.88
Married	51.2	51.8		50.2	43.5	
Living with a partner	11.6	13.6		9.1	8.3	
Single	33.7	26.4		27.6	34.1	
Separated/divorced	2.3	3.6		8.1	9.9	
Widowed	1.2	4.5		5.1	4.2	
Education	$N = 85$	$N = 110$	11.255**	$N = 297$	$N = 384$	10.19*
Grade school or less	8.2	10.9		0.30	0.8	
Some high school	29.4	37.3		1.3	1.3	
High school graduate	8.2	10.0		8.8	12.8	
Some college	12.9	18.2		31.3	38.3	
College graduate	24.7	20.0		42.4	35.9	
Postgraduate work	16.5	3.6		15.8	10.9	
Employment status	$N = 86$	$N = 110$	7.309**	$N = 297$	$N = 383$	0.97
Employed, full time	65.1	54.5		60.9	64.5	
Employed, part time	23.3	18.2		17.2	16.2	
Not employed	11.6	27.3		21.9	19.3	
Dual income household	$N = 86$	$N = 110$	3.788	$N = 297$	$N = 384$	0.156
Yes	55.8	41.8		51.2	47.9	
No	44.2	58.2		48.8	52.1	
Household income	$N = 77$	$N = 99$	7.399	$N = 290$	$N = 378$	9.31*
Under US$20,000	18.2	23.2		12.1	12.7	
US$20,000 to US$29,999	22.1	28.3		13.1	16.4	
US$30,000 to US$39,999	22.1	28.3		15.5	19.0	
US$40,000 to US$54,999	15.6	8.1		14.8	17.5	
US$55,000 to US$69,999	13.0	5.1		14.8	14.3	
US$70,000 or more	9.1	7.1		29.7	20.1	

*Significant difference at the 0.10 level; **significant difference at the 0.05 level.

aged between 25 and 44, more highly educated and employed are more likely to be familiar with GM food. In the USA respondents that are college graduates and have a higher household income are more likely to be familiar with GM food.

In both Ireland and the USA respondents that are familiar with GM food are more likely to read nutrition and ingredient labels (Tables 2.14 and 2.15). Organic food is more likely to have been purchased in the past year by respondents familiar with GM food in both countries (Table 2.16). Therefore, it is not surprising that respondents in Ireland and the US that are familiar with GM food are more likely to feel that mandatory labelling of GM food is very important (Table 2.17). Familiar respondents in both countries are more likely to purchase GM food products than respondents that are not familiar with GM food (Table 2.18).

Table 2.14. Nutrition label readership.

	Ireland			US		
	Familiar (%) (*N* = 86)	Not familiar (%) (*N* = 110)	Chi square	Familiar (%) (*N* = 297)	Not familiar (%) (*N* = 384)	Chi square
Very often	39.5	19.1	12.726**	54.2	42.2	15.62**
Somewhat often	24.4	31.8		30.6	31.0	
Not very often	24.9	29.1		10.4	18.2	
Not at all	8.1	20.0		4.8	8.6	

**Significant difference at the 0.05 level.

Table 2.15. Ingredient label readership.

	Ireland			US		
	Familiar (%) (*N* = 86)	Not familiar (%) (*N* = 110)	Chi square	Familiar (%) (*N* = 297)	Not familiar (%) (*N* = 384)	Chi square
Very often	32.6	14.7	11.443**	50.2	29.1	34.64**
Somewhat often	31.4	32.1		27.6	33.2	
Not very often	26.7	33.0		8.3	25.4	
Not at all	9.3	19.3		2.1	12.3	

**Significant difference at the 0.05 level.

Table 2.16. Past year organic food purchase.

	Ireland			US		
	Familiar (%) (*N* = 86)	Not familiar (%) (*N* = 110)	Chi square	Familiar (%) (*N* = 297)	Not familiar (%) (*N* = 384)	Chi square
Yes	68.2	40.9	14.367**	75.1	63.9	13.73**
No	31.8	59.1		24.9	35.1	

**Significant difference at the 0.05 level.

Table 2.17. Government imposition of mandatory labelling of GM foods.

	Ireland			US		
	Familiar (%) ($N = 86$)	Not familiar (%) ($N = 110$)	Chi square	Familiar (%) ($N = 297$)	Not familiar (%) ($N = 384$)	Chi square
Very important	81.4	61.8	11.746**	56.9	48.7	11.38**
Somewhat important	18.6	30.9		22.2	34.0	
Not very important	0	6.4	13.5	11.3		
Not at all important	0	0.9		7.4	6.0	

**Significant difference at the 0.05 level.

Table 2.18. Likelihood of purchasing GM foods.

	Ireland			US		
	Familiar (%) ($N = 86$)	Not familiar (%) ($N = 110$)	Chi square	Familiar (%) ($N = 297$)	Not familiar (%) ($N = 384$)	Chi square
Definitely	8.1	0	19.515**	12.2	2.6	36.30**
Probably	26.7	13.6		31.1	26.2	
Maybe	20.9	27.3		31.8	42.0	
Probably not	18.6	36.4		14.5	22.3	
Definitely not	25.6	22.7		10.5	6.8	

**Significant difference at the 0.05 level.

Likelihood of Purchasing GM Foods

Respondents in each country were examined based on two subgroups: those that are likely to purchase a GM food product compared with those that are not likely to purchase a GM food product. Respondents that are identified as *purchase GMO* indicated that they would definitely or probably purchase a food product that has been GM. Respondents that are identified as *not purchase* indicated that they would maybe, probably not, or definitely not purchase a food product that has been GM. The Irish respondent who is likely to purchase GM food is more highly educated. The US respondent who is likely to purchase GM food is more likely to be male, single, separated or divorced, and from a single income household (Table 2.19).

Readership of nutrition or ingredient labels when purchasing food products is not related to the probability that a respondent in Ireland or the US will purchase a GM food product (Tables 2.20 and 2.21). In Ireland past year purchasers of organic food are more likely to purchase GM food. However, in the USA past year purchasing of organic food is not related to the purchase probability of GM food (Table 2.22). Respondents in Ireland and the USA who are likely to purchase a GM food product are more likely to be familiar with GM food products than those that are not likely to purchase GM food products (Table 2.23).

Attitudes towards mandatory labelling of GM food are not related to the willingness to purchase a GM food product in Ireland. Mandatory labelling of GM food is less important to likely purchasers of GM food product in the USA. However, a majority, 73%, of respondents in the USA who are likely to purchase GM food indicated that mandatory labelling of GM food is important (Table 2.24).

Table 2.19. Demographics.

	Ireland			US		
	Purchase GMO (%)	Non-purchase (%)	Chi square	Purchase GMO (%)	Non-purchase (%)	Chi square
Sex	$N = 45$	$N = 152$	0.650	$N = 239$	$N = 437$	13.408**
Female	51.1	57.9		42.3	57.0	
Male	48.9	42.1		57.7	43.0	
Age	$N = 45$	$N = 151$	5.431	$N = 239$	$N = 439$	10.456*
Under 20 years	0	3.3		3.3	2.7	
20 to 24 years	13.3	19.2		16.7	12.1	
25 to 44 years	55.6	38.4		36.4	39.9	
45 to 54 years	20.0	27.2		19.2	22.6	
55 to 59 years	6.7	6.0		12.1	6.8	
60+ years	4.4	6.0		12.1	15.9	
Marital status	$N = 45$	$N = 152$	2.156	$N = 239$	$N = 440$	8.742*
Married	55.6	50.0		43.1	48.2	
Living with a partner	13.3	12.5		7.5	9.3	
Single	28.9	30.3		36.8	28.6	
Separated/divorced	2.2	3.3		10.0	8.2	
Widowed	0	3.9		2.5	5.7	
Education	$N = 45$	$N = 151$	11.780**	$N = 239$	$N = 440$	5.292
Grade school or less	11.1	9.3		0.4	0.7	
Some high school	26.7	36.4		1.3	1.4	
High school graduate	2.2	11.3		8.4	12.5	
Some college	15.6	15.9		38.1	33.6	
College graduate	24.4	21.2		36.4	39.8	
Postgraduate work	20.0	6.0		15.5	12.0	
Employment status	$N = 45$	$N = 152$	1.876	$N = 239$	$N = 439$	3.893
Employed, full time	62.2	57.9		67.8	60.1	
Employed, part time	24.4	19.7		14.2	18.0	
Not employed	13.3	22.4		18.0	21.9	
Dual income household	$N = 45$	$N = 152$	1.437	$N = 239$	$N = 440$	3.078*
Yes	55.6	45.4		44.8	51.8	
No	44.4	54.6		55.2	48.2	
Household income	$N = 40$	$N = 137$	4.534	$N = 236$	$N = 430$	2.795
Under US$20,000	20.0	21.9		13.1	12.1	
US$20,000 to US$29,999	17.5	27.7		14.8	15.1	
US$30,000 to US$39,999	30.0	24.1		17.4	17.7	
US$40,000 to US$54,999	17.5	9.5		16.9	15.8	
US$55,000 to US$69,999	5.0	9.5		11.4	15.8	
US$70,000 or more	10.0	7.3		26.3	23.5	

*Significant difference at the 0.10 level; **significant difference at the 0.05 level.

Table 2.20. Nutrition label readership.

	Ireland			US		
	Purchase GMO (%)	Non-purchase (%)	Chi square	Purchase GMO (%)	Non-purchase (%)	Chi square
	N = 45	*N* = 152		*N* = 239	*N* = 440	
Very often	35.6	25.7		43.1	50.2	
Somewhat often	26.7	28.9		33.1	29.5	
Not very often	28.9	28.9		16.7	13.4	
Not at all	8.9	16.4	2.629	7.1	6.8	3.465

Table 2.21. Ingredient label readership.

	Ireland			US		
	Purchase GMO (%)	Non-purchase (%)	Chi square	Purchase GMO (%)	Non-purchase (%)	Chi square
	N = 45	*N* = 152		*N* = 239	*N* = 440	
Very often	20.0	23.2		33.5	41.3	
Somewhat often	37.8	29.8		33.1	29.9	
Not very often	28.9	30.5		23.4	19.9	
Not at all	13.3	15.9	1.328	10.0	8.9	4.113

Table 2.22. Past year organic food purchase.

	Ireland			US		
	Purchase GMO (%)	Non-purchase (%)	Chi square	Purchase GMO (%)	Non-purchase (%)	Chi square
	N = 45	*N* = 151		*N* = 238	*N* = 437	
Yes	62.2	49.7		71.4	68.0	
No	37.8	50.3	2.191*	28.6	32.0	0.868

*Significant difference at the 0.10 level.

Table 2.23. Familiarity with GM foods.

	Ireland			US		
	Purchase GMO (%)	Non-purchase (%)	Chi square	Purchase GMO (%)	Non-purchase (%)	Chi square
	N = 45	*N* = 152		*N* = 238	*N* = 439	
Very familiar	4.4	15.3	14.4**	9.7	5.7	16.74**
Somewhat familiar	62.2	31.6		44.1	32.6	
Not very familiar	24.4	63.45		34.0	41.9	
Not at all familiar	8.9	19.8		12.2	19.8	

**Significant difference at the 0.05 level.

Table 2.24. Government imposition of mandatory labelling of GM foods.

	Ireland			US		
	Purchase GMO (%)	Non-purchase (%)	Chi square	Purchase GMO (%)	Non-purchase (%)	Chi square
	$N = 45$	$N = 151$		$N = 239$	$N = 438$	
Very important	68.9	71.1	3.730	41.4	58.2	25.5**
Somewhat important	26.7	25.0		31.4	27.6	
Not very important	2.2	3.9		15.9	10.0	
Not at all important	2.2	0		11.3	4.1	

**Significant at the 0.05 level.

Conclusions

This research shows that there is a similar level of familiarity with GM food in Ireland and the USA. Approximately 43% of respondents in both countries indicated that they were familiar with GM food. However, familiar Irish consumers are aware from more sources than the US consumer. The Irish consumers appear to be more interested in GM food because they were more likely to indicate that they had discussions with family, friends and colleagues concerning GM food. More highly educated respondents in both countries are more likely to be familiar with GM food.

Most consumers in both countries indicated that government imposition of mandatory labelling is important, 95% in Ireland and 81% in the US. The more familiar US and Irish consumers are with GM food, the more likely they are to indicate that mandatory labelling is very important. Further, almost three-quarters of US respondents who are likely to purchase a GM food product indicated that mandatory labelling is important. The European Union has imposed legislation requiring the mandatory labelling of food products containing GM ingredients. It appears that efforts in the US by legislators to require labelling are appropriate. In California, there was an attempt to qualify an initiative for the November 2000 ballot that would state that the people of California wish for labelling of genetically engineered food. At the federal level, a bipartisan group of 20 sponsors introduced a bill on 9 November 1999 to require a label on food containing GMOs.

A minority of consumers in each country said that they were likely to purchase a GM food product. However, the more familiar respondents in Ireland and the US are with GM food, the more likely they are to purchase a GM food product. Therefore, familiarity with GM food seems to increase positive attitudes towards it. Educational programmes concerning the process of producing GM food to increase consumers' familiarity are likely to generate more positive attitudes.

Familiar respondents in both countries described GM food as: 'grown in the US, improves the production of food', and 'is made by splicing genes from one plant or organism to another'. The three characteristics describing GM foods the least were: 'is the same as a hybrid, has superior taste', and is 'harmful to consumers'. There is a difference in attitudes between the Irish consumer and consumers in the US towards GM food. The familiar US respondents perceived GM food to have neutral or positive attributes. The Irish consumer attributed more negative attributes to GM food. Further, they are more likely to indicate that mandatory labelling is important and less likely to purchase a GM food product.

Attitudes were examined for differences over a three-month time period in both Ireland and the USA. Attitudes did not differ between the two time periods. Additional research will be conducted in October 2000 to determine if familiarity and attitudes concerning GM food have remained constant or changed over a 1-year period.

References

Central Statistics Office (1999) Home page, principal statistics. www.cso.ie/principalstat/pristat2.html

US Bureau of the Census (1991) *State and Metropolitan Data*, US Bureau of the Census.

Wolf, M., Amspacher, W. and Noyes, B. (2000) *A Marketing Research Study of San Luis Obispo / Five Cites Area Residents' Attitudes Regarding the Local Viticulture and Wine Industry.* California State University Agricultural Research Initiative.

3 Differences in Public Acceptance between Generic and Premium Branded GM Food Products: an Analytical Model

Annelies Verdurme, Xavier Gellynck, Jacques Viaene and Wim Verbeke
Faculty of Agriculture and Applied Biological Sciences, Department of Agricultural Economics, Division of Agro-Marketing, Ghent University, Ghent, Belgium

Problem Definition and Objective

Biotechnology is an enormously expanding discipline with various applications in the medical, agricultural and food sector. The use of genetic modification technology in food production is only one example, nevertheless a very important one. It offers *new perspectives* for product development, cost reduction and environmental protection. According to the White Paper, modern biotechnology includes an impulse for competitiveness and economic growth (European Commission, 1994). However, biotechnology in general and genetically modified (GM) food in particular came recently to the forefront of public attention. Abundant questions are being asked about safety for human health and for the environment, ethics, free consumer choice, socio-economic and legal issues.

Public support is especially low for biotechnology applications in the food sector. Europeans show more concern and resistance towards GM food compared with other parts of the world (Table 3.1). These consumer concerns seriously jeopardize the future market success of modern biotechnology products, including GM food products (EFB Task Group on Public Perceptions of Biotechnology, 1998, 1999). Consequently, gaining insight into consumer beliefs, attitudes and behavioural intentions concerning GM food is essential.[1] Against this background, the objective of this chapter is twofold:

Table 3.1. Willingness to eat GM food (%). Example: GM fruit with improved taste. (From International Food Information Council, 1999; Einseidel, 2000; INRA, 2000; Shimbun, 2000.)

Willingness to eat GM food (%)			
Europe	25	Japan	38
Canada	34	USA	54

- explore consumer attitudes and behavioural intentions towards GM food as well as their determinants;
- verify whether or not indications of differences in determinants of consumer attitudes and behavioural intentions exist between generic and branded GM food products.

The chapter starts with a description of the research methodology. Next, findings in literature are confronted using results of qualitative market research. The chapter focuses on determinants of both consumer attitudes and behavioural intentions towards GM food. Hereby, a distinction is made between

[1] Research financed by Flanders Interuniversity Institute for Biotechnology (VIB) in the frame of the Technology Assessment Programme.

generic and premium branded GM food products. Throughout the chapter, food products without a brand name or with a private label are considered as generic food products, while food products of well-known top brands are considered as premium branded food products. Finally, some conclusions and topics for further research are formulated.

Research Methodology

Theoretical framework

The formation of beliefs, attitudes and behaviours may occur directly or indirectly (Mowen, 1993). In the case of direct formation, a belief, attitude or behaviour is created independent of each other. In the case of indirect formation, belief, attitudes and behaviours build on each other to create hierarchies, that is hierarchies of effects. Depending on the type of purchase process, four hierarchies or orders in which beliefs, attitudes and behaviours occur, can be distinguished (Table 3.2). The various hierarchies of effects are idealized representations of consumer buying behaviour. No matter what the purchase process, consumers are likely to have vague beliefs and attitudes about a product before buying it. However, these different orders in the sequence of consumer behaviour provide a feel for the relative emphasis of beliefs, attitudes and behaviour within the different purchase processes.

Since to date only a few GM food products have been available on the food market, the chapter focuses on behavioural intentions rather than on effective behaviour. *Behavioural intention* is defined as 'the determination of a consumer to engage in some act, such as purchasing a product', a GM food product in this research (Mowen, 1993).

Data collection

Based on a study of literature, both consumer attitudes and behavioural intentions towards GM food are identified as well as the determinants. Therefore, several international scientific journals, reports, books, some web sites and a few popular magazines and newspapers were consulted.

To explore what determines consumer attitudes and behavioural intentions towards GM food in Belgium and, more specifically, in the Flemish region, *qualitative market research* was conducted during the period September 1999 – February 2000. It comprised focus group discussions with consumers and in-depth interviews with experts, such as representatives from public authorities, political parties, scientific world, industries, press, consumer and environmental organizations. In total

Table 3.2. Purchase processes and their possible hierarchies of effects.

Purchase process	Hierarchy of effects	Description
High involvement	Standard learning hierarchy[a]:	1. Investigating the product to learn about its attributes.
	Beliefs → affect → behaviour	2. Using this information to form attitudes towards it.
		3. If attitudes are positive, the product is purchased.
Low involvement	Low-involvement hierarchy[b]:	1. Superficial investigation of the product.
	Beliefs → behaviour → affect	2. Purchase of the product.
		3. Formation of attitudes towards it.
Experiential	Experiential hierarchy[c]:	1. Strong affective response to the product.
	Affect → behaviour → beliefs	2. Product purchase.
		3. Development of beliefs to justify the buying act.
Behavioural influence	Behavioural influence hierarchy:	1. Strong situational factors give rise to product purchase.
	Behaviour → beliefs → affect	2. Attitudes or beliefs form after the buying act.
	Behaviour → affect → beliefs	

[a] Ray, 1973; [b] Krugman, 1961; Lavidge and Steiner, 1961; Olshavsky and Granbois, 1979; [c] De Bruicker, 1979.

Table 3.3. GM food applications discussed in the focus groups.

Generic GM food products
environmentally friendly pig meat
broccoli preventive against cancer
bread made from wheat able to grow under dry weather conditions
Premium branded GM food products
chips made from potatoes resistant to a fungus
candy bar made from soybeans resistant to the herbicide Roundup
coffee made from plants able to grow under dry weather conditions

six focus group discussions were organized. Each focus group consisted of six to nine participants or consumers. The following socio-demographic criteria were taken into consideration when selecting participants:

- Sex: all participants are women responsible for food purchases within their family.
- Age: between 25 and 50-years-old.
- Profession: groups are composed of both housewives and women with a part-time or a full-time job outside the home.

A topic list served as a guide for the discussions. Each focus group discussion took about 3 hours and was recorded on videotape to make analysis afterwards possible. The topics were presented to the respondents according to the 'funnel technique', where the interview starts with a discussion about the subject in a very broad context (food in general) and as the discussion proceeds the general frame is gradually narrowed (GM technology and GM food) down until the core subject (GM food products) is reached. However, in half of the focus groups, the GM technology was discussed after presenting six concrete GM food products to the respondents (Table 3.3). Three of these food products were generic and three premium branded food products.

This approach enables verification of whether or not indications exist that attitude and behavioural intentions differ between the technology as such and its derived products. By making the distinction between generic and premium branded GM food products, the identification of indications about differences in consumer attitudes and behavioural intentions between these two types of food products becomes possible.

In the frame of the in-depth interviews, 18 experts were interviewed. These experts were chosen among well-known representatives of involved parties and authors of articles recently published in journals, proceedings, the Internet, magazines and newspapers. Each in-depth interview took about 2 hours during which major topics regarding GM and GM food were discussed: consumer attitudes and behavioural intentions, risks and benefits, legislation and control, communication strategies and future development.

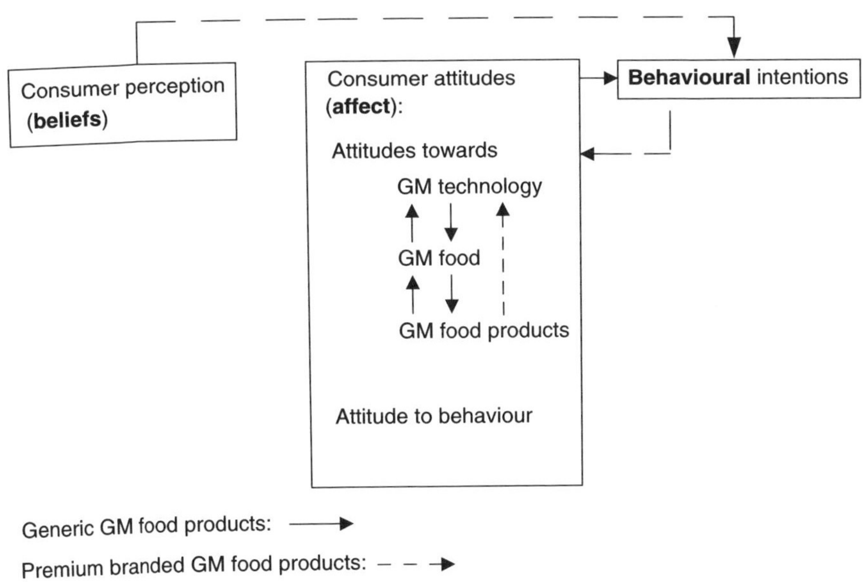

Fig. 3.1. Analytical model.

Results

Based on the results from literature study, focus group discussions with consumers and in-depth interviews with experts, an *analytical model* that describes consumer attitudes and behavioural intentions towards GM food as well as the determinants, was developed (Fig. 3.1). In the model, a distinction is made between generic and premium branded GM food products.

Consumer attitudes towards GM

By evaluating consumer attitudes towards GM, three levels can be distinguished (Fig. 3.2). The levels range from the rather abstract and distant 'technology' level, to the 'application' level and the final and more concrete 'product' level. Objections are not to GM as a *technology*, but focus rather on the field and the organism involved (Frewer *et al.*, 1997). The application generating the most objections will determine the attitude towards the technology.

At the *application* level, a distinction is made between two aspects, namely the field and organism involved. Public support for applications in the medical field such as the detection of hereditary diseases is high while extremely low for food applications (INRA, 2000). This difference in public support according to the application field was also illustrated during the focus group discussions and can be explained as follows:

- When people are ill and risk death, they want to recover no matter how, even through GM medicine. However, according to the focus group discussions, some consumers fear that genetic modification will give rise to new diseases instead of curing the existing ones.
- Until the 1980s, food was regarded to be essential for human growth and strength in most parts of Western Europe. Market saturation, efforts to sensitize the public regarding diet-related diseases (e.g. cancer) as well as food scandals such as hormones, BSE and the dioxin crisis changed this view and food became more associated with human health and shape. Nowadays, some food products such as meat are even regarded as health threatening.
- Because of food scandals such as bovine spongiform encephalopathy (BSE) and the dioxin crisis, people are much more aware about the link between health on the one hand and the quality and quantity of food consumed on the other hand. Food also has an important socio-cultural function (e.g. traditional birthday cake, local food products). Consequently, food produced according to conventional methods is preferred

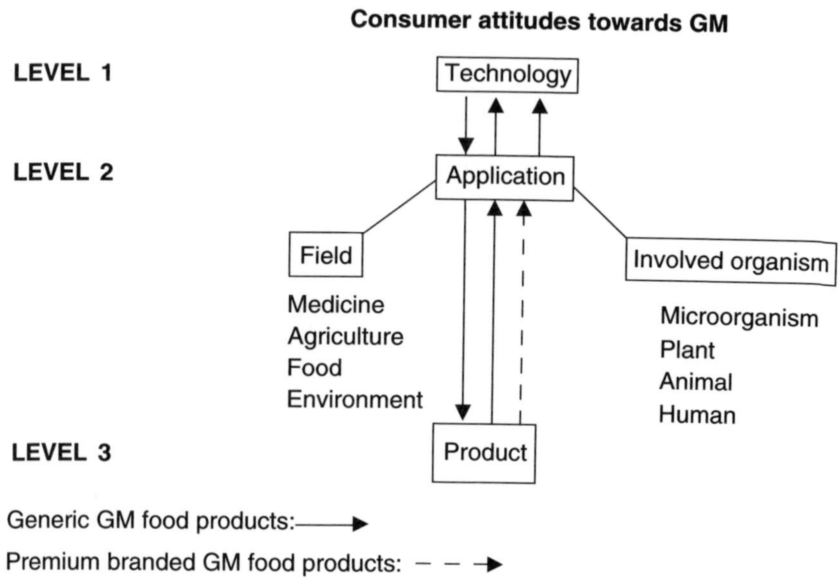

Fig. 3.2. Three levels of consumer attitudes towards GM.

above GM food products, which are perceived as artificial and a potential danger to traditional values.

Just as in the case of the application field, differences in public support exist according to the organism involved. GM in which microorganisms or plants are involved, provokes less disapproval compared with animals and certainly humans (INRA, 2000). In contrast to plants and microorganisms, animals are considered to have feelings and to be more closely related to humans. It makes the step towards GM humans smaller. Consumers express the fear that the technology will be abused, especially in relation to humans (e.g. parents choosing looks and sex of children).

Both findings in literature (Bredahl and Grunert, 1998; Bredahl, 2000) and results from the focus group discussions indicate that consumers reject concrete GM food *products* based on negative attitudes towards the technology and its application in food production. These attitudes are highly emotional, describing the technology as unnatural, evil and uncontrollable. As a result, respondents merely concentrate on the perceived risks and not on the benefits when discussing the generic GM food products. While this relation holds for the focus groups where the discussion started with the technology, a case-by-case evaluation takes place in the groups where the discussion started with concrete examples. Hereby, benefits are confronted with risks and the perceived balance between these two determines consumers' attitude towards the product. In this way, indications exist that the more emotional attitudes are replaced by a more rational approach in the case of concrete examples. Therefore, communication regarding GM food should focus on concrete food products instead of on the technology and GM food in general.

Major determinants of consumer attitudes towards generic GM food products

Consumer attitudes towards generic GM food products (affect) are determined by the perception of risks and benefits regarding GM (beliefs), which in turn is based on general attitudes and knowledge (Viaene *et al.*, 1999a; Bredahl, 2000) (Fig. 3.3). Perceived consumer benefits may partly compensate for believed risks (Frewer *et al.*, 1999). Therefore, not only are personal effects taken into consideration but also those for loved ones, future generations and the environment.

Risks consist of hazard and outrage. Hazard is the actual risk and answers the question 'What may cause damage to whom and how much?'. Outrage represents loathing, panic and perceptions of evil, injustice and exploitation. It corresponds with ethical objections (European Commission, 1998a). If the public is outraged, a greater risk is perceived, even if the hazard does not reflect a real danger

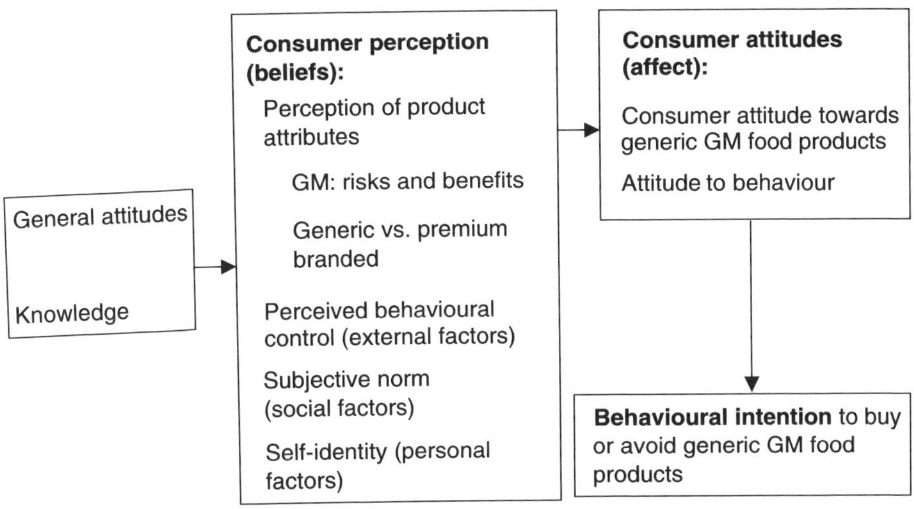

Fig. 3.3. Determinants of consumer attitudes and behavioural intentions towards generic GM food products.

(European Commission, 1998b). Consumer concerns focus on potential hazards for human health (own, others) and for the environment. Ethical objections relate to issues of animal welfare, a dislike of tampering with nature/food and societal injustice such as problems in the developing world (Frewer, 1998a; Aubrée *et al.,* 1999; Nuffield Council on Bioethics, 1999). During the discussions, consumers declared that GM food is unhealthy and worked up. Future generations will lose the notion of the natural origin of food. They even compared the use of GM technology in the production of food with the use of hormones, which are both considered as crossing ethically acceptable limits.

Benefits comprise both producer and consumer benefits. At the consumer level, personal (e.g. improved taste), environmental (e.g. less pesticide use) and societal (e.g. solving world hunger) benefits are distinguished. Products with consumer benefits are more readily accepted than products that merely benefit the producer (European Commission, 1998b). However, the current GM products offer few consumer benefits (Isenterant, 1999). The focus group discussions made clear that most consumers are aware of this fact and, as a result, renounce GM food. However, even when the use of GM technology in food production offers a solution to major environmental or societal problems, alternative solutions such as organic farming or a redistribution of purchasing power and resources are preferred over the use of GM. It has to be stressed that the research focuses on a saturated food market where consumers can afford such an attitude.

General attitudes comprise several major attitude domains such as general attitudes towards nature/the environment, technology/science, food, health and trust in public authorities and in industries (Frewer *et al.*, 1997; Cantley and Miyamura, 1999; Bredahl, 2000). The *Eurobarometer* demonstrates that enthusiasm about modern biotechnology (including GM) is low, although the majority of Europeans are not technophobes (INRA, 2000). This study also reveals a strong decline in public trust compared with 1996. The focus group discussions confirm these results. Recent food scandals (e.g. BSE, dioxin) have severely damaged consumer trust in public authorities and industries (Viaene *et al.*, 1999b). More transparency in the risk assessment procedure and increased public participation in the decision making process regarding the further development of GM (food) could restore this trust which is considered to be essential for consumer acceptance of GM (food) (Frewer, 1998b; Frewer *et al.*, 1999).

Through its impact on the perception of risks and benefits, *knowledge* about GM also determines consumer attitudes towards generic GM food products. In 2000, Europeans' awareness and knowledge of GM was still low, despite the widespread controversy and increasing press coverage since the previous survey in 1996 (INRA, 2000). A knowledge deficit arouses consumer uncertainty and leads eventually to the overall rejection of GM (Aubrée *et al.*, 1999; Scholderer and Balderjahn, 1999). Consequently, increasing the knowledge of GM by the provision of information may reduce consumers' reluctance. However, several research studies reveal that consumer knowledge about GM is not necessarily positively correlated with consumer acceptance (International Food Information Council, 1999; Nuffield Council on Bioethics, 1999; Viaene *et al.*, 1999b). In contrast, providing information tends to intensify prior attitudes rather than change them (Frewer *et al.*, 1999). Moreover, consumer attitudes on the 'technology' and 'application' level are characterized by a great deal of emotion and little reason, which make these attitudes difficult to alter through pure knowledge-oriented and rational information.

Determinants of behavioural intentions towards generic GM food products

Behavioural intentions to either buy or avoid generic GM food products are determined by consumers' *attitude to behaviour* (affect) (Fig. 3.3). Attitude to behaviour is the attitude a consumer holds towards performing a certain behaviour (Ajzen and Fishbein, 1980), in this case, buying generic GM food products (Bredahl *et al.*, 1998). It is influenced by consumer attitudes towards generic GM food products (affect) and by social (subjective norm), personal (self-identity) and external (perceived behavioural control) factors (beliefs).

The *subjective norm* deals with consumers' motivation to perform the behaviour of which important others (e.g. loved ones) are believed to approve (Mowen, 1993). In this research, the behaviour corresponds with purchasing or avoiding generic GM food products. *Self-identity* is the way an individual regards him or herself. Consumers tend to affirm and bolster the self-image through specific buying intentions, including those towards generic

GM food products (Sparks and Shepherd, 1992). Ajzen (1985) introduced the term 'perceived behavioural control' in the 'Theory of Planned Behaviour'. Applied to generic GM food products, this term covers the effect of various external factors such as time, availability and recognition (labelling), which consumers believe to influence the degree of personal choice to buy or avoid generic GM food products (Sparks *et al.*, 1995; Bredahl *et al.*, 1998). A free food choice (or at least the perception of it) is essential for consumer acceptance of generic GM food products (Robinson, 1997; Viaene *et al.*, 1999b). In the absence of a free food choice, consumers feel coerced into acceptance and, as a result, consumers adopt a reluctant attitude towards the products. The actual labelling policy does not increase consumer perception of being able to decide freely since only food products containing more than 1% of GM ingredients at the moment of purchase must carry a label.

Determinants of consumer attitudes and behavioural intentions towards premium branded GM food products

The determinants of consumer attitudes and behavioural intentions towards premium branded GM foods are supposed to differ considerably from those of generic GM food products. When discussing the premium branded food products in the focus groups, consumers became more tolerant towards GM at all three levels. Premium brands are believed to stand for *high quality* and *tradition* (traditional values). The premium brand is considered as being fully responsible for the safety and soundness of the food products, not the consumer or the one preparing the food as in the case of generic products. Therefore, it is suggested that in the case of premium branded food products, the consumer perception of premium brands (beliefs) determine behavioural intentions towards premium branded GM food products (Fig. 3.1). Based on positive beliefs about a premium brand, consumers (intend to) buy (GM) food products of the premium brand first and adapt their attitudes towards GM (affect) afterwards in order to justify their behaviour(al intentions). Therefore, a strong brand may render the modified nature of the food product less or even irrelevant for consumers (brand loyalty).

However, the inverse hypothesis that the unfavourable and negative atmosphere surrounding GM (food) is too strong to be influenced by a premium brand's positive image is still to be tested. In that case, a negative attitude towards GM will contribute to the behaviour(al intention) to avoid all food products of a premium brand associated with GM.

It must be noted that in the analysis of the focus group discussions, the differences in consumer acceptance between generic and premium branded GM food products are linked to the consumer perception of premium brands. However, all generic food products discussed were fresh products, while all the premium branded ones were processed. Therefore, the differences in consumer acceptance could also be linked to the fact that fresh food products are perceived as a bigger threat to human health than processed food products because no 'dilution effects' as a result of the processing occur.

Conclusions and Further Research

Based on literature study, focus group discussions with consumers and in-depth interviews with representatives of various parties involved with the issue of public acceptance of GM, determinants of consumer attitudes and behavioural intentions towards GM food were investigated. The results show a clear difference between generic and premium branded GM food products.

In general, three levels of consumer attitudes towards GM can be distinguished: the 'technology' level, the 'application' level and the 'product' level. Attitudes towards the technology are negative and highly emotional. Its application in food production receives little consumer support, especially where animals are involved.

With regard to *generic* GM food products, consumer attitudes (affect) are determined by the perception of risks and benefits (beliefs), which in turn is based on general attitudes and knowledge. Risks comprise potential hazards for human health (own, others) and for the environment on the one hand and ethical concerns related to animal welfare, nature and the developing world on the other hand. Benefits consist of consumer (personal, environmental or societal) and producer benefits. Personal consumer benefits are valued the highest and producer benefits the lowest. General attitudes such as trust in public authorities and industries appear to play a major role in consumer attitudes. Consumers' knowledge about GM is low. The precise relation

Table 3.4. Experimental design of the survey.

Order	The technology first, GM food products next	GM food products first, the technology next	Total
First premium branded food products	250	250	500
First generic food products	250	250	500
Total	500	500	1000

between knowledge and attitude remains unclear. However, results suggest that providing pure knowledge-oriented rational information about GM will not alter prior negative attitudes towards it. Behavioural intentions towards generic GM food products are determined by the attitude to behaviour (buying or avoiding GM food products) (affect) which in turn is influenced by consumer attitudes towards generic GM food products (affect) and by social (subjective norm), personal (self-identity) and external (perceived behavioural control) factors (beliefs).

With regard to *premium branded* GM food products, behavioural intentions are determined by the consumer perception of premium brands (beliefs). Consumers (intend to) buy (GM) food products of a specific premium brand first and adapt their attitudes towards GM (affect) afterwards in order to justify their behaviour(al intentions). As a result, a premium brand may render the modified nature of the food less or even irrelevant for consumers (brand loyalty).

Based on the findings in literature and on the results of the focus group discussions, the following hypotheses can be developed:

- Consumer attitudes towards GM technology are much more negative than towards concrete GM food products;
- Among GM food products, premium branded products are more readily accepted than generic ones.

In order to verify these hypotheses, a quantitative market research study was organized in July 2000. In the research, 1000 consumers have been interviewed on the basis of a structured questionnaire and according to an experimental design (Table 3.4). The experiment consists of discussing the technology before or after concrete GM food products. Moreover, a distinction is made between generic and premium branded GM food products.

References

Ajzen, I. (1985) From intentions to actions: theory of planned behaviour. In: Kuhl, J. and Beckmann, J. (eds) *Action Control – From Cognition to Behaviour.* Berlin: Springer-Verlag, pp. 11–39.

Ajzen, I. and Fishbein, M. (1980) *Understanding Attitudes and Predicting Social Behaviour.* Prentice-Hall, Englewood Cliffs, New Jersey.

Aubrée, H., Diard, S., Garcia, S., Jaegle, S., Lunel, C., Mauvielle, V., Messager, O. and Rade, V. (1999) *Organismes génétiquement modifiés et filière agro-alimentaire.* Etude de filière, 4ème année, 6ème promotion, INSFA, Remmes, France.

Bredahl, L. (2000) *Determinants of Consumer Attitudes and Purchase Intentions with Regard to Genetically Modified Foods – Results of a Cross-national Survey,* working paper No. 69. MAPP, Aarhus, Denmark.

Bredahl, L. and Grunert, G. (1998) *Consumers' Cognitions with regard to Genetically Modified Foods – Results of Qualitative Study in Four Countries,* working paper No. 59. MAPP, Aarhus, Denmark.

Bredahl, L., Grunert, G. and Frewer, L.J. (1998) *Consumers Attitudes and Decision-making with regard to Genetically Engineered Food Products – a Review of the Literature and a Presentation of Models for Future Research,* working paper No. 52. MAPP, Aarhus, Denmark.

Cantley, M. and Miyamura, Y. (1999) Alimentation, technologie, réglementation et confiance des consommateurs. *l'Observateur de l'OCDE,* 216, 21–25.

De Bruicker, F.S. (1979) An appraisal of low-involvement consumer information processing. In: Maloney, J. and Silverman, B. (eds) *Attitude Research Plays for High Stakes.* American Marketing Association, pp. 112–130.

EFB Task Group on Public Perceptions of Biotechnology (1998) *Biotechnology in the Public Sphere. A European Sourcebook.* Science Museum, London.

EFB Task Group on Public Perceptions of Biotechnology (1999) Ethical aspects of agricultural biotechnology. Working Document for the European Conference, 16–19 May 1999, Oviedo, Spain.

Einseidel, E. (2000) Biotechnology and the Canadian public: 1997 and 2000. Graduate Program in Communications, University of Calgary, Calgary, Canada.

European Commission (1994) *Growth, Competitiveness, Employment: the Challenges and Ways Forward into the 21st Century.* White Paper, Office of Official Publications of the European Communities, Luxembourg.

European Commission, DG XII (1998a) *Looking at the Biotechnology Consumer.* Office of Official Publications of the European Communities, Luxembourg.

European Commission, DG XII (1998b) *Strategies for Communication on Food Biotechnology.* Office of Official Publications of the European Communities, Luxembourg.

Frewer, L.J. (1998a) Consumer opinion on transgenic animals. *AgBiotech News and Information* 10, 213–216.

Frewer, L.J. (1998b) Perception of risks and benefits and consumer decision making regarding genetically modified food. In: *Proceedings of AIR-CAT 5th Plenary Meeting: Effective Communication and GMO-foods.* Vienna, Austria, 11–14 June 1998, pp. 15–17.

Frewer, L.J., Hedderley, C., Howard, C. and Shepherd, R. (1997) 'Objection' mapping in determining group and individual concerns regarding genetic engineering: risk, benefit and ethics. *Agriculture and Human Values* 14, 67–79.

Frewer, L.J., Howard, C., Hedderley, D. and Shepherd, R. (1999) Reactions to information about genetic engineering: impact of source characteristics, perceived personal relevance and persuasiveness. *Public Understanding of Science* 8, 35–50.

INRA (Europe) (2000) *Eurobarometer 52.1 – the Europeans and Modern Biotechnology.* European Co-ordination Office, Brussels.

International Food Information Council (1999) U.S. consumer attitudes towards food biotechnology. Reports from the Wirthlin Group Quorum Surveys in March 1997, February 1999 and October 1999.

Isenterant, D. (1999) Genetische modificatie in voeding: wat en waarom? In: *Proceedings of the Symposium 'Genetische Modificatie en de Agro-voedingssector'* Flanders Expo, Ghent, Belgium, 19–23 April 1999, pp. 6–14.

Krugman, H. (1961) The impact of television advertising: learning without involvement. *Public Opinion Quarterly* 29, 59–62.

Lavidge, R. and Steiner, G. (1961) A model for predictive measurements of advertising effectiveness. *Journal of Marketing* 25, 59–62.

Mowen, J.C. (1993) *Consumer Behaviour.* MacMillan Publishing, New York.

Nuffield Council on Bioethics (1999) *Genetically Modified Crops: the Ethical and Social Issues,* fifth report published in May 1999.

Olshavsky, R.W. and Granbois, D.H. (1979) Consumer decision making – fact or fiction? *Journal of Consumer Research* 6, 93–100.

Ray, M. (1973) Marketing communications and the hierarchy-of-effects. In: Clarke, P. (ed.) *New Models for Mass Communications.* Sage Publications, Beverly Hills, California, pp. 147–176.

Robinson, C. (1997) Genetically modified foods and consumer choice. *Trends in Food Science & Technology* 8, 84–88.

Scholderer, J. and Balderjahn, I. (1999) Consumer information strategies for genetically modified food products. *Proceedings of 28th EMAC Conference Marketing and Competition in the Information Age.* European Marketing Academy, Brussels.

Shimbun, Y. (2000) Poll: More than 60% reluctant to eat GM foods. *The Daily Yomiuri,* 12 August 2000.

Sparks, P. and Shepherd, R. (1992) Self-identity and the theory of planned behaviour. *Social Psychology Quarterly* 22, 388–399.

Sparks, P., Shepherd, R. and Frewer, L.J. (1995) Assessing and structuring attitudes towards the use of gene technology in food production: the role of perceived ethical obligation. *Basic and Applied Social Psychology* 16, 267–285.

Viaene, J., Gellynck, X., Verbeke, W. and Verdurme, A. (1999a) Consumer attitudes towards genetic engineered food and development of a segmented communication policy. In: *Scientific Report VIB 1999,* pp. 197–200.

Viaene, J., Gellynck, X., Verbeke, W. and Verdurme, A. (1999b) Determinants of consumer attitudes towards GM food: similarities and differences between empirical findings in Flanders and abroad. In: *Verhandelingen van de FLTBW, naar Aanleiding van de 48ste Post-Universitaire Onderwijsdag,* no. 39. Faculty of Agricultural and Applied Biological Sciences, Ghent University, Ghent, Belgium, 1 December 1999, pp. 57–72.

4 Is European Consumers' Refusal of GM Food a Serious Obstacle or a Transient Fashion?

Claus-Hennig Hanf[1] and Andreas Böcker[2]
[1]*Department of Agricultural Economics, Christian-Albrechts-University, Kiel, Germany;* [2]*Department of Agricultural Economics and Market Research, Justus-Liebig-University, Giessen, Germany*

Introduction

In recent years, several genetically modified (GM) crop varieties have been introduced. In less than 5 years, almost half of the US acreage of cotton, soybeans and maize has been planted with GM varieties. In many other countries, the adoption of GM products also started well (Franks, 1999: 566). In the seeding season 2000, however, the adoption seems to have been interrupted, and even in the US hesitation has been observed.

This stagnation has its origin in the extensive rejection of GM food by consumers in European countries and in Japan. The EU responded to the massive protests of lobbying groups by introducing a compulsory labelling system for all GM food. The introduction of the labelling obligation caused an almost complete stop of imports of GM food to Europe and became a serious import barrier. There are two reasons for this:

- Many of the exporting countries do not possess respective labelling systems, at least not obligatory ones, so that no separate trading channels for GM products are developed. In consequence, the exporting firms cannot give a warranty that no GM food is mixed in and therefore the firms are forced to label more or less all products as 'GM' food.
- The vast majority of European consumers are not willing to accept GM food even if the modified products are somewhat cheaper.

US merchants and producers have not taken very seriously these reservations of European and especially Northern European consumers. They rather relied on their persuasive power and on the fact that European consumers have always eventually followed the development in North America with a more or less distinct time lag (Connor, 1994). However, the confidence in an only brief delay of adoption seems to have vanished in the past months. Recent investigations of consumers' behaviour came to the conclusion that the resistance against GM food has even increased whereas an increasing acceptance of the 'white gene technology' in the medical domain has been observed. In this context, two questions arise:

1. Is the rejection of GM food by the European consumer a serious peril for agricultural genetic engineering, or is it only a fashion that will vanish after a short time?
2. What can be done to increase the likelihood of GM food being broadly accepted as fast as possible?

In the following, we will address the above-mentioned questions. At first, we describe a simple model of consumer behaviour that can be employed to capture the relevant elements of food purchase decisions. This model and some empirical evidence is then used to approach the question of persistence and duration of the non-acceptance period. Finally, the conditions and possibilities of breaching the buying 'boycott' are discussed.

Consumer Behaviour

Occasionally, the European consumer is characterized as extremely anxious, emotional or even irrational. Although the widely held opinion that the consumer trusts only advocacy groups like Greenpeace, and deeply mistrusts governments and scientists, is exaggerated to some degree, this picture is not completely wrong. For example, Frewer *et al.* (1996: 479 ff.) point out that both government and business representatives are only moderately trusted, if their statements are perceived to be made in obvious self-interest. Here, we treat consumers' behaviour in the case of GM food as rational decision making where rationality is defined considering that information is costly and information processing is effort and time consuming. Hence, it is assumed that consumers act rationally in the sense of bounded rationality.

Despite the intense and ongoing public debate on genetic engineering in food and agriculture, most consumers probably still feel inadequately informed about this topic. For example, Weiss (1997: 427) reports that 84% of 700 adults who participated in a consumer survey in Vienna felt that way. The same study revealed that consumers learn about GM food mainly from television (73%) and newspapers and magazines (41%), while more sophisticated sources such as books (5%) or lectures (4%) play only a minor role. Television and newspaper reports, however, tend to have been rather extreme, that is, mainly negative, and biased in that very seldom are both risks and benefits reported. Owing to this bias, the complexity of the topic and the general lack of adequate information at the individual level, it is likely that consumers tend to perceive the additional risk to be rather high. This belief is closely connected with the experience that new and 'revolutionary' technologies (e.g. the pesticide DDT) often caused unexpected negative outcomes, although the majority of scientists and politicians had stated this would not be. The prevailing assumption of European consumers that GM food is connected with the possibility of hazard is a logical consequence of the widely distributed technological pessimism. Based on a cross-national survey carried out in Denmark, Germany, the UK and Italy, Bredahl (2000) found that '(a)cross countries, attitudes towards genetic modification in food production were deeply embedded in more general attitudes held by consumers, in particular attitude towards nature and attitude towards technology'.

The negative attitude towards GM food means no absolute rejection, but a serious hindrance for adoption. A very simple model of adoption can be described as follows. First, assume that GM technology is targeted at cost reduction only so that there is no difference in direct consumer benefits between traditional and GM food, except that GM food will be less expensive. Consumers will then choose the GM variant, if the following holds:

$$PR_{\text{trad}} - PR_{\text{GM}} > P(GM) \times D_{\text{GM}} \qquad (4.1)$$

That is, for GM food to be bought, the price difference between traditional (PR_{trad}) and GM foods (PR_{GM}) must exceed the perceived risk of GM food consumption. In this sense, the price difference may serve as a premium for taking a risk (Dnes, 1996), which is assumed to be individually calculated by multiplying the expected damage of that consumption activity (D_{GM}) with the individually perceived hazard probability $P(GM)$.

Now assume that the consumer expects a different utility or benefit from GM food (U_{GM}) than from the traditional variant (U_{trad}). Since overall utility maximization is equivalent to maximizing utility per money unit spent, the decision criterion for buying GM food can be generalized to:

$$\frac{U_{\text{GM}}}{PR_{\text{GM}}} - \frac{U_{\text{trad}}}{PR_{\text{trad}}} > P(GM) \times \frac{D_{\text{GM}}}{PR_{\text{GM}}} \qquad (4.2)$$

From Equation 4.2 three possibilities for overcoming European consumers' scepticism can be inferred:

- Reduce the subjectively perceived high damage expectation, through reducing $P(GM)$, D_{GM} or both.
- Lower prices PR_{GM}, which means to transfer cost reductions from primary producers to consumers.
- Increase perceived benefits of GM foods by emphasizing improvements in existing or development of new product attributes.

We will argue in the following that it is rather difficult to bring GM food into a competitive position in Europe by product development even if GM food possesses advantages over non-GM food. The main reason is the obligatory labelling.

Expected damage from GM food

The perceived probability of occurrence of a danger from GM food is probably very small but much overestimated. There are several reasons for this assumption. The literature frequently refers to the fact that consumers' perception of food risks deviates considerably from the judgement of experts (Wiegand and von Braun, 1994). Furthermore, Lichtenstein *et al.* (1978) show that there exists a general tendency to overestimate small risks, while high risks are often underestimated. Wiegand and von Braun (1994) adduce further reasons for an overestimation of small probabilities; for example, if the risk is taken involuntarily, or if the effects of the risk factor are widely unknown (Lowrance, 1976: 87). Both these aspects are typical for GM food. Consequently, one should expect the perceived subjective probability, although still very small, to be many times larger than the average expert judgement.

Furthermore, a number of psychometric studies have shown that food safety related risk perception is dominated by the perceived severity of a hazard, thus generally reducing the impact of probability information. For example, Slovic *et al.* (1980), Sparks and Shepherd (1994) and Fife-Shaw and Rowe (1996) found that the most important component of the perception process was related to the severity of a hazard. The second most important component or factor related to familiarity and awareness of a hazard. Characteristics related to the probability or number of people being exposed, however, were loaded on the factor that was only third most important.

The strong role of hazard severity in individual risk perception is strengthened further by the fact that food shopping is usually done by one person for the entire household. A potential hazard is thus not restricted to the buyer him or herself, but is instead multiplied. Herrmann *et al.* (1997: 518), for example, found that the presence of young children in the household led to reduced apple consumption during the Alar[1] crisis in the late 1980s. Closely related is the finding of Fife-Shaw and Rowe (1996), that the perception of hazard severity also includes concern for others, especially vulnerable groups, but also future generations. This element of hazard severity must be taken very seriously in the case of GM food, as consumers have learned by the mass media that the environmental consequences of a 'bad case' – if it happens – may be catastrophic and long lasting, and may therefore be very expensive to resolve. Although industry has reacted to such environmental issues by increasing the precision of DNA alteration and developing measures to reduce the risk of cross pollination (Franks, 1999: 575), these efforts and the corresponding progress are difficult to communicate to laypersons.

Altogether, owing to the overestimation bias of both hazard severity and probability, the perceived risk of GM crops is probably much too large at present to be easily overcome by marketing measures within a few years. But within the EU, considerable differences between single countries can be observed. For example, an international consumer survey by Bredahl (2000) showed that in Denmark and Germany about one-third of the sample had a strong intention to avoid purchasing GM food. The British consumers were slightly more willing to purchase GM products (yogurt and beer) and the Italian generally least hesitant. We can only speculate about the reasons for these national differences in the attitude towards GM food. In several studies it has been reported that Italians are less pessimistic about the consequences of technical development on the environment. This general attitude may have affected the assessment of expected damage and/or perceived probability in such a way that a less pronounced rejection resulted.

Competition by prices

A study about the willingness to pay for GM and traditional food conducted at the University of Kiel (Gath, 1998) came to the result that the price difference between these two types of food must be considerable, before a recognizable market share can be expected for GM food. About one quarter of consumers expressed that they would not buy GM food, even if it cost less than half the price of ordinary food. As only very few test persons asserted that they would buy GM food without any hesitation,

[1] Alar is a pesticide brand that was widely applied in apple production, until it was shown to cause severe health problems.

the majority proved to be price elastic. This majority of consumers, however, would only buy the GM variant if it was about 25–50% cheaper. The results, which are based on contingent valuation, were almost the same for cheese, yogurt, vegetables and fruit juice, and are broadly in accordance with the findings of the international study reported by Bredahl *et al.* (1998) and Bredahl (2000). For Germany and Denmark, this study revealed that every third (fourth) person declared that he or she would try under all conditions to avoid buying GM yogurt (beer). For both products and both countries, the share of participants indicating an extreme purchase intention was also very low, between 1 and 3%. Thus, again, the majority of consumers can be expected to react to some price and quality change, so in the next step, the necessary magnitude of that change will now be discussed.

Luehrs (1987) carried out an investigation of the local price distribution and price variation in a medium sized city (300,000 inhabitants) in Germany. She investigated about 25 branded food products and 12 generic products in 11 retail chains over a period of 30 weeks. Luehrs reports an average weekly price difference of about 40% for branded products and of about 60% for generic products. Out of the total of 37 prices the share of those that had changed from one observation to the other varied between 2.2 and 7.0%. While this seems to suggest rather stable prices over time, a more detailed analysis showed considerable differences with respect to retail chains. For 5 of the 11 retail chains, the size of the price changes were above 20%, and four of these belonged to those retail chains which changed prices relatively often. These investigation results can be interpreted in the sense that consumers are used to price differences and price variations in the range of 20 or 25%. From this, we infer that a price change, in this case a reduction, must exceed this range in order to induce the price elastic majority of consumers to buy GM food.

A price reduction of at least 25%, however, is unrealistic, if the price reduction is exclusively based on a respective cost reduction in agriculture. In processed food, the cost share of the agricultural raw material is seldom larger than 25%. But also in the trade of fresh fruits and vegetables, it seems to be rather unrealistic to assume that the cost savings in agriculture alone allow a sufficient purchase incentive to be offered to consumers in order to succeed with labelled GM food.

This cost reduction becomes even more unlikely, if – in addition to costly segregating and labelling requirements – GM crops are burdened by further constraints, which national monitoring organizations have the authority to require for commercial licences. Franks (1999: 575 ff.) discusses this for the UK, where the corresponding body, the Advisory Committee on Releases to the Environment, may do so in regions where valuable genetic resources may be put at risk from commercial release. Such additional constraints might range from agronomic practices, such as planting barrier crops or maintaining reproductive isolation distances between crops and valuable wild gene pools, to the prohibition of planting GM crops. The consequence is in any case a cost increase relative to conventional or traditional production techniques.

Creation of superior quality

An alternative starting point to overcome consumers' hesitation over GM food is to be seen in the enlargement of the utility difference between GM and non-GM food. This presupposes a considerable improvement of the food quality. An often expressed hope of the advocates of GM food is that the second generation of GM food will be likely to concentrate on the improvement of food properties that are positively valued by consumers, or create food with additional desirable characteristics. One example of quality improvement is enrichment of grain with vitamins and mineral substances (Erbersdobler, 1999); an example of a completely new property is the elimination of allergenic proteins in rice varieties.

Many of those new products are likely to have a nutritional and purchase value that considerably exceeds the value and price of the conventional produce by far more than 25%. However, an increment of the nutritional value and of the attributed utility by 50 or more per cent does not guarantee an increase of demand for the respective GM food, at least not in the case of processed food. Almost all of the food quality improvements can be realized either by improving the basic agricultural raw material or by technological progress in food processing. If a food improvement becomes known for which a sufficiently large buyer segment exists which is willing to pay, the respective food will be constructed by supplementing, mixing, extraction, exchanging and other processing activities (Böcker *et al.*, 1997).

Breeding a new raw material is only one possibility for producing a functional food; food processing often offers many. It is different with agricultural products that are dedicated to fresh consumption or are at least consumed without any serious processing (only cooked or fried). Here, food processing is not such a close substitute for breeding. However, this does not necessarily provide GM products with a very large competitive advantage over conventional fresh food in developed countries with high income levels. Consumers can and do afford to feed themselves on a plethora of food products. If they need or believe they need a functional food they can choose how they provide themselves with the respective function. The individual utility increase which results from a single specific food item is thereby strongly limited.

Summary and Concluding Remarks

The above considerations make it clear that it is not very likely that even a new generation of improved GM food will raise the individual consumer utility sufficiently in order to gain a substantial competitive advantage over conventional food stuffs. The level of consumer acceptance is not only low at present. These strong attitudes towards, that is, mainly against, GM foods are firmly nested in higher-order attitudes or beliefs (Bredahl, 2000), so that low acceptance can be expected to prevail in the future.

The only way this obstacle could be effectively overcome is by actual and, of course, positive product experience. For this to happen, however, a substantial increase in consumer benefit, that is value or utility per money unit, would have to be delivered by new products. In mass markets, where the only benefit has to be seen in potentially reduced prices with no additional value created by genetic engineering, we have argued that this is highly unlikely. And even in cases where additional value is created, substitute food processing technologies exist that can – at least partially – mimic the benefits, thus reducing the expected competitive advantage of GM. We therefore argue, that GM food will remain restricted to high priced niche markets in the near future. From there, supported by the spreading news of positive product experience, it might increasingly enter into the mass markets and thus gain in market share.

References

Böcker, A., Hanf, C.-H. and Turss, O. (1997) *Interdependenzen der Produktdifferenzierung innerhalb der Nahrungsmittelkette. Diskussionsbeiträge* No. 6, 192 S. Institut für Agrarökonomie, Christian-Albrechts-Universität zu Kiel.

Bredahl, L. (2000) *Determinants of Attitudes and Purchase Intentions with Regard to Genetically Modified Foods – Results of a Cross-National Survey*, MAPP Working Paper No. 69. MAPP, Arhus, Denmark.

Bredahl, L., Grunert, K. and Frewer, L. (1998) *Determinants of Attitudes and Purchase Intentions with Regard to Genetically Modified Foods – A Review of the Literature and Presentation of Models for Future Research*, MAPP Working Paper No. 52. MAPP, Arhus, Denmark.

Connor, J.M. (1994) North America as a precursor of changes in Western-European food-purchasing patterns. *European Review of Agricultural Economics* 21, 155–173.

Dnes, A.W. (1996) An economic analysis of the BSE scare. *Scottish Journal of Political Economy* 43, 343–348.

Erbersdobler, H. (1999) Gesunde Ernährung mit Novel Food (healthy nutrition and novel food). *Schriftenreihe der Agrar- und Ernährungswissenschaftlichen Fakultät der Universität Kiel, Vorträge zur Hochschultagung* 88, 209–218.

Fife-Shaw, C. and Rowe, G. (1996) Public perceptions of everyday food hazards: a psychometric study. *Risk Analysis* 16, 487–500.

Franks, J.R. (1999) The status and prospects for genetically modified crops in Europe. *Food Policy* 24, 565–584.

Frewer, L.J., Howard, C., Hedderley, D. and Shepherd, R. (1996) What determines trust in information about food-related risks? Underlying psychological constructs. *Risk Analysis* 16, 473–486.

Gath, M. (1998) Der einfluß der kennzeichnung auf die verbraucherakzeptanz gentechnisch veränderter lebensmittel (Impact of labeling on consumer acceptance of genetically modified food). Master thesis, Kiel University, Germany.

Herrmann, R., Warland, R.H. and Sterngold, A. (1997) Who reacts to food safety scares?: examining the alar crisis. *Agribusiness* 13, 511–520.

Lichtenstein, S., Slovic, P., Fischhoff, B., Layman, M. and Combs, B. (1978) Judged frequency of lethal events. *Journal of Experimental Psychology, Human Learning and Memory* 4, 551–578.

Lowrance, W.W. (1976) *Of Acceptable Risk: Science and Determination of Safety*. W. Kaufman Inc., Los Altos, California.

Luehrs, F. (1987) Lokale Preisschwankungen bei Verbrauchsgütern. *Diplomarbeit.* Universitaet Kiel, Institut for Agaroekonomie, Kiel.

Slovic, P., Fischhoff, B. and Lichtenstein, S. (1980) Facts and fears: understanding perceived risk. In: Schwing, R.C. and Albers, W.S. (eds) *Societal Risk Assessment: How Safe is Safe Enough?* Plenum, New York, pp. 181–214.

Sparks, P. and Shepherd, R. (1994) Public perceptions of the potential hazards associated with food production and food consumption: An empirical study. *Risk Analysis* 14, 799–806.

Weiss, E.W. (1997) Sicherheitsbeurteilung neuartiger Lebensmittel (safety evaluation of novel foods). *Ernährung/Nutrition* 21, 426–429.

Wiegand, G. and von Braun, J. (1994) Zur ökonomik von schadstoffen in lebensmitteln. *Agrarwirtschaft* 43, 257–307.

5 Estimates of Willingness to Pay a Premium for Non-GM Foods: a Survey

Catherine A. Mendenhall[1] and Robert E. Evenson[2]
[1]Environmental Health Sciences, Department of Epidemiology and Public Health, and [2]Economic Growth Center, Department of Economics, Yale University, New Haven, Connecticut, USA

Introduction

The genetically modified (GM) food debate has been intensifying with time. Proponents of the new technologies speak of making healthy foods even more nutritious, reducing the need for toxic chemicals, and increasing the yield of plants, all helping to feed the Earth's expanding population. In March 2000, the Food and Agriculture Organization (FAO) issued its first public statement on genetically modified organisms (GMOs), with the message that genetic engineering could help to increase output in agriculture, forestry and fisheries. Genetically engineering food may dramatically change the way farming is done. For example, Frommer *et al.* (1999) have found that varieties of a single plant species, such as barley or tomato, exhibit a high degree of variation in salt tolerance, suggesting that only a few mutations in several key transporter or regulatory proteins could confer salt tolerance on salt-sensitive plants. These plants could change agriculture patterns by allowing agriculture production where previously it was not possible. The potential for GM crops to change the way the world is fed may be large.

But fears of GM foods have also been growing. The FAO also issued words of warning in its statement: 'caution must be exercised in order to reduce the risk of transferring toxins from one life form to another, of creating new toxins or of transferring allergenic compounds from one species to another, which could result in unexpected allergic reactions'.

Since the public's perception of GM foods is currently unknown, there is also a lack of knowledge regarding the current demand for non-GM crops relative to GM crops. Currently, the premium for non-GM soybeans is around 3–10% (in the USA; estimates as large as 50% have been reported in Japan). This indicates that food processors are demanding non-GM crops and are willing to pay a premium in order to obtain them. Estimating the actual demand is necessary for preventing the possibility of market disruption.

It is possible that the demand for non-GM foods is sizeable. For example, Marks & Spencer issued a statement in July 1999 stating that it would no longer sell GM foods, and would change suppliers and processors to make this a reality. In August 1999, Gerber and H.J. Heinz baby-food makers rejected the use of modified food ingredients. Whole Foods Market, the largest natural foods chain in the US, currently requires suppliers of its house brands to certify that their products are GM-free.

Demand for non-GM maize could possibly reach 28% of the entire market:

1. Currently, maize processors use 3.2% of the maize market. Supposing that the entire industry were to demand non-GM maize, around 3% of the maize market would be non-GM.
2. If the sweetener industry were also to demand non-GM maize, 6% of the market could potentially be non-GM.

3. Exported maize may have to be certified non-GM. Up to 19% of the market could move to non-GM maize in response (Kansas Corn Growers Association, 2000).

As of 3 March 2000, GM foods were declared unable to be considered 'organic' under federal guidelines. The USDA revised its national standard requirements for organic food in response to a flood of negative publicity. The industry had repeatedly requested national standards, understanding that without guidelines, there is nothing to back up the claim that a product is organic. Consumers were questioning what the organic label really meant, and whether it was worth paying more for food designated as organic.

Determining what factors influence consumers' risk perceptions towards GM foods, and working up a profile for the individual who is likely to pay a premium for certified non-GM foods will aid in estimating the overall demand for non-GM crops. Grobe *et al.*'s (1999) study of consumer risk perception profiles regarding recombinant bovine growth hormone (rbGH) contends that self-protective behaviour, defined as (i) seeking assurance that milk purchased came from a non-treated herd; or (ii) changing milk consumption levels, is significantly affected by household size and identification with environmental groups. Grobe also declares that identifying and understanding these profiles would enable risk communicators to design more effective risk communication strategies.

The goal of this chapter is to determine what factors influence consumers' risk perceptions towards GM foods, to determine the profile for the individual who is likely to pay a premium for certified non-GM foods, and to estimate the demand for non-GM crops.

Methods

Public acceptance of GM foods can be measured either by analysing the strategy, attitude and political weight applied by the stakeholders in the GM controversy in their attempt to sway public opinion, or by conducting a representative survey among the general public. Since the number of stakeholders in the GM controversy is too great (governmental institutions, business, non-governmental organizations (NGOs) and churches, and international NGOs, among others) the second strategy was chosen for the present study.

In order to examine the public's acceptance level of GM food, a standardized questionnaire was created. The first questions concentrated on the individual's concern about GM foods for their health, the environment and their children. The next section addressed the individual's willingness to distinguish between GM and non-GM foods, and their willingness to pay a premium for non-GM foods. The third section focused on statements with regard to the individual's profile, with general questions regarding age, income and current food purchasing habits, in addition to benchmarking the individual's concern about his or her health in comparison to his or her peers. Respondents were asked specifically whether they had heard or read anything about the GM controversy. Sixty per cent of the 54 individuals surveyed had heard or read something about GM technology. This awareness level was higher than expected, but may have been due to the media-saturated city in which the survey was conducted.

Respondents who were not aware of GM technology were provided with a neutral description, given that the framing of the information could influence the respondent's answers to the questionnaire.

The respondent could indicate to what extent he or she agreed with a certain statement by applying a scale of marks from strongly agree to strongly disagree. A telemarketed survey of random New Haven, Connecticut, residences was conducted. Every fifth number in the New Haven phone book was dialled. Interviews were conducted with the person identified as a household resident 'who is 18-years-old or older, and responsible for the household's food purchasing decisions'.

Results

Data consisted of 54 completed surveys, gathered from individuals in the city of New Haven. Initial contact was made with 131 individuals and the response rate for the entire frame sample was 41%. Of these surveys, the average respondent was 32-years-old, and had a median income of approximately US$38,000. The average household size was 2.1. Race and sex were not asked.

Table 5.1. Chi-square analysis (*P* values).

	Concern about safety of GM foods	Prefer non-GM foods in general	Willingness to pay a premium for non-GM foods	Feel that GM foods should be labelled	Concerned about environmental impact of GMOs
Concern about health	0.058**	0.031*	0.031*	0.356	0.026*
Exercise	0.058**	0.228	0.905	0.182	0.427
Smoking	0.232	0.720	0.646	0.290	0.996
Age	0.087**	0.636	0.348	0.921	0.270
Currently buying fat-free foods	0.197	0.410	0.646	0.324	0.065**
Currently buying organic foods	0.025*	0.519	0.041*	0.712	0.050*
Education	0.164	0.244	0.636	0.469	0.464
Number of people in household	0.795	0.602	0.038*	0.781	0.014*
Having children	0.922	0.806	0.216	0.600	0.532
Income	0.459	0.252	0.953	0.605	0.657

* Significant; **marginally significant.

Variables

Individuals were questioned on *safety* (how they interpreted the safety of consuming GM foods), *premium* (their willingness to pay a premium for certified non-GM foods), *label* (their opinion on whether GM foods should be labelled), *environment* (whether they were concerned about GM foods' impact on the environment) and *prefer* (preferring non-GM foods in general if given a choice).

Variables reflecting personal health influences relevant to food purchases were *concern* (concern about health in relation to peers), *food* (already buying fat-free or low cholesterol foods) and *organic* (already purchasing at least some organic foods). Variables reflecting social and cultural influences were *education, smoking* and *exercise* (exercising regularly).

Variables that were more inherited and less related to individual attitudes were *household size, children* (presence of children in the household) and *age*. A person's perceived sense of control was thought to be correlated with their income status, so *income* (annual household income) was used as a proxy.

Statistics

The objective of the descriptive data analysis was to investigate which segments of the general public are concerned about the safety of GM foods, and if those concerned would be willing to pay a premium for non-GM foods. Statistical tests were performed in SAS. Table 5.1 reports the results of 2-way cross tabulations of the endogenous variables using the chi-square test statistic. Table 5.2 reports comparable tests between endogenous variables.

Perceived safety of GM foods

Overall, 26% of those surveyed indicated that they were very concerned about the safety of GM food. Over 50% said that they were somewhat concerned,

Table 5.2. Chi-square analysis, dependent variables (*P* values).

	Willingness to pay a premium for non-GM foods	Feel that GM foods should be labelled	Concerned about environmental impact of GMOs	Prefer non-GM foods in general
Concern about safety of GM foods	0.022*	0.258	0.0001*	0.0001*
Willingness to pay a premium for non-GM foods	—	0.571	0.131	0.096
Feel that GM foods should be labelled	—	—	0.205	0.003*
Concerned about environmental impact of GMOs	—	—	—	0.0001*

* Significant; ** marginally significant.

and the remaining 24% were divided with 5% being somewhat unconcerned about GM safety and 19% not at all concerned.

The marginal effect of age (P = 0.087) indicates that those who were younger are less inclined to be concerned about the safety of GM foods, and perceived less of a health risk than those who were older. Individuals who are willing to pay a premium for non-GM foods are also significantly more likely to be concerned about the safety of GM foods (P = 0.022), with concern increasing with increasing willingness to pay. Safety is also a function of concern about health (*concern*, P = 0.058). Those individuals that are more concerned about their heath in general are more concerned about the safety of GM foods.

Using chi-square analyses, it was determined that those who currently buy fat-free foods are not more likely to have concerns over GM foods (P = 0.197, marginal significance with t-test, P = 0.076), and those who buy organic are more likely to report that they are concerned (P = 0.025). Individuals who exercise have a moderately elevated level of concern (marginal significance, P = 0.058). Concern over the safety of GM foods was not related to smoking (P = 0.232), level of education (P = 0.164), number of people in the household (P = 0.795), having children (P = 0.922) or income (P = 0.459).

Preferring non-GM foods

Fifty per cent of those surveyed indicated that they would be very likely or somewhat likely to purchase non-GM foods if they cost up to 20% more than GM foods. Surprisingly, individuals with two members in the household are most likely to pay more for non-GM foods (P = 0.038). Individuals who currently buy organic foods are also those who are more inclined to buy certified non-GM foods at a premium (P = 0.041). Similarly, those currently inclined to be concerned about their health are more likely to pay up to 20% more for non-GM foods (P = 0.031).

Willingness to pay a premium for non-GM foods was not significantly related to exercising (P = 0.905), smoking (P = 0.646), level of education (P = 0.636) or income (P = 0.0953).

Labelling

Eighty-two per cent of those surveyed strongly believe that the foods made with GM ingredients should be labelled. Ten per cent indicated that they somewhat believe labelling should be mandatory. Overall, over 90% of those surveyed feel that some kind of labelling for GM foods should be required.

Feeling that GM foods should be labelled was not significantly related to having children (P = 0.600), concern about health (P = 0.356), exercising (P = 0.182), smoking (P = 0.29), currently buying fat-free foods (P = 0.324) or currently buying organic foods (P = 0.712).

Impact on the environment

If a person was concerned about the safety of GM foods, they were also likely to be concerned about the impact of GM foods on the environment (P = 0.0001). Those who were concerned about their health were also concerned about the environmental impact of GM foods (P = 0.026), and individuals who currently bought fat-free foods or organic foods were also significantly more likely to be concerned (P = 0.065 and 0.050, respectively). Concern declined with increasing numbers of people in the household (P = 0.014), and was not related to exercise, smoking, age, education level, having children or income.

Discussion

Limitations of the study

This is undoubtedly a pilot study. The results and conclusions stated should be interpreted with caution, due to the small sample size and the possibility of bias by non-response. In other words, the individuals that responded to the survey may feel more strongly about the subject matter, making the survey results unrepresentative of the country as a whole. This survey should also be repeated on a national level, for the racial and social economic status of individuals in New Haven is not necessarily similar to that of the rest of the US.

Also, if an individual states that he or she would be willing to pay a premium, it does not necessarily translate into the individual actually paying a premium when faced with the decision. Methods to deal with this phenomenon should be employed when repeating this survey.

Perceptions of GM foods

Over 75% of the individuals surveyed indicated that they were at least somewhat concerned about the safety of GM foods, and over 90% felt that there should be some form of labelling distinguishing food containing GM ingredients from non-GM food. This is not surprising; after all, the European Union just passed legislation requiring a label for all foods containing more than 1% of GM ingredients.

Feeling concerned about GM foods was not a function of education. However, those individuals who were already concerned about what they were eating, such as the individuals who already bought organic foods, were more likely to be concerned. Participating in an exercise programme or already paying attention to their health in general also increased the level of concern. In summary, the concerned consumer was health conscious. Individuals with larger households or with children were not as likely to be concerned, perhaps due to financial constraints or other, more pressing, issues.

Willingness to pay a premium for non-GM foods was not a function of income level or education. Instead, it was related to the current level of concern about the individual's health and current buying practices. If the person was already eating organic food (and therefore currently paying a premium for organic food), he or she was willing to pay a premium for certified non-GM food, too. An individual's willingness to pay a premium decreased with increasing household size. Again, this may be due to financial constraints or more pressing issues concerning the household.

Older people were more concerned about the safety of GM foods and the possible environmental impacts of GMOs on the environment, but this concern did not translate into willingness to pay a premium for non-GM foods. This appears to be unrelated to financial constraints; in this study, age did not correlate with income ($P = 0.218$).

The profile for the consumer who had concern for the environmental impacts of growing GM crops was specific, and was similar to the consumer who was willing to pay a premium for non-GM food. Both had concerns about their health, were already concerned about the food they consumed, and were less likely to be concerned when there were more individuals in the household.

In conclusion, it appears that the individuals surveyed were concerned about GM food, and that they would like to know if GM ingredients were in their food. The USDA and FDA should look towards instituting labelling requirements for GM foods. Being proactive instead of reactive in its labelling legislation would probably be looked upon favourably by the general public.

The establishment of willingness to pay a premium for non-GM foods means change for the agricultural supply chain: separate fields, separate harvesting procedures and possibly equipment, separation of GM and non-GM in transportation, processing, packaging and marketing. There is an increase in cost to the farmer for keeping the GM and non-GM separate, but this is not currently being reflected in the prices for non-GM vs. GM crops.

At the farm level: an analysis of costs

The high level of concern and willingness to pay a premium indicated by the consumer shows that there is room for a non-GM market, and that a premium can be charged for non-GM foods. This is also verified by the recent creation of a niche market for non-GM crops. However, one area that has not shown much change during the creation of this market is the premium awarded to the farmer for certified non-GM crops. If the current model is not adjusted, farmers will not receive enough to justify growing non-GM crops.

Typically, Americans spend 10% of their income on food, but that money does not go directly to the farmer. Today, farmers get about 20 cents or less of each food dollar. The remaining 80 cents goes towards transportation and marketing of the product, including packaging, labour and advertising.

The Florida Department of Agriculture and Consumer Services has shown that there is a large discrepancy between the prices paid to the farmer for fresh fruits and vegetables and the prices paid at the store. For the winter and spring of 1996 to 1997, Florida retail prices for most popular commodities were 300–600% above the price paid to the growers. The bottom line is that a 0.6 cent premium kg^{-1} for non-GM maize is not going to translate into that large a price increase at the consumer level, and if 50% of individuals are willing to pay up to a 10% premium at the consumer level, the amount of price increase at the farmer level can be calculated.

For example, the current farm price for sweetcorn is 11 cents kg^{-1}. The retail price is 24 cents kg^{-1}, so farmers are getting 46% of the retail price. If 50% of consumers are willing to spend 10% more, or to pay 26 cents kg^{-1}, we can calculate the price premium the farmers could obtain. At the retail level 10% translates to 1/0.46 (0.10) = 21%. Everything held constant, the price to the farmer could be approximately 13 cents, or a 21% premium. If the percentage spread for farm to retailer is to be held constant (though this may vary, it is currently 216.66%), the price to the farmer would be approximately 12 cents, or an 8.3% premium. Either way, this is more than the 3% premium farmers are currently receiving for non-GM maize.

Out of a US$3.50 box of Wheaties, the farmer receives 3 cents,[1] or 0.8% of the total. The price spread is US$3.47, and the percentage spread is 11,666.66%. If Kellogg's was to charge a 10% premium for certified non-GM Wheaties, or US$3.85, holding other expenses constant, the farmer could conceivably be earning 10% = 1/0.008 (0.10) = 1250% of the 3 cents, or 37.5 cents per non-GM Wheaties box. In other words, a 10% premium at the retail level translates into an increasing percentage along the supply chain. If the farmers are not given an increase in the premium paid for non-GM crops, they may not be able to produce them and the opportunity for this new market to thrive will be lost.

In a final example, margarine currently sells for around 90 cents kg^{-1}, and is 80% soybean oil. Soybean oil is priced at 9.45 cents kg^{-1}, which means that 7.6 cents from each kg of margarine (or 8.4% of the price of margarine) goes to the farmer. If the price of certified non-GM margarine was raised 10% to 99 cents, it could translate into a 10% = 1/0.084 (0.10) = 120% premium on certified non-GM soybeans, or 11.34 cents kg^{-1}.

Policy recommendations

With the current level of interest in non-GM food, the US Government should develop guidelines for certification. Labelling the non-GM foods as opposed to the GM foods will be a way to inform the customer without attaching a negative stigma to food or food products containing GM ingredients.

Those strongly concerned about GM foods are already taking precautionary measures, such as buying organic food, and will pay a premium to have non-GM food when the opportunity to do so is offered to them.

References

Barrett, J. and Segerson, K. (1996) Prevention and treatment in food safety: an analysis of conceptual issues. In: Caswell, J. (ed.) *The Economics of Reducing Health Risk from Food.* Food Marketing Policy Center, Storrs, Connecticut, pp. 19–43.

Caswell, J.A. and Mojduszka, E.M. (1996) *Using Informational Labeling to Influence the Market for Quality in Food Products,* NE-165 Working paper No. 43. Department of Resource Economics, University of Massachusetts, Amherst, Massachusetts.

Douthitt, R., Zepeda, L. and Grobe, D. (1996) *Comparison of National and Poor Households: Results of a Survey of Consumer Knowledge and Risk Perceptions of Food-related Biotechnologies,* special report No. 68. Institute for Research on Poverty, Madison, Wisconsin.

Economic Research Service, USDA (2000) Impacts of adopting genetically engineered crops in the United States. [online article] www.econ.ag.gov/whatsnew/issues/gmo/

Frommer, W.B., Ludewig, U. and Rentsch, D. (1999) Taking transgenic plants with a pinch of salt [comment]. *Science* 285(5431), 1222–1223.

Grobe, D., Douthitt, R. and Zepeda, L. (1996a) *Exploring Consumers' Risk Perceptions of Recombinant Bovine Growth Hormone and Recombinant Porcine Growth Hormone by Income and Gender: A Focus Group Study.* Institute for Research on Poverty, Madison, Wisconsin.

Grobe, D., Douthitt, R. and Zepeda, L. (1996b) *Measuring Consumer Knowledge and Risk Perceptions of Food-related Biotechnologies: Survey Instrument.* Institute for Research on Poverty, Madison, Wisconsin.

Grobe, D., Douthitt, R. and Zepeda, L. (1999) Consumer risk perception profiles regarding recombinant bovine growth hormone (rbGH). *Journal of Consumer Affairs* 33(2), 254–275.

John, E.L., Rayor, L.S. and Carter, M.E. (1999) Transgenic pollen harms monarch larvae. *Nature* 399, 214.

[1] www.wheatonline.com/nrtext/price1026.txt

Kansas Corn Growers Association (2000) The many uses of corn. www.ksgrains.com/corn/uses.html

Kristal, A.R., Levy, L., Patterson, R.E., Li, S.S. and White, E. (1998). Trends in food label use associated with new nutrition labeling regulations. *American Journal of Public Health* 88(8), 1212–1215.

Luoma, J.R. (2000) Pandora's Pantry. *Mojo Wire* [serial online] January–February. www.bsd.mojones.com/mother_jones/JF00/pandora.html

Marietta, A.B., Welshimer, K.J. and Anderson, S.L. (1999) Knowledge, attitudes, and behaviors of college students regarding the 1990 Nutrition Labeling Education Act food labels. *Journal of the American Dietetic Association* 99(4), 445–449.

Marra, M., Carlson, G. and Hubbell, B. (1997) Economic impacts of the first crop biotechnologies, [online presentation] December. www.ag-econ.ncsu.edu/faculty/marra/FirstCrop/sld018.htm

Myhr, A. and Traavik, T. (1999) The precautionary principle applied to deliberate release of genetically modified organisms. *Microbial Ecology in Health and Disease* 11(2), 65–74.

Nestle, M. (1996) Allergies to transgenic foods – questions of policy [editorial; comment]. *New England Journal of Medicine* 334(11), 726–728.

6 A Consumer-based Approach towards New Product Development through Biotechnology in the Agro-food Sector

Nickolas M. Spetsidis and Günter Schamel
Faculty of Agriculture and Horticulture, Institute of Economic and Social Sciences, Humboldt University of Berlin, Berlin, Germany

Introduction

The development and application of biotechnology is set to have revolutionizing effects on the agro-food sector. Its influence is wide ranging: from enabling new evolutionary insights in living organisms to the redesigning of crop plants and the development of new rational targets for the most common food products. Biotechnology is the toolbox available to introduce innovative ways of realizing biological processes, and to provide opportunities not only to improve production processes but also to develop new approaches to control quality and improve food production.

Having made massive investments in biotechnology, companies are continuously capitalizing on new products. Product development using this technology in the agro-food sector has today become a business endeavour. Companies have often failed to establish close coordination between R&D and commercial success (McElroy, 1999). As technology develops rapidly the main focus of product development activities has been to keep pace with technology and neglect market performance. The additional value that technology delivers was mainly focused on specific targets and not dispersed to all the actors involved. On the other hand, consumers do not seem willing to accept new technologies employed in food production (Hamstra, 1995), which, in addition, do not deliver advantages on consumption and benefit the producers only. Thus, consumers appear to be more reluctant to accept novel products derived from biotechnology. What appears to be vital is for the producers to shift from this technology-push situation more into market orientation, by tracing and satisfying consumer needs.

The core idea of this chapter is that continuous research into the needs of all the actors involved can be a valuable guide to the new product development (NPD) process. An assessment of the technology is carried out in terms of consumers benefits delivery. Screening areas of application within the agro-food sector, we identify what genetic manipulation technology is able to deliver to consumers (final or intermediate). This is illustrated by a match of consumer needs with certain product features derived from the technology. For this illustration we use the methodology of the 'house of quality' (HoQ) using the process of quality function deployment (QFD). Later, the benefits identified are linked with relevant aspects of consumers' acceptance of genetically modified (GM) foods in order to assess the benefits realization.

Our analysis moves along three dimensions:

1. *The technology itself*: the nature of biotechnology has the potential to provide specific product features. We analyse the technological developments in plant and food biotechnology and demonstrate what it could offer in terms of concrete product features.

2. *The substrate of technology*, referring to the product and its relation to consumers. This will be illustrated using the process of QFD and the HoQ methodology. We do not start filling in the consumer needs, rather the product characteristics that particular techniques provide (roof of the house). Later, we match the product characteristics with identified consumer (either the intermediate or end user) needs (consumer 'wants'). Eventually, the output is a translation of product features into consumer benefits.

3. *The consumers' reaction towards biotechnology.* Consumers appear reluctant to accept the use of genetic modification for food production. The analysis focuses on those concerns against the technology itself despite the benefits that it is able to provide. Since we identify the additional value that technology delivers, we investigate the conditions under which consumers can realize this value.

The analysis begins with an introduction to the concept of consumer-oriented NPD and then we provide an overview of the QFD process. We adapt the relevant issues of QFD and of consumer-oriented NPD to genetic modification for food production and outline some key issues related to consumer orientation. Further, we shift to technology and summarize the current state of affairs of genetic manipulation with regard to plants and food production. We transpose the technology to the consumers' side by a translation of product features generated by the deployment of technology into consumer benefits using the HoQ. The goal is to identify the additional value and investigate under which conditions this is realized by the consumer. At the end of the chapter we confront the relevant issues of QFD and of consumer-oriented NPD with key findings of a survey conducted using specific product scenarios with German consumers and derive some main conclusions.

Principles of Consumer-oriented New Product Development

Introduction

NPD is widely recognized as a crucial activity for most companies. Long-term survival is increasingly dependent on the ability of companies to develop and successfully introduce new products on to the marketplace. It is commonly argued that a substantial share of the expected profits will come from products which are not currently on the market (van Trijp and Steenkamp, 1998). New products launched on the market can provide significant rewards in cases of success and heavy penalties in cases of failures. It is generally accepted in marketing literature that, from the total amount of new ideas turned to new products, only one-third can realize successful commercialization. Rapidly changing technologies, heightened competition and the dynamic nature of consumer needs shape an environment characterized by constant turbulence both on the technology and on the market side (Cooper, 1996; van Trijp and Steenkamp, 1998). This environment means NPD is an extremely risky but also a challenging activity.

The high rates of failure and the high associated costs have given rise to a significant amount of research over the determinants of NPD success. The ultimate goal is to identify elements of success relevant to the internal structure of the organization as well as to its external position and strategy within the industry and towards the competition. Success depends among other factors (Urban and Hauser, 1993) on the rate that the new products successfully address identified consumer needs and at the same time surpass the competition. Consumer-oriented NPD takes consumers needs as a starting point for the product development process and the product and production technology as a derivative thereof (van Trijp and Steenkamp, 1998). Under this concept technology is nothing more than the tool used to make a product that consumers want. Technological developments have their effect on the products as they generate opportunities to fulfil existing consumer needs more efficiently through new or improved products (van Kleef, 1999, Wageningen Research University, personal cooperation and collaboration). In essence, it is the company's ability to exploit its technological capabilities to fulfil carefully selected market opportunities by satisfying identified consumer needs (van Trijp, 1999).

The NPD process needs strong links and cooperation between the functional groups in order for it to be established within the organization for the generation of new ideas that will be translated into new products. This is a complex and cross-functional process (Baker and Hart, 1999) which involves the active participation of all the parties of interest, namely marketing, manufacturing, suppliers and customers (Scott, 1998). It requires a 'balanced' approach in which marketing and R&D

collaborate and share responsibilities (van Trijp and Steenkamp, 1998). The borders are not firm, rather a dynamic interrelationship prevails involving marketing moving more towards the manufacturing process and R&D to taking more account of the external market activities (Scott, 1998). The key to success is how those links operate in order to effectively translate marketing inputs into desired product characteristics. Communication among the functional groups turns out to be the cornerstone of effectiveness in product development (Moenaert *et al.*, 1994; Cooper, 1996; Griffin and Hauser, 1996; Scott, 1998; van Trijp and Steenkamp, 1998).

Quality function deployment and the house of quality

The level of integration of different functional groups for the product development process is heavily influenced by the strategic orientation of the company. Gatignon and Xuereb (1997) define technology-oriented firms as 'firms with the ability and willingness to acquire substantial technological background and use it in the product development'. The company develops the capabilities to exploit its technical knowledge to provide new solutions that meet market demands. On the other hand, Kohli and Jarowski (1990; Jarowski *et al.*, 1993) define market orientation of firms as 'the *organisationwide* generation and dissemination of market intelligence related to current and future customer needs across the departments and the *organisationwide* responsiveness to it'. In this case the company has the mission to identify, analyse and exploit knowledge about consumer needs and use it as its core competency in the whole product development process (Gatignon and Xuereb, 1997).

Marketing and R&D responsibilities are neither static nor independent and cannot be analysed separately (Griffin and Hauser, 1996). Recent evidence suggests that the likelihood of product development success is enhanced if marketing and R&D staff members have high levels of communication (Griffin and Hauser, 1993). However, marketing literature and empirical evidence indicate that disharmony between R&D and marketing has been the rule rather than the exception (Griffin and Hauser, 1996; Scott, 1998; Temponi *et al.*, 1999). Marketing and R&D have different levels of abstraction in realization of consumer needs (van Trijp and Steenkamp, 1998). The level of communication and cooperation between the groups is a crucial issue in the product development process.

One process for enhancing effectiveness in communication and cooperation is QFD. It is the mechanism that translates the 'consumer voice' into technical terminology and helps functional groups

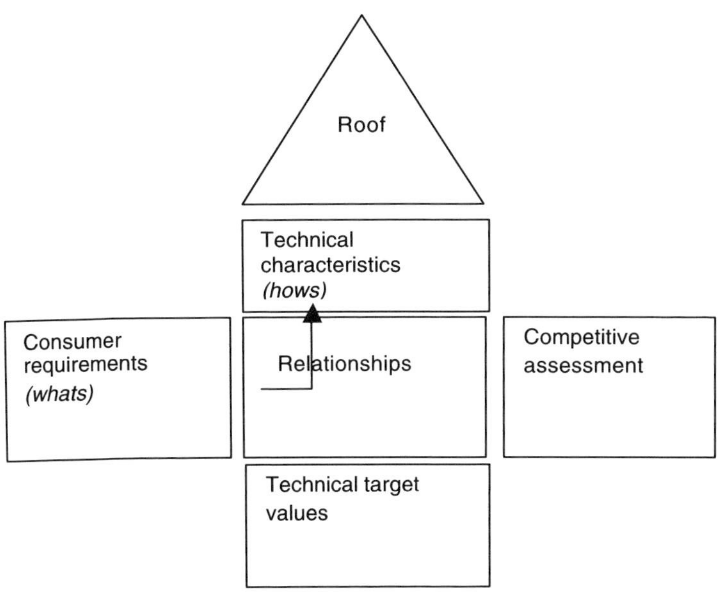

Fig. 6.1. House of quality.

to make key trade-offs between what consumers want and what the company is able to deliver (Griffin and Hauser, 1996). QFD reduces the internal barriers between marketing and R&D providing the means to share and utilize information. The HoQ is the tool used for this communication (Hauser and Clausing, 1998). It is a careful statement of consumer wants for the item being developed, followed by a stipulation of technologies that can be used to achieve desired product characteristics (Crawford, 1996; Govers, 1996; Hauser and Clausing, 1998). HoQ is illustrated in Fig. 6.1.

The HoQ begins with filling in the *consumer requirements* ('whats'). This is a structured list of what the consumer 'wants' using a familiar consumer terminology. The consumer in this case is not only the end user but also regulators and intermediate users. *Competitiveness assessment* illustrates the customer's perception of how the company's product fulfils perceived needs relative to the competition. This comparison may identify opportunities for product improvement, or sources of competitive advantage. At the top of the HoQ, the product is analysed in technical terminology. In the *technical characteristics* ('hows') section, the product attributes are assigned to physical measurement units. Thus, the *roof* represents the technical and functional interrelations of the product features. In this part, necessary engineering trade-offs are demonstrated. The association of consumer demands towards the product and its technical and functional features takes place in the central 'room' of the HoQ, where the impact of the technical response against the consumer needs is measured and evaluated. At the bottom, the *technical target values* include a summary of the basic technical data that have been generated from the interpretation of the information from the main matrix of the HoQ. In addition, the assessment of the competitor's technical performance indicates margins of possible advantage. Technical target features indicate the priorities for the product development process (Hauser, 1993; Cohen, 1995; Temponi *et al.*, 1999).

New product development using biotechnology

Companies involved with biotechnology can be characterized as predominantly technology oriented. The focus of R&D activities is primarily on product and process development and not on successful commercialization. The process for a successful new product development has been largely neglected in favour of more research into the basic levels of science (Wheelwright, 1994; Giusepin, 1999, Vlaardigen, The Netherlands, personal communication).

Companies utilizing biotechnology within the agro-food sector could be also considered as 'high-tech companies' as they are mainly innovative, using offensive technological and marketing strategies to achieve high levels of product innovation (Nyström, 1990). They are an example of organizations that need to innovate under a turbulent environment where technology and the market are subject to constant change. Since the knowledge basis used is constantly changing, transformations to the scientific substrate surrounding the industry also take place (van Vliet, 1998). New technologies rapidly follow one another, either replacing the existing ones, or generating new paths for development. Therefore, the technology push is the result of new production opportunities arising from the evolution of the technology.

The innovation process has been mainly focused on improving R&D performance, measured by the number of patents developed or acquired (Giusepin, 1999, personal communication), or on the generation of products with novel features (McElroy, 1999). The challenge lies with how the product development process is capable of bridging the gap between what the technology delivers and what consumers really want. Increasing competition in production technology and heightening market demands urge companies to realize that rapid and successful NPD should have top priority (Wheelwright, 1994; McElroy, 1999). Rapid and high-quality product development can be a key strategic advantage for a competitive strategy (Wheelwright, 1994). Superior product development process refers to how rapidly and effectively new ideas can be translated into product features desired by consumers.

Genetic Manipulation Technology: Dynamics and Potential

The current state of affairs

Biotechnology can be defined as 'the set of technologies using living organisms or parts of them to make or modify products to improve plants or animals or to develop microorganisms for specific uses' (Darf, 1990). It is the utilization of technologies to explore and exploit the biological potential of living organisms to improve processes or to introduce new methods of production. Genetic manipulation technology and recombinant DNA technology embrace all the technologies and their derivative techniques applied to implant foreign agents into the genetic structure of an organism or to modify its existing genetic sequence in order to manipulate target characteristics[1] (Persley, 1992; Roller and Harlander, 1998). Therefore, biotechnology has the potential to offer cost-efficient ways of producing an array of novel or value-added products.

Genetic manipulation technology is of great significance for agriculture. The world market for transgenic plants in 1999 was US$3.5 billion with a planted area of almost 40 Mha (*The Economist*, 2000; ISAAA, 2000). The application of genetic manipulation initially pursued the same path as the conventional breeding and cultivation methods. The difference was that with biotechnology the time needed to achieve the desired results was dramatically decreased (Persley, 1992). The primary objective of research was to isolate single genes, transform and introduce them to the desired crops (Darf, 1990; Mazur, 1999; McElroy, 1999; www.EuropaBio.be). The so-called 'first generation' of genetically modified organisms (GMOs) are products with improved input agronomic traits such as high yield, resistance to pests and diseases, herbicide tolerance and yield stability. Their value in generating revenues was by reducing the costs of production (Mazur, 1999; McElroy, 1999). Advances in technologies for sequencing the genetic material of organisms allowed the development of techniques to incorporate more than one gene in order to manipulate and control more complex traits: the 'quality traits' (second generation of products) (Rekonski, 1997; Haseltine, 1998; Roller and Harlander, 1998; Mazur, 1999; McElroy, 1999).

The key technological driver behind the latest advances is 'genomics'. Its potential arises as the advances in genetic sequencing have provided the tools to identify and investigate groups of genes contributing to a biochemical function or trait. Genomics technology identifies the location, impact and function of genes affecting certain traits of interest. The constant comprehension of genetic information by improved genetic sequencing will be the guide to monitor the ways in which genes interact to produce traits of interest. The challenge is to identify genes that safely and effectively generate commercial opportunities in agriculture (Mazur, 1999). The most important agronomic properties are genetically complex. The agronomic properties generate additional value by changing the properties of the final product to meet sophisticated consumer demands such as combined agronomic traits or quality properties of final products (McElroy, 1999; www.EuropaBio.be).

The task for the future will be not only to control one or two genes, but also to understand and manipulate whole physicochemical and physiological procedures (Renkonski, 1997; Mazur, 1999). Developments in sequencing technologies can provide a lot of information about the association of genes with corresponding traits, but say nothing about their functions in the whole organism (functional genomics) (Haseltine, 1998; Mazur, 1999; Galperin and Koonin, 2000). The path to take is 'bioinformatics' which involves the successful management and coordination of huge quantities of information in conjunction with the use of information technology to achieve the desired output (Mazur, 1999; Thomson, 1999).

Technological advances will give rise to more radical changes, which will further shorten the product development process (McElroy, 1999; Hodgson, 2000). Genetic manipulation allows new combinations of genetic material to improve microorganisms for tailor-made products of interest. It is now possible to isolate particular genes coding for enzymes and introduce them into microorganisms used in food production (e.g. chymosin derived from the bacteria *Kluyveromyces lactis*). It is also possible to intervene in the genetic material of enzymes or other microorganisms (bacteria), which are often selected as a host to express particular genes or proteins. Yeast can produce enzymes, precursors for chemicals, or specific proteins. Lactic acid bacteria may also secrete proteins that serve as natural anti-microbial compounds. Moulds are increasingly being chosen as large-scale producers of enzymes (Unilever, 1995). Modification of basic ingredients can lead to tailor-made functional components. It has been estimated that 50% of all industrial enzymes have already been genetically manipulated (Roller and

[1] The terms biotechnology and genetic manipulation technology are used interchangeably to denote technologies which lie beyond classical genetics.

Harlander, 1998). This specific performance achieved by genetic engineering technologies has attracted the industry's interest in producing probiotics and extracting bioactive components used for specific purposes (Belem, 1999). Biotechnology is the toolbox available to boost sales of food product categories such as functional foods or functional pharmaceutical foods (Belem, 1999; Goldberg, 1999), to rediscover food products (Belem, 1999) or to improve the value of the existing ones by enhancing incremental properties (Roller and Harlander, 1998).

Biotechnology deployment for food production

The use of genetic manipulation for food production is mainly focused on improving quality in several ways. The improvement of product quality, and the reduction of production time and costs illustrate some of the main targets of biotechnology (Frewer *et al.*, 1997a; Roller and Harlander, 1998). The area of applications can be wide. In this section some selected examples are given to indicate how the technology is employed to deliver additional value. Extensive analysis can be found in the tables in the Appendix.

The objective of improved product quality includes targets like fatty acid metabolism, nutrient metabolism, elimination of undesired properties, extended shelf-life, improved colour and processing quality (www.EuropaBio.be). The first application was in the tomato with a delayed ripening process (the FlavrSavr® tomato) introduced by Calgene. Zeneca also developed tomatoes with retarded ripening properties suitable for tomato pulp. Today, tomatoes with increased levels of provitamin A have been successfully developed to enhance nutritional value especially for vulnerable consumers (Römer *et al*,. 2000). Another significant achievement is the higher level of provitamin A in rice.

The manipulation of lactic acid bacteria for fermented food products such as buttermilk, yogurt, cheese, sausage or sauerkraut was also one of the initial applications of biotechnology in foods (Wymer, 2000). Some products have already been commercialized, such as the recombinant chymosin for cheese, yogurts with modified bacteria (*Lactobacillus bulgaricus* and *Streptococcus thermophilus*) to prevent the so-called 'post-fermentation acidification', and bacteria that overproduce alanine from lactose to derive cheese or yogurt with improved or novel flavours (Wymer, 2000). There have been recent attempts to modify essential components of milk for low-lactose content for lactose intolerant persons. This application ensures adequate intake of Ca and at the same time allergenic reactions can be avoided (Whitelaw, 1999).

Biotechnology deployment in agriculture

Application of biotechnology in plant production was initiated for plant breeding. A thorough understanding of the relevant biochemical reactions and their control points led to new opportunities to induce and utilize genetic variation. Genetic engineering techniques increase the precision of alterations that can be made, by increasing the precision of the definition of changes needed to achieve a desired phenotype (Persley, 1992). Agronomic properties are a key target of application. Herbicide tolerance as well as resistance to fungi, insects, nematodes and viruses was the primary research target. Increased nutrient uptake and efficient nutrient exploitation is another significant application. As technologies develop, the focus is on complex traits controlled by more than one gene or on complex and combined agronomic properties (e.g. high yields of a given energy storage compound) (Mazur, 1999; McElroy, 1999; www.EuropaBio.be). An extensive description of the application of technology in agriculture is given in Tables A1 and A2 in the Appendix.

Benefits Delivery and Consumer Reaction

Introduction

Biotechnology has turned out to be of paramount importance for the companies in the agro-food sector. Its application targets are to enhance an array of properties in food products in order to improve their functionality or even to create entirely new ones. Genetic manipulation offers the capacity to combine a bundle of different characteristics in one product. Technology is accelerating the pace at which new products are in the pipeline to deliver additional value to consumers. Food or life science companies increasingly realize that successful commercialization is going to be a crucial factor for long-term profitability and growth. Success will also depend on the degree of consumer acceptance whereby the key is benefits realization. There is strong evidence that the commercial success of new genetically modified (GM) products will depend on public acceptance and the benefits that new products will deliver to all actors involved in the chain (Wheelwright 1994; Mazur, 1995, 1999; Rekonski, 1997; McElroy, 1999).

Consumers do not value products *per se;* rather they value the benefits they deliver on consumption (Grunert *et al.*, 1997). Consumer theory considers each product to be a bundle of physical characteristics/technical features generated by the production technology. The new product then is a bundle of concrete attributes which the company offers for superior benefit delivery (van Trijp and Steenkamp, 1998). Kaul and Rao (1995) distinguish product attributes and product characteristics. Product attributes are 'the dimensions that define consumers perceptions relative to the product', whereas product characteristics are 'the various physical features that define the product'. Product attributes are either concrete or abstract though still directly related to the product's characteristics (Steenkamp, 1997). Consumers attach to each product an assortment of attributes, which differentiate it from alternatives. Product features influence the formation of product attributes and consumers will choose the products that offer superior benefits according to their perception of these attributes. Superior benefit delivery is the domain within which a product is able to fulfil the perceived consumer needs better than another. The need is merely an 'unsatisfied' condition of a consumer, which leads him or her to an action that will improve this condition (Steenkamp, 1997; Sheth *et al.*, 1999). The realized benefits are the outcome of the product's use (Steenkamp, 1997).

This section aims to screen and analyse the benefits that genetic manipulation technology is able to deliver in terms of product characteristics. Despite the benefits that technology provides, significant concerns are raised by consumers against the technology and ultimately against the products. Later, we review the relevant literature and confront it with key findings from our study of German consumers. The survey included specific scenarios of products (yogurt, vegetables and functional foods) derived with the help of biotechnology, and tested the consumers' reactions towards concrete benefits delivery.

Consumer requirements

There has been extensive research to investigate consumer needs with regard to foods. The decision-making process and consumers' intention to purchase foods has been thoroughly investigated and several models have been developed to describe this process. Consumers today are more critical about the technologies and processes used to produce foods (Hamstra, 1995). Innovations stemming from the employment of new technologies for food production are confronted with suspicion when they convey a certain level of uncertainty relevant to consumption. Consumers' concerns and requirements may vary from a simple food ingredient used, to the nature and scope of the technology used. Research evidence indicates that the majority of European consumers perceive new technologies as unnecessary for food production (Hamstra, 1995; European Commission, 1997; Frewer *et al.,* 1997; Wohl, 1998). Additional technologies employed in the food chain are perceived not to improve products but worsen their quality (European Commission, 1997). The findings of the survey conducted in Germany outline this issue. The majority of respondents believed that new technologies employed in food production tend to worsen the quality of foods.

A large amount of research suggests that consumers attach a significant weight to health issues. Health is judged as one of the most important product attributes that consumers attach to food (Hamstra, 1991, 1995; Fuller, 1994; Ter Hofstede *et al.,* 1996; Grunert *et al.,* 1997; Bredahl *et al.*, 1998; van Trijp and Steenkamp, 1998; Katz, 1999). Relevant studies indicate that 'health' is related to food products without or with only minimal levels of additives (Hamstra, 1991, 1995; Bredahl *et al.,* 1998), or with products supplemented with probiotics, or with low fat content (van Trijp and Steenkamp, 1998).

Concerns about safety are also significant. Food safety is related to the hazards of food consumption either in the short run via allergenic reactions or in the long run through impact on the health of future generations (Fuller, 1994; Hamstra, 1995; Bredahl *et al.*, 1998; Wohl, 1998). Food safety can also include aspects of security in the sense of trust for the product (Bredahl *et al.,* 1998).

The taste appeal is also an element that influences consumers' acceptance of food products. If the product does not taste properly it is not likely to be purchased. The issue of 'taste' includes some dimensions which are implicitly required such as rich flavour (Katz, 1999), rich or smooth texture (Fuller, 1994), or low fat content and wholesomeness (Bredahl *et al.,* 1998). Taste is also an element of the quality perception that consumers hold (Fuller, 1994; Ter Hofstede *et al.*, 1996; Steenkamp, 1997; van Trijp and Steenkamp, 1998).

Convenience is another factor which influences consumers' acceptance of foods. Convenience is related to the perceived ease of use and processing the product (Hamstra, 1995). It can be related to long shelf-life (Hamstra, 1995; van Trijp and Steenkamp, 1998), a long storage time (Unilever, 1995) or the ease of preparation, handling and cooking (Ter Hofstede *et al.*, 1996; Steenkamp, 1997; Katz, 1999).

Translation of product features into consumer benefits

To outline the role of genetic manipulation technology in the NPD process, more insight is required about the benefits that technology delivers to the different actors involved. Concrete product features derived from genetic manipulation are identified, and translated into concrete customer benefits (using consumers' terminology). Initially, we define which genetic manipulation technique is used, and explain how this technique is applied. Later we translate product features derived into product attributes. Thus, product attributes can be translated into benefits for each actor (farmer, producer, retailer and end user consumer) using its own terminology. This process can be considered as using the HoQ the other way round. The consumer needs are not the starting point, rather the product features derived from genetic manipulation. Data about the products' physical characteristics are similar to the data at the top of the HoQ. The attributes delivered resemble the left part of the house. The final result is an assessment of what technology is able to deliver to different types of consumers. This analysis is illustrated in Tables A1–A3 in the Appendix.

Plant production

Table A1 shows that application of genetic manipulation in plant production often leads to improvements of agronomic traits (like herbicide, insecticide, fungal or bacterial resistance). This contributes in certain cases to the improvement of the quality of the final product. Soya, maize or potatoes that are resistant to devastating pests (e.g. Colorado beetle, certain weeds) can provide effective crop protection, resulting in fewer losses. Other advantages are cost reduction and enhanced yields. Grains that are herbicide tolerant lower the management costs. Lower cost contributes to more efficient production. The product entering the market may be of better quality, since it may contain less chemical residues from herbicides and will probably be of a lower price. Genetically manipulated tomatoes, which are easy to handle during the postharvest process, provide benefits for both producer and consumer.

Food production

Genetic manipulation technology contributes to the improvement of processing traits in a direct or indirect way (Tables A2 and A3). GM tomatoes (through a thicker skin or slow ripening) provide benefits for the producer, since they can be stored for a longer time. This results in fewer losses. Transport costs can be reduced as well. The quality of the product remains changeless for a longer time. This offers convenience to the producer for storing and processing the product. Longer storage time also benefits retailers and consumers.

Genetic manipulation offers more controlled fermentation processes, using raw materials that are more stable and are of superior performance. GM enzymes or other starter microorganisms (bacteria) can be considered as purer ingredients since the need for additives can be eliminated. Application of technology to oil derivatives targets production of oil with improved taste characteristics for the consumer. Chymosin is also a necessary ingredient for cheese production, which can be produced with the help of genetically modified enzymes. Chymosin is purer in the sense that larger amounts of enzymes are produced compared with the traditional method.

Advances in technology are essential for food product development. New technologies may provide more efficient solutions to production problems or offer the means for more efficient production. Genetic manipulation technology can be considered as a 'clever way' to improve processes of production and the characteristics of the products. Technology is applied to intermediate products to improve the processes and provide ways for more 'green production' since naturally modified ingredients are used instead of chemical ones (Giuseppin, 1999, personal communication).

Genetic manipulation technology provides products with additional health benefits and improved taste, and offers ingredients with specific function (Roller and Harlander, 1998), at the same time reducing undesirable effects of specific components (Giuseppin, 1999, personal communication). With respect to quality of production, genetic manipulation may reduce the need for chemical additives through tailor-made microorganisms which provide a wide range of specific product attributes for specific target groups (Giuseppin, 1999, personal communication).

Consumer acceptance

One issue that is significant for inducing consumers' acceptance of new technologies for food production is benefits realization (Hamstra, 1995). Many of the products introduced so far may carry little or no benefits to consumers. When consumers have been asked about the scope and the potential advantages of the deployment of biotechnology in foods, the advantage for the producer was by far the most common response (European Commission, 1997; Frewer *et al.*, 1997b; Bredahl, 1998; Sheehy *et al.*, 1998; *The Economist*, 2000). To date, the complexity of the technology prevented consumers from understanding the attributes that technology provides. This results in an unequal relationship between producer and consumer benefits where the consumer only sees producer benefits (Frewer *et al.*, 1997; Sheehy *et al.*, 1998).

In general consumers in Europe do not seem willing to accept the use of genetic modification in foods. When they are confronted with the option to avoid GM products, they would probably do so (Bredahl, 1998, 2000). Consumers show their unwillingness to accept GM products which provide less perceived additional value than that currently available from the existing ones (Hamstra, 1995; Bredahl *et al.*, 1998; Sheehy *et al.*, 1998). In the study conducted with German consumers the product scenario with a GM yogurt (with GM bacteria) gained the least acceptance, as it was perceived that it has the same function as the traditional one but is more hazardous for health.

It has been verified that consumers' behaviour towards the technology and products is shaped by a risk–benefit evaluation of the technology overall coupled with the evaluation of risks and benefits of the products. In addition some general attitudes held by consumers play an influential role. It has been well documented that a 'risk averse' attitude towards technologies in foods can stimulate reservations to technological innovations (Hamstra, 1995; European Commission, 1997; Bredahl, 2000). Moreover, attitudes towards environment and nature (Hamstra, 1995; Frewer *et al.*, 1997; Kuznesof *et al.*, 1997; Bredahl, 2000), perceived trust in regulators and public agencies (Frewer *et al.*, 1996; European Commission, 1997; Bredahl, 2000) and price sensitivity (Zechendorf, 1994; Kuznesof *et al.*, 1997; Bredahl, 2000) can also be elements of the general attitudes consumers have. Bredahl, in a series of studies (Bredahl, 1998, 2000), analysed consumers' attitude formation towards genetic manipulation technology in foods and suggested a behavioural model in which the determinants of attitudes are the perceived risks and benefits trade-offs related to product and technology. Moreover, the perceived process risks and benefits influence the attitude towards the process of production. However, empirical evidence shows that a clear distinction of attitudes between technology and the product could not be inferred (Bredahl, 2000).

Biotechnology is a relatively new, complex and constantly changing technology surrounded by a level of uncertainty about the long-term implications. Complexity and uncertainty are issues that need time for the consumers to comprehend and ultimately evaluate the pros and cons (Menrad *et al.*, 1998; Sheehy *et al.*, 1998; Biefang, 1999). This is a major factor influencing attitudes formation (Biefang, 1999). This aspect is confirmed by the findings of the survey of German consumers. Biotechnology was perceived as not 'a natural way of production' which involves a high degree of risk. In this case, consumers have the perception of being involved in involuntary, man-made risks which might have chronic and disastrous impacts on human health among other things (Biefang, 1999). Taking all these considerations into account, the perception of risk includes many other factors which lie beyond the domain of technical and scientific risks (Wohl, 1998; Biefang, 1999). Risk theory suggests that activities or technologies judged as highly risky are considered low in benefits (Wohl, 1998; Biefang, 1999). This is probably true in the case of biotechnology.

There is also strong evidence by research that the acceptance of genetically manipulated products can be assessed on a case by case basis. If we take the example of 'rapid method of production' using biotechnology, this was perceived as a positive one for the cheese products (Hamstra, 1995; Frewer *et al.*, 1997), but not when it was referred to in yogurt or beer (Bredahl, 1998). This argument could be verified by the pattern with which the German consumers responded to different product scenarios stemming from the same application of technology. When the scenarios represented explicitly visible benefits (functional foods) then there were more positive responses than for scenarios representing weak or vague consumer benefits (the case of yogurt or vegetables). Recall that the value of visible benefit delivery is a key issue.

High levels of consumer awareness are shown in most relevant studies. Information about the technology is highly desired and accepted. Clear correlation between the level of knowledge and the corresponding levels of acceptance could not be inferred. Findings by the European Commission (1997) show that increased information does not seem to influence the attitudes consumers hold, but enforces the existing ones. Some authors claim, and evidence exists, that although consumers show high levels of awareness, the levels of actual knowledge about the technology appear to be relatively low (Hamstra, 1995; European Commission, 1997; Sheehy *et al.*, 1998). Sheehy *et al.* (1998) explain this situation and suggest that biotechnology is new and complex, thus not evident to the casual consumer by the final product. This is an argument justifying consumers' demand for labelled products (Hoban, 1997, 1999). Consumers seek to be informed and ask for the right of an informed choice based on actual information-based labelling. The vast majority of German consumers in our study (over 50%) are willing to accept GM foods in the market under the condition of clear labelling.

Conclusions

This chapter aimed to analyse the topic of new product development with the help of genetic manipulation technology in the agro-food sector. The analysis moved along three dimensions. One was to review the literature on some basic principles of NPD using a consumer-oriented approach. Application of genetic manipulation technology offers the opportunities to improve processes of production or to deliver an entirely new array of products. With regard to this aspect, we screened technology in terms of its potential in agriculture and food production, and identified what it is able to deliver in terms of concrete benefits to all actors involved. Biotechnology has the potential to offer additional value and the consumer emerges as a bottleneck for this additional value delivery. Along this line, we ascertain some crucial factors that influence consumers' acceptance of GM products.

Utilization of genetic manipulation technology for NPD is set to take place under a turbulent environment both on the supply side (producers need to be adapted to a constantly changing technological environment) as well as on the demand side (as consumer concerns and preferences are subject to constant change). Companies need to have a formal and structured NPD procedure where the consumer needs and 'wants' emerge as a crucial element. Formalized procedures such as the QFD can be appropriate to improve the effective translation of consumer needs into physical product features desired by consumers.

Genetic manipulation in food production is employed to enhance quality by improving metabolic pathways, by eliminating undesired properties, by extending the shelf-life of products or by offering improved quality in the process of production. Micro ingredients are targets of technology in order to introduce tailor-made products for specific cases. Agriculture is one of the main areas of biotechnology application. The main target initially was to improve breeding, later to introduce crops with desired traits of interest. Nowadays the focus of research and production is to shift from simple traits to more complex reactions coded by interacting genetic sources (genes). Overall, genetic manipulation technology has significant potential to provide additional value to products, which presumably can benefit the consumer, either the end user or the intermediate. Exhibits in Tables A1, A2 and A3 represent only a fraction of the whole picture, which is constantly growing.

Although the benefits of biotechnology can be obvious, consumers tend to be increasingly critical of GM foods. In general, they are inclined to be more demanding, and become more concerned about the process of production and the scope of technology used. With regard to biotechnology, a number of factors are identified which influence the levels of consumer acceptance. The literature suggests that a determinant of acceptance is the attitudes the people hold towards the technology and accordingly the product. Attitudes can also be influenced by the consumers' subjective perception of the actual product characteristics. Perceived risks and benefits have a significant impact in evaluating the products and technology overall. The benefits delivery is a crucial factor for determining the risks–benefits trade-off. Concrete, specific and affordable benefits delivered on consumption can significantly raise acceptance of GM foods.

Ultimately, product developers within the agro-food sector need to adapt themselves to an environment where the consumer benefits appear to be the bottleneck for added value delivery. Consumer benefit delivery needs to be taken very seriously into account during the early stages of the product development process. At the same time adequate information should be communicated by the producer to the public, either by clear labelling or by other risk-communication activities about the actual risks and benefits that the new products convey. Product development using biotechnology in food and plant production needs to be associated with additional inputs, which are essential for successful commercialization. Those include an in-depth intelligence of what consumers already know and believe about food biotechnology, as well as recognition that risk is not only a technical concept but also a social construct. Those aspects suggest that success in the market would also have to include an in-depth communication strategy which would gradually include the whole spectrum of issues surrounding food biotechnology (Biefang, 1999). All these arguments constitute the need for producers to shift from a technology-push situation more to a consumer-oriented product development, where radical innovations in technology can be successfully transformed into real additional value realized by consumers.

References and Background Reading

Baker, M. and Hart, S. (eds) (1999) *Product Strategy and Management.* Prentice-Hall, London.

Begemann, B.D. (1997) Competitive strategies of biotechnology firms. Implications for US agriculture. *Journal of Agriculture and Applied Economics* 29, 118–119.

Belem, M.A.F. (1999) Application of biotechnology in the product development of nutraceuticals in Canada. *Trends in Food Science & Technology* 10, 101–106.

Biefang, S. (ed.) (1999) *Strategies for Communication on Food Biotechnology,* EUR 18600EN. European Communities, Office for Official Publications of the European Communities, Luxembourg.

Bredahl, L. (1998) *Consumer Cognition with regard to Genetically Modified Foods. Results of a Qualitative Study in Four Countries,* working paper No. 59. Center for Market Surveillance Research and Strategy for the Food Sector, The Aarhus School of Business, Denmark.

Bredahl, L. (2000) *Determinants of Consumer Attitudes and Purchase Intentions with Regard to Genetically Modified Foods: Results of a Cross-national Survey,* working paper No. 69. Center for Market Surveillance Research and Strategy for the Food Sector, The Aarhus School of Business, Denmark.

Bredahl, L., Grunnert, R.G. and Frewer, L. (1998) *Consumer Attitudes and Decision-making with regard to Genetically Engineered Food Products: a Review of the Literature and a Presentation of Models for Future Research,* working paper No. 52. Center for Market Surveillance Research and Strategy for the Food Sector, The Aarhus School of Business, Denmark.

Bucio, L.J., Vega, M.O., Garcia, A.G. and Estrella, H.L. (2000) Enhanced phosphorus uptake in transgenic tobacco plants that overproduce citrate. *Nature Biotechnology* 18, 450–453.

Cohen, L. (ed.) (1995) *Quality Function Deployment.* Addison-Wesley, New York.

Cook, J.D., Emptage, L.S., Miller, F.W., Rauch, S. and Ruiz-Funes, J.-P.M. (1997) Food biotechnology: can you afford to be left out? *McKinsey Quarterly* 3, 78–89.

Cooper, R.G. (1996) Overhauling the new product process. *Industrial Marketing Management* 25, 465–482.

Crawford, C.M. (ed.) (1996) *New Products Management,* 5th edn. Irwin/McGraw-Hill, Michigan.

Darf, P.J. (1990) Plant production: introduction. In: Persley, G. (ed.) *Agriculture Biotechnology: Opportunities for International Development. Biotechnology in Agriculture* No. 2. CAB International for The World Bank, International Service For International Agricultural Research, CAB International, Wallingford, UK.

The Economist (2000) Blech. *The Economist* 354, 75.

European Commission (1997) *Eurobarometer* 46.1. *The Europeans and Modern Biotechnology.* European Commission Directorate General XII, Science Research and Development, Office for Official Publications, Luxembourg.

European Commission (1998) *Looking at the Biotechnology Consumer* EUR 18492 EN. Directorate-General Science, Research and Development, Office for Official Publications, Luxembourg.

Frewer, L.J., Howard, C., Hedderley, D. and Shepherd, R. (1997a) Consumer attitudes towards different food-processing technologies used in cheese production: the influence of consumer benefit. *Food Quality and Preference* 8(4), 271–280.

Frewer, L.J., Howard, C. and Shepherd, R. (1997b) Public concerns in the United Kingdom about general and specific applications of genetic engineering: risks, benefit, and ethics. *Science Technology and Human Values* 22(1), 98–124.

Fuller, G.W. (ed.) (1994) *New Food Product Development. From Concept to Marketplace.* CRC Press, Boca Raton, Florida.

Galperin, M.Y. and Koonin, E.V. (2000) Who's your neighbor? New computational approaches for functional genomics. *Nature Biotechnology* 18.

Gatignon, H. and Xuereb, J.-M. (1997) Strategic orientation of the firm and new product performance. *Journal of Marketing Research* Feb, XXXIV, 77–90.

Goldberg, R. (1999) The business of agriceuticals. *Nature Biotechnology* 17, BV5–BV7.

Govers, C.P.M. (1996) What and how about quality function deployment (QFD). *International Journal of Production Economics* 46, 47, 575–585.

Griffin, A. and Hauser, J. (1993) The voice of the customer. *Marketing Science* 12(1), 1–27.

Griffin, A. and Hauser, J. (1996) Integrating R&D and marketing: a review and analysis of the literature. *Journal of Product Innovation Management* 13, 191–215.

Grunert, K., Harmsen, H., Hartvig, H., Larsen, E. and Sorensen, S.B. (1997) New areas in agricultural and food marketing. In: Wierenga, B., van Tilburg, A., Grunert, K., Steenkamp, J.-B.E.M. and Wedel, M. (eds) *Agricultural Marketing and Consumer Behavior in a Changing World.* Kluwer Academic Publishers, Boston, Massachusetts.

Hamstra, A.M. (1991) *Biotechnology in Foodstuffs: Towards a Model of Consumer Acceptance.* The SWOKA Institute, The Hague.

Hamstra, A.M. (1995) *Consumer Acceptance Model for Food Biotechnology: Final Report.* The SWOKA Institute, The Hague.

Haseltine, W.A. (1998) The power of genomics to transform the biotechnology industry. *Nature Biotechnology* 16 (Suppl.), 25–27.

Hauser, J.R. (1993) How Puritan-Bennet used the house of quality. *Sloan Management Review* Spring, 61–70.

Hauser, J. and Clausing, D. (1998) The house of quality. *Harvard Business Review* May–June, 63–73.

Hoban, J. (1997) Consumer acceptance of biotechnology: an international perspective. *Nature Biotechnology* 15, 232–234.

Hoban, J. (1999) Public perceptions and understanding of agricultural biotechnology. Economic perspectives: an electronic journal of the US Department of State. www.usia.gov/journals/journals.html

Hodgson, J. (2000) Crystal gazing the new biotechnologies. *Nature Biotechnology* 18, 29–31.

ISAAA (2000) Global review of commercialised transgenic crops. www.isaaa.org

Jarowski, B.J., Ajay, K. and Kohlli, A.K. (1993) Market orientation: antecedents and consequences. *Journal of Marketing* 57(3), 53–79.

Johne, F.A. and Snelson, P.A. (1998) Success factors in product innovation: a selective review of the literature. *Journal of Product Innovation Management* 5, 114–118.

Katz, F. (1999) Top product development trend in Europe. *Food Technology* 53(1), 38–42.

Kaul, A. and Rao, V.R. (1995) Research for product positioning and design decisions: An integrative review. *International Journal of Research in Marketing*, 293–320.

Kohli, A.K., Ajay, K. and Jarowski, B.J. (1990) Market orientation: the construct research propositions and managerial implications. *Journal of Marketing* 54(1), 1–20.

Kuznesof, S., Gofton, L. and Ritson, L. (1997) *Consumer Acceptability of Biotechnology in Relation to Food Products, with Special Reference to Farmed Fish.* European Communities, Office for Official Publications of the European Communities, Luxembourg.

McElroy, D. (1999) Moving agbiotech-downstream. *Nature Biotechnology* 17, 1071–1074.

Mazur, B. (1995) Commercializing the products of plant biotechnology. *Trends in Biotechnology* 13, 319–323.

Mazur, B. (1999) Technology issues in plant development. *Nature Biotechnology Supplement* 17, BV9–BV10.

Menrad, K., Kochatzky, K., Maßfeller, S. and Strauß, E. (1998) *Communicating Genetic Engineering in the Agro-Food Sector to the Public*, EUR 18358EN. European Communities, Office for Official Publications of the European Communities, Luxembourg.

Millar, H. (1999) Consumer expectations and openness. *Nature Biotechnology Supplement* 17, BV39–BV40.

Moenaert, R.K., Souder, W.E., De Meyer, A. and Deschoolmeester, D. (1994) R&D-marketing integration mechanisms, communication flows and innovation success. *Journal of Product Innovation Management* 11, 31–45.

Nyström, H. (ed.) (1990) *Technological Market Innovation.* John Wiley & Sons, Chichester, UK.

Persley, G.J. (ed.) (1992) *Beyond Mendel's Garden: Biotechnology in the Service of World Agriculture.* CAB International for the World Bank, Wallingford, UK.

Renkonski, M.A. (1997) Marketing strategies of biotechnology firms: implications for U.S. agriculture. *Journal of Agricultural and Applied Economics* 29(1), 123–131.

Roller, S. and Harlander, S. (eds) (1998) *Genetic Modification in the Food Industry.* Chapman & Hall, London.

Römer, S., Fraser, P.D., Kiano, J.W., Shipton, C.A.,

Misawa, N., Schuch, W. and Bramley, P.M. (2000) Elevation of the provitamin A content of transgenic tomato plants. *Nature Biotechnology* 18, 666–669.

Saguy, S. and Moskowitz, H.R. (1999) Integrating the consumer into new product development. *Food Technology* 53(8), 67–73.

Saviotti, P.P. (1998) Industrial structure and the dynamics of knowledge generation in biotechnology. In: *Biotechnology and Competitive Advantage.* Jacqueline Senker, Bath.

Scott, G.M. (1998) The new age of new product development: are we there yet? *R&D Management* 28(4) 225–236.

Sheehy, H., Legault, M. and Ireland, D. (1998) Consumers and biotechnology: a synopsis of survey and focus group research. *Journal of Consumer Policy* 21, 359–385.

Sheth, J.N., Mittal, B. and Newman, B.I. (eds) (1999) *Customer Behavior: Consumer Behavior and Beyond.* The Dryden Press, Dryden, Ontario.

Steenkamp, J.-B.E.M. (1997) Dynamic in consumer behavior with respect to agricultural and food products. In: Wierenga, B., van Tilburg, A., Grunert, K., Steenkamp, J.-B.E.M. and Wedel, M. (eds) *Agricultural Marketing and Consumer Behavior in a Changing World.* Kluwer Academic Publishers, Boston, Massachusetts.

Temponi, C., Yen, J. and Tiao, A.W. (1999) House of quality: a fuzzy logic-based requirements analysis. *European Journal of Operational Research* 117, 340–354.

Ter Hofstede, F., Steenkamp, J.-B.E.M. and van Trijp, J.C.M. (1996) *Pan European Study of Vegetables and Fruits,* September. Unilever, The Netherlands.

Thomson, J.D. (1999) Shortcuts from gene sequence to function. *Nature Biotechnology* 17, 1158–1159.

Unilever (1995) *Unilever and Biotechnology; Merging Science and Consumer Benefits.* Unilever Corporate Relations, The Netherlands.

Urban, G.L, Hauser, J.R. (eds) (1993) *Design and Marketing of New Products*, 2nd edn. Prentice-Hall, London.

van Trijp, H.J.C.M. (1999) Consumer Understanding as Inspiration for Innovation, Nordic Conference, Nordic Food Conference, Norway.

van Trijp, H.J.C.M. and Steenkamp, J.-B.E.M. (1998) Consumer-oriented new product development: principles and practice. In: Jongen, W.M.F. and Meulenberg, M.T.G. (eds) *Innovation of the Food Production Systems.* Wageningen, 37–66.

van Vliet, R. (1998) Introduction. In: Senker, I. and van Vliet, R. (eds) *Biotechnology and Competitive Advantage.* Edward Elgar Publishing Ltd, USA.

Wheelwright, S.M. (1994) Commercializing biotech products. *Bio/Technology* 12, 877–880.

Whitelaw, B. (1999) Towards designer milk. *Nature Biotechnology* 17, 135–136.

Wohl, J.B. (1998) Consumers decision-making and risk perception regarding foods produced with biotechnology. *Journal of Consumer Policy* 21, 387–404.

Wymer, P. (ed.) (2000) *The Application of Genetically Modified Lactic Bacteria in Food Products in Europe,* the reports of a workshop. Cambridge Biochemical Consultants, Vlaardigen, The Netherlands.

Zechendorf, B. (1994) What the public thinks about biotechnology. *Bio/Technology* 12, 870–875.

Appendix: Translation of Product Features derived with the help of Genetic Manipulation Technology in Attributes and Benefits. Examples related to Plant, Food and Dairy Production.

Table A1. Translation of product features derived from genetic engineering techniques, into attributes and benefits: examples related to plant production.

Biotechnological feature, possibility	How applied in product	Attribute derived by new feature	Benefits
Implantation of genes	Insertion of gene (*Cry* III) to make *potatoes* to produce toxin to kill Colorado beetle	Colorado-resistant potatoes	Beetle control → less chemical residues → environmentally friendly (P, C) Beetle control → crop protection → fewer losses → cost reduction and yield enhancement (F)
Isolation and implantation of genes	Isolation and implantation of genes in *potatoes* to increase solid content	Starch-modified potatoes	Increased carbohydrate content → more solid content in potatoes → reduced cooking time → convenience (C) Increased solid content → better cooking performance → convenience (C) Increased carbohydrate content → low calories → prevent overweight → health (C) No postharvest damage → reduce transport costs (F, R)
Implantation of toxic proteins	Insecticide (borer) proteins from *Bt* implanted in *cereals* in order to make them resistant to insects	Insect-resistant cereals	Crop protection → less damage → low cost → better product quality in marginal areas (F)
Enzymes coded by gene parts	Enzymes made herbicide-resistant cereals by gene-coding	Herbicide-resistant cereals	Lower herbicide amount to plants → environmentally friendly (P, C) Lower chemical residues to the plant → improve quality → more natural product → health (C) Pre-emergence applications → prevent devastation → lower damage → lower costs → enhance yields → more efficient production (F)
Implantation of bacterial gene	To moderate the carotenoid biosynthesis	Increased production of provitamin A in tomatoes	Increased production of β-carotene → tomatoes with more provitamin A → more nutrients → enhance health (C)
Modification of the tobacco genome	To enhance phosphorus uptake	Plants which utilize phosphorus more efficiently	P is more available to the plant → less P is wasted → plants able to grow in difficult soil conditions → lower costs on fertilizers (F) Less P wasted → less soil pollution (C)
Implantation of genes	Seed storage protein 'Beta phaseolin' implanted in *rice*	Increased protein level rice	Increased protein level → more nutritious rice (C)

Modification of gluteins	Modify glutein content in wheat for *bread* making	Glutein-rich wheat	Improved glutein content of wheat → improves baking process → better products (more/less fluffy) Improved glutein content of wheat → improves flavour → different products for different tastes and uses (for toasts, for meals) (R,C)
Modification of yeast	Manipulation of amilolytic yeast used in *beer* production	Beer with features derived from a variety of yeast	Natural ingredient → environmentally friendly (P, C) Natural ingredient → lower content of artificial ingredients → quality enhancement → more 'natural' product (C) More controlled fermentation → cost saving → lower prices (P, C)
Cloning of genes, including coding and modification	Modification of sodium dioxide in *beer* fermentation	More stable flavour	Stable fermentation process → more controlled fermentation → cost saving (P) Stable fermentation process → quality enhancement (P, C)
Implantation of genes	Insertion of gene compartments from bacteria virus from wild species (*Canola*)	Alter oil composition, high lauric acid	Expand use of plant oils in soap and food products → improved quality of plant oil in cosmetics (C)
Modification of genes	Implantation of gene to alter oleic acids in *soya*	Oil with high linoleic acid, high oleic and palmitic acid Reduced transfatty acids	Replacement of hydrogenated acids → improved quality of soya oil (C) Replacement of hydrogenated acids → enhances healthiness (C)
Modification of genes coding for useful amino acids	Manipulation of genes coding for biochemical reactions	Increase production of amino acids	Plants used as sources of amino acids useful for drugs → lower costs of produ ction (P) Plants used as sources of amino acids useful for drugs → more natural substances → more efficient drugs → 'Pharming' (C)
Modification of genes	Modification of lectin level in *legumes*	Lowered lectin level	Avoid allergic reactions → stay healthy (C) Product for different types of consumers → serve specific consumers (allergenic ones) (P)

(F), farmer; (R), retailer; (C), consumer; (P), producer.

Table A2. Translation of product features, derived from genetic engineering techniques, into attributes and benefits: examples related to vegetables and fruits.

Biotechnology feature possibility	How applied	Attribute derived by the new feature	Benefits
Implantation of foreign agents	Implantation of bacteria and virus to alter pectin in *tomatoes*	Thicker skin tomatoes	Enhanced process value → possibility to store longer → lower transport costs → lower losses (P, R) Stable quality → reduced energy costs → reduced cost of paste (P) → lower prices of paste (C) Prevent softening of skin → better cooking performance (C) Longer shelf-life → convenience (C, R) Improved flavour → superior taste (C)
Manipulation of genes	Identification and modification of TOM 41 gene in *tomatoes*	Delayed ripening tomatoes	Longer (storage) shelf-life → convenience (R, C) Improved flavour → improved taste → improved cooking performance (C) Resistant to postharvest pathogens → improved quality → lower product loses (F, P, R)
Selection of strains of potatoes to produce extra amounts of thaumatin (sweetener)	Identification and implantation of relevant genes to host plants	Plants to produce thaumatin	Natural non-carbohydrate-containing sweetener → suitable for dietetic products (diabetes) → promote health → avoid allergies (C) Not only potatoes to produce Th. → expansion of this cultivation outside Africa (P)
Manipulation of genes	Modification of relative gene in *strawberries*	Long-life strawberries	Longer shelf-life → convenience (C, R) Longer storage life → fewer losses to the market (P, R)

(F), farmer; (R), retailer; (C), consumer; (P), producer.

Table A3. Translation of product features, derived from genetic engineering techniques, into attributes and benefits: examples related to dairy products.

Biotechnology feature possibility	How applied	Attribute derived by the new feature	Benefits
Selecting and producing enzymes from bacteria	Selecting and producing chymosin from bacteria for *cheese* production	Vegetable cheese, made with pure enzymes	Rapid production → lower cost → faster to the market → competitive advantage (P) Lower loss of enzymes → cheaper ingredients → efficient production (P) Cheaper ingredients → lower production costs → lower prices for the same high quality (C) Not derived from animals anymore → vegetarian cheese → animal welfare (C)
Modification of bacteria	*Yogurt* microbes contain modified gene	Avoid 'post-fermentation acidification'	Natural product (no additives) → healthy (C) Longer to keep → more convenient → avoid acidification (R, C)
Modification of autolytic starter bacteria	Modify bacteria	More controlled fermentation	Faster production → more controllable → lower costs (P) Green dairy process (P, C)
Modify gene expression (low lactose transgenic livestock)	Intervene in gene function to produce lactose	Low lactose milk with nutritive properties	No allergenic reactions to lactose → different products for different types of consumers (P, C) Low lactose milk with high Ca levels → enhance health for lactose allergens by normal Ca intakes (C)

(F), farmer; (R), retailer; (C), consumer; (P), producer.

7 The Impact of Bovine Somatotropin on Farm Profits

Loren W. Tauer
Cornell University, Ithaca, New York, USA

Introduction

The compound bovine somatotropin (bST) has been commercially available to US dairy producers since February 1994. Before being approved for sale, bST was subjected to years of investigation and testing. Given the large production response per cow that most of these tests reported, bST was generally projected to be profitable for dairy farmers, with estimates often exceeding US$100 year^{-1} per cow (Butler, 1992), although some projected little or no profit (Marion and Wills, 1990). Now that bST has been used by US farmers for a number of years, it is possible to estimate their actual production and profit responses.

Bovine somatotropin is a hormone produced by the dairy cow that regulates milk production. The genetic material for this compound has been isolated and is produced by recombinant biotechnology. This recombinant-produced bovine somatotropin can then be injected into the dairy cow to augment her naturally produced levels of this hormone, enhancing milk production, but requiring additional feed and other inputs to increase milk production. Monsanto (www.monsanto.com/dairy/) is currently the only US supplier of recombinant bST under the registered tradename POSILAC. As of 11 May 1999, Monsanto stated that 13,000 dairy producers were using bST, and of the nearly 9 million dairy cows in the US, approximately 30% of the cows are in herds that are supplemented with POSILAC.

Tauer and Knoblauch (1997) used data from the same 259 New York producers in 1993 and 1994 to estimate the impact of bST on milk production per cow and return above variable cost per cow. bST was not available in 1993, but one-third of these farmers used bST in 1994. The use of bST had a positive and statistically significant impact on average production per cow ($\alpha = 0.01$), but the profit effect, although positive and large, was not statistically different from zero ($\alpha = 0.14$).

Stefanides and Tauer (1999) also analysed the production and profit effects using the same data source, but included data from 1995, resulting in a panel data set of 211 farms. They corrected for self-selection bias by using the two-step Heckman approach, and estimated a probit adoption function for each year (Greene, 1997). They likewise found a statistically significant and positive effect on milk production per cow from the use of bST, but found that the impact of bST on profits was statistically zero. They suggest that farmers may still be learning how to use bST profitably, or that such a large number of farmers are using bST, including those getting a low return, that the average farm is not making a profit from its use.

Nonetheless, there may still be a subset of farmers earning a positive return from the use of bST. These farmers may either have been the most

(eds V. Santaniello, R.E. Evenson and D. Zilberman)

effective learners or have a unique position or characteristic to profit from bST. This chapter implements that concept using 4 years of bST-use data and measures the impact by different types of farmers. Observations from 138 New York dairy farms are available. Since farmers displayed various historical use patterns over the 4 years of bST availability, those patterns are used to accommodate any self-selection bias. Likewise, since data were available from 1993 before bST was available, farms are sorted based on managerial abilities displayed that year, and the impact of bST in later years is estimated by management level.

Methods and Model

The data comprise a group of farms, some of which use or have used bST. The intent is to determine whether the use of bST increases profits. A profit function was estimated where profit is a function of output and input prices, other exogenous determining variables, and the use or non-use of bST. The data available are from an ongoing farm business analysis programme, and actual or expected prices were not collected, except for an implicit milk price computed by dividing milk revenue by milk sold. Published price data are available for only a few inputs, and those that exist are collected and reported at the state level, resulting in no variability across farm observations. Four years of data also provides little temporal price variability. Because of these data limitations, prices are not included in the profit function except for the price of milk.

Included in the profit function was a dummy variable representing the use or non-use of bST. However, the potential exists for self-selection bias since the farmers themselves determined their use or non-use of bST. Farmers that use bST may be more or less profitable as a group even without the use of bST. The result is that the error term on the profit function may be correlated with this treatment effect. There are a number of remedies to this data limitation, most of which involve the estimation of a separate equation explaining the selection decision, and using the prediction from that equation as an instrumental variable (Davidson and MacKinnon, 1993). Stefanides and Tauer (1999) used Heckman's two-step estimation procedure and found insignificant evidence of bST selection bias in an earlier use of this data source. This result may be due to the fact that all the farms in the data set are relatively well managed. Hence, the selection of bST may not be related to a management performance variable such as profit.

Measuring the impact of an event on welfare is pervasive in economic research. Examples include estimating the returns to education or to membership in labour unions. If data are available for a number of years, and individuals have histories of membership and non-membership in an event, then these historical patterns can be used to estimate the returns to individuals exclusive of membership in the event, resulting in a better estimate of the return to the event (Card, 1996). The dairy data used here include farmers that ceased using bST after initially using bST, and farmers who delayed their adoption of bST. Modelling the error structure as dependent on these various bST use histories alleviates the correlation between the error term and the use of bST. That approach was used in the empirical results reported below.

The profit w_{it} of farm i in period t is specified as:

$$w_{it} = a_t + \beta_t x_{it} + \delta\, u_{it} + v_{it} \qquad (7.1)$$

where u_{it} denotes the use of bST of farm i in period t and δ is the impact of bST use on profit; x_{it} represents observed exogenous variables for farm i and time t and β_t is the impact of those variables on profit at time t; a_t is the intercept for time t; and v_{it} is a residual component of profit. This residual profit can be separated into a permanent farm η_i component and a transitory (error) component as:

$$v_{it} = \eta_i + \varepsilon_{it} \qquad (7.2)$$

It is assumed that bST will only affect the permanent component so that $E(u_{it}\varepsilon_{it}) = 0$.

Following Card (1996), the permanent component of profit for farm i is specified as a function of a bST use history dummy variable u_{ih}, where h defines some historical bST use pattern, and a vector of exogenous variables x, as:

$$\eta_i = \sum_{h=1}^{H} u_{ih}\phi_h + \lambda x_{it} + \xi_i \qquad (7.3)$$

where ξ_i is an independent error term. The variable u_{ih} models various patterns of bST use and non-use over time for the specific farm i, ϕ_h is the impact of that specific bST use history on farm profit, and λ is the impact of the exogenous variables on farm profit.

The data consist of four separate years of bST

use or no bST use, so 16 possible combinations of annual use and non-use are possible. Many of these combinations are null or sparse. Many farms either have used bST for the entire 4 years or have not used bST in any of the 4 years. bST use is coded and is translated into a usage history as $h = kl$; $k = 0,1$; $l = 0,1$; where 0 is non-usage and 1 is bST usage during the first 2 years k, or during the last two years l. bST usage history is then coded as the four-member set $h = \{00, 01, 10, 11\}$ with membership of {50, 11, 15, 62}.

The profit Equation 7.1 can also be modified to use this bST use history as:

$$w_{it} = a_t + \beta_t x_{it} + \delta u_{ikl} + v_{it} \qquad (7.4)$$

where $\delta = 0$ by definition if bST is not used in period t. Then inserting Equation 7.3 into 7.2 into 7.4, produces a system of four equations, one equation for each year:

$$w_{i1} = a_1 + (\beta_1 + \lambda)\, x_{i1} + (\delta + \phi_{10})u_{i10} + \phi_{01}u_{i01} + (\delta + \phi_{11})u_{i11} + \xi_i + \varepsilon_{i1}$$

$$w_{i2} = a_2 + (\beta_2 + \lambda)\, x_{i2} + (\delta + \phi_{10})u_{i10} + \phi_{01}u_{i01} + (\delta + \phi_{11})u_{i11} + \xi_i + \varepsilon_{i2}$$

$$w_{i3} = a_3 + (\beta_3 + \lambda)\, x_{i3} + \phi_{10}u_{i10} + (\delta + \phi_{01})u_{i01} + (\delta + \phi_{11})u_{i11} + \xi_i + \varepsilon_{i3}$$

$$w_{i4} = a_4 + (\beta_4 + \lambda)\, x_{i4} + \phi_{10}u_{i10} + (\delta + \phi_{01})u_{i01} + (\delta + \phi_{11})u_{i11} + \xi_i + \varepsilon_{i4}$$

With this specification, λ represents the impact of general skill on profits, δ measures the impact of bST on profits when bST is used, while ϕ_{kl} represents the impact of the various use histories on the profit for any period t. If bST users would have larger profits even if bST is not used, then the impact of that is estimated separately as ϕ_{11} rather than from using bST. The parameter ϕ_{01} represents the impact on profits for later users of bST even if they are not using bST during years 1 or 2, and conversely ϕ_{10} represents the base profits of later bST users even if bST is not used.

This model specifies that the impact of bST on profits as measured by δ is not influenced by bST use history. Thus, the profit from bST for those farms that tried but discontinued the use of bST is the same as those farms that continuously used bST, and those farms that waited to use bST. Readers may find this tenuous. Many would believe that farms may have ceased using bST because those farms found bST unprofitable, while farmers who found bST profitable would have continued to use bST. However, modelling separate bST effects (separate δ's) for three bST use patterns, and including the bST history variables, would lead to a singular equation system without the ability to estimate separately modelled δ's. As a remedy, the bST parameter δ, and use history parameter ϕs can be combined to measure bST profitability, with the understanding that any inherent greater profitability of continuous bST farms that would have occurred even without the use of bST might be erroneously included as the return to bST. Combining coefficients may be appropriate since Stefanides and Tauer found no evidence of self-selection bias in a 2-year version of this data set. The estimate of δ might then be interpreted as a lower bound on bST return from the modelled equation, and the estimate $(\delta + \phi_{11})$ might be considered as an upper bound.

This system of four equations was jointly estimated with cross-equation restrictions imposed. The error structure consists of a term specified by farm across years, ξ_i, as well as an error unique to each period, ε_{it}. These errors were estimated by seemingly unrelated regression which allows for this contemporaneous correlation.

This system approach was also used to model and estimate the change in milk production per cow from the use of bST. The dependent variables w_{it} were replaced with milk production changes per cow.

Initial model estimates using the Wald test determined that unique a or $\beta + \lambda$ vectors by year did not exist. As such, a common intercept and beta $(\beta + \lambda)$ vector coefficients were estimated across equations.

Previous studies using earlier years of this data source concluded that bST has no impact on profits (Tauer and Knoblauch, 1997; Stefanides and Tauer, 1999). It is difficult to imagine that farmers would continue to use bST if it failed to generate a profit, although as Stefanides and Tauer (1999) suggest, the impact on milk output is unmistakable, possibly making it difficult for farmers to ascertain the true profitability of bST. Yet, it might be possible that a subset of bST users may be finding that product profitable. Various authors have conjectured that bST may only be profitable for the better managed farms (Marion and Wills, 1990). This was tested by dividing the sample into groups, based on individual farm performance in 1993 before bST was available.

Data

The data were from the New York Dairy Farm Business Summary Program (Knoblauch and Putnam, 1998). This is a record collection and analysis project primarily meant to assist dairy farmers in managing their operations. Farmers receive a business analysis of their farm and benchmark performance measures from combined participants. Farm analysis is done temporally if farmers participate annually. For the 5-year period of 1993–1997, a total of 138 farms participated every year. This provides the data set for the empirical analysis with data from 1993 used to sort farms, and bST impacts measured with data from 1994 to 1997.

This is not a random sample. It represents a population of farmers that actively participate in agricultural extension and research programmes. It would be tenuous to make inferences for the general population of dairy farms. The farms in this sample are larger on average than New York dairy farms and they experience higher levels of production per cow. Yet, there is a significant amount of hetereogeneity in the data. The smallest farm has only 29 cows, while the largest has over 2000 cows. The average number of cows is 165 (standard deviation of 236).

Variable specification is consistent with the annual Dairy Farm Business Summary (DFBS) Report (Knoblauch and Putnam, 1998). Production per cow is the total milk sold for the year divided by the reported average number of dairy cows. Short-run profit is measured as milk receipts minus the operating costs of producing milk. Operating cost includes variable costs, and excludes fixed cost items such as depreciation. If bST does increase farm revenue, farmers may use that additional revenue to purchase additional equipment not necessary for bST use, increasing computed depreciation, which would mute the measured impact of bST on profits. To be included in this published data set, milk receipts must constitute at least 90% of total farm receipts, yet some culled cows, calves, and excess feed were sold each year. Those receipt items were subtracted from total operating costs to estimate the operating costs of producing milk. This calculation assumes that the cost of producing products other than milk was equal to the value of those products. Although not necessarily true, detailed cost accounts were not collected by enterprise or receipt source.

Many of these farms are multiple owned operations, mostly parent–child. This was coded as 1 if multiple owned, 0 otherwise. Since there are multiple operators, age and education of the first manager were used. Education was coded as 1 if more than a high school education, 0 otherwise. Age was measured in years. Milking system was coded as a 1 if parlour, 0 if stanchion. Milk price was computed as gross milk receipts divided by the quantity of milk sold during the year. Finally, cows are the average number of cows the farm reported for the year. The average and standard deviation of these data are reported in Table 7.1.

As stated in the model section, farmers were coded as either using bST or not using bST during the first 2 years and then the last 2 years of bST availability. The DFBS surveys for each of the 4 years asked farmers to indicate their use of bST in one of five categories as (0) not used at all, (1) stopped using it during the year, (2) used on less than 25% of the herd, (3) used on 25–75% of the herd, or (4) used on more than 75% of the herd. Most responses over the period were in categories 0 and 3. Very few farms indicated that they used it on more than 75% of the herd, and if they did during one of the years, the other years they typically dropped back to using it on 25–75% of the herd. Likewise, few farms used it on less than 25% of the

Table 7.1. Definition of variables.

Variables	Definition	Mean 1993	SD 1993
SRPCOW	Short-run profit per cow	US$603	US$325
MILKCOW	Milk production per cow (kg)	8577	1180
COWS	Average number of cows	165	236
MILKPR	Milk price per cwt	US$13.20	US$0.52
AGE	Age of principal owner in years	46	10
BUS_ORG	Business organization, 1 if multiple owner	0.41	0.04
EDUC	Education, 1 if more than high school	0.54	0.04
MILK_SYS	Milking system, 1 if parlour	0.53	0.04

N = 138 observations.

herd, and if they did during any year, other years their use was usually 25–75% of the herd.

This bST use coding has limited informational content. Although most of these farms are DHIA (Dairy Herd Improvement Association) members, that organization does not code bST use on individual cow records, so neither age nor production level of individually treated cows was known. This lack of detailed bST management information precludes analysis on bST use tactics, which may be complex and unique by farm. Farmers using bST must believe that it is profitable on their farms. As such, farms were simply sorted into bST users and non-users.

Empirical Results

The system regression of return over operating costs per cow for all 4 years that bST was available, using the entire sample of 138 farms, shows that the impact of bST on profits was statistically not different from zero (Table 7.2). The profit estimate from bST use was US\$15.88 year^{-1} per cow, but the t-value testing whether this estimated coefficient was statistically different from zero is only 0.32. Farmers that used bST continuously for each of the 4 years experienced an overall numerical increase in profits per cow of US\$44.92 compared with farmers who did not use bST during any of the 4-year period. This US\$44.92 represents a bST use impact of US\$15.88 and a base profit impact of \$29.04. As discussed earlier, the value of US\$44.92 might be viewed as an upper bound to the return from continuous bST use while the value of US\$15.88 can be viewed as the lower bound estimate. However, a Wald test that the sum of δ and ϕ_{11} (US\$44.92) is different from zero, produced a chi-square value of only 0.69, failing to reject the null hypothesis, and accepting the alternative hypothesis that continuous bST users are not making money from using bST. Likewise, although the estimate of the combined bST and base return to those who stopped using bST

Table 7.2. Impact of bST on milk production per cow and short-run profit per cow; full sample.

Independent variables (standard error of each estimate is in parentheses below each coefficient estimate)	Dependent variable	
	Δ (MILKCOW)(kg)	SRPCOW (US\$)
bST use (δ)	297	15.88
	(101)	(50)
Continuous bST(ϕ_{11})	393	29.04
	(164)	(74)
bST use/drop(ϕ_{10})	191	−43.77
	(193)	(80)
bST wait/use(ϕ_{01})	184	−11.99
	(211)	(92)
AGE	9	10.88
	(36)	(17)
AGE^2	−0.10	−0.131
	(0.39)	(0.18)
EDUC	55	38.79
	(95)	(42)
BUS_ORG	167	133.96
	(96)	(43)
MILK_SYS	−28	−188.46
	(100)	(44)
MILKPR	−88	70.88
	(22)	(12)
Constant	819	−613.87
	(964)	(438)
Wald tests		
Ho: $\delta + \phi_{11} = 0$ chi-square (prob)	28.46 (0.00)	0.69 (0.41)
Ho: $\delta + \phi_{10} = 0$ chi-square (prob)	7.02 (0.01)	0.13 (0.72)
Ho: $\delta + \phi_{01} = 0$ chi-square (prob)	4.77 (0.03)	0.00 (0.96)

was a negative US$28, and the estimate of the combined bST and base return to the wait/use sequence was a positive US$4, neither of these estimates was statistically different from zero by Wald tests.

Although bST does not appear to be profitable on average for the types of farmers this sample represents, bST does increase milk output per cow. Knoblauch and Putnam (1998) report that DFBS farms using bST increased milk sold per cow from 9187 kg in 1993, before bST was available, to 10,469 kg in 1997. In contrast, farms not using bST sold 7905 kg per cow in 1993 and 7778 kg in 1997. This pattern suggests that the difference in milk sales each year from the base year of 1993 rather than level values of milk sold per cow should be defined as the regressand. The difference in milk produced per cow from the base year of 1993 was regressed on the same independent variables used in the profit equation. The bST impact was estimated to be 296 kg year^{-1}, and with a t-value of 2.93, was statistically different from zero. This is the increase in the herd average and includes cows that may not have been treated with bST at all during the period. Farmers that used bST continuously for each of the 4 years experienced an increase in milk production per cow of 689 kg year^{-1} compared with farmers who did not use bST during any of the 4-year period (Table 7.2). This 689 kg year^{-1} represents a bST impact of 296 kg year^{-1} and a base impact of 393 kg year^{-1}. Farms that used bST during the early part of the 4-year period, but discontinued during the last part of the 4-year period, experienced an overall response of 486 kg year^{-1}, while farmers that waited to use bST for the second half of the 4-year period experienced an overall response of 480 kg year^{-1}.

These results support the earlier estimates of Tauer and Knoblauch (1997), and Stefanides and Tauer (1999) that bST has a measurable and significant impact on output per cow, but has no statistically significant impact on profits per cow. With more experience, bST may spawn a larger milk output response, supporting the assertion that learning has occurred, but that learning only vaguely appears to translate into greater profits, since the bST use coefficients on the profit equation have high relative standard errors. Yet, some individual producers might be making money from using bST. To identify these farmers, the next section divides the sample into sub-samples based on various metrics used to measure management and performance among dairy farmers, to determine if better managers are making money from the use of bST.

Empirical Results of Sub-samples

One measure of managerial performance is profit. Return above operating costs is the statistic used to measure the impact of bST on profits per cow, and is used to sort farms into one of three managerial levels. Return above operating cost from 1993, the year before bST was available, is used to sort the 138 farms into three groups of 46 farms from lowest to the highest return.

Since profit per cow is transient by year, with few benchmarks for comparisons, two other proxies for management were used to sort the farms. One measure is milk sold per cow. It is generally acknowledged that better managed farms have higher output per cow. It is also a variable that is easily measured and commonly collected. It does have limitations. Production per cow may be low not only because of poor feeding and managing of the cow, but also because of low genetic potential of the herd. Although the selection and use of inferior genetics might be viewed as the consequence of poor management, the herd might be optimally managed subject to that genetic resource. This has implications for measuring the impact of bST, since test results have shown that even cows with mediocre genetics can respond to the use of bST if they are fed and managed optimally (Patton and Heald, 1992). Since bST is known to increase output per cow, the 138 farms were sorted into three groups based on milk per cow from the year 1993 when bST was not yet available. The group ranges were 5463–8153 kg per cow for the low group, 8159–9157 kg for the middle group, and 9161–11,088 kg for the high group.

Although not a direct measure of management, many believe that larger farms on average are better managed, or at least have the opportunity to benefit the most from the use of bST. Although small farms may be able to provide more individual attention to cows, correct feeding is important to effectively use bST, and larger farms are more apt to have state-of-the-art feeding facilities. This has policy implications since if larger farms benefit more from bST, in a competitive environment there will be additional pressure on the ability of the small dairy farm to survive (Tauer, 1992). Size as measured by the number of dairy cows is also an easily measured and reported statistic. Here the farms were again separated into three equal size groups of 46 farms based upon the average number of dairy cows during the year 1993. The groups' ranges were 29–71 cows for the small farm, 71–139 cows for the medium farm, and 139 to over 2000 cows for the large farm size, although most of the large farms in the sample had fewer than 250 cows.

The regression results, reported in Table 7.3 by sorting by profitability in 1993, show that the use of bST has no statistical impact on profits per cow regardless of the previous (inherent) profitability of the farm. Although the low profit per cow farms experienced a positive numerical return to bST (US$49.47), and the use of bST on higher profit per cow farms generated negative profits, the standard errors on these estimates are so large that statistically one might conclude that the effect is zero profit response across all profit levels. Combining the bST impact coefficient with the various bST use history coefficients produces combined coefficients which are also not statistically different from zero, as determined by the individual Wald tests in Table 7.3, except for possibly the high managed farms which make US$124 per cow using bST ($\alpha = 0.20$).

The regression results from the sort based on milk production per cow in 1993 exhibits an interesting pattern (Table 7.4). On the low production per cow farms, bST use increases milk production per cow by 419 kg year^{-1}, statistically different from zero. On the middle production per cow farms, bST use increases milk production per cow by 575 kg year^{-1}, again statistically different from zero. But on the high production farms, bST use appears to have no statistical impact on production per cow. These results support previous experimental results that low producing cows respond to bST (Patton and Heald, 1992). If base production effects are folded into the bST impact coefficients, then Wald tests show that all continuous bST users experience positive output effects from bST use, even on high production farms. On low production farms, that is

Table 7.3. Impact of bST on short-run profit per cow; sample sorted by profitability.

Independent variables (standard error of each estimate is in parentheses below each coefficient estimate)	Profit per cow (US$)		
	Low	Middle	High
bST use (δ)	49.47	−27.13	−11.68
	(84)	(73)	(97)
Continuous bST(ϕ_{11})	−67.04	−8.56	135.43
	(114)	(86)	(137)
bST use/drop(ϕ_{10})	−17.49	−114.71	24.21
	(97)	(83)	(151)
bST wait/use(ϕ_{01})	50.31	−48.78	27.77
	(123)	(71)	(231)
AGE	16.17	−7.47	−41.35
	(21)	(25)	(32)
AGE^2	−0.21	0.05	0.41
	(0.22)	(0.26)	(0.32)
EDUC	84.17	73.6	40.58
	(59)	(38)	(76)
BUS_ORG	43.47	38.61	59.33
	(58)	(47)	(77)
MILK_SYS	21.21	−12.91	−252.23
	(68)	(45)	(76)
MILKPR	102.23	37.01	42.29
	(21)	(15)	(21)
Constant	−1349	192	1193
	(594)	(619)	(785)
Wald tests			
Ho: $\delta + \phi_{11} = 0$ chi-square (prob)	0.06 (0.81)	0.58 (0.44)	1.67 (0.20)
Ho: $\delta + \phi_{10} = 0$ chi-square (prob)	0.12 (0.73)	3.55 (0.06)	0.01 (0.93)
Ho: $\delta + \phi_{01} = 0$ chi-square (prob)	0.64 (0.42)	0.95 (0.33)	0.01 (0.94)

1165 kg year^{-1}; on middle production farms, that is 449 kg year^{-1}; and on high production farms, that is 1070 kg year^{-1}. Interestingly, only on the low production farms are the combined bST and history use coefficients statistically different from zero for farms that discontinued bST use or waited to start the use of bST. On middle and high production farms, these combined coefficients were not statistically different from zero.

Regressions of returns over operating costs per cow for the three (1993) production levels show a statistically zero profit bST response for both the low production farms and high production farms, but a US$176.10 profit effect on the middle production farms (Table 7.5). So although it appears that low production farms generated an output response from bST use, that did not translate into profits. That was not the case on the middle production farms that turned their positive bST output response into profits. High output production farms experienced neither an output nor a profit response from using bST. Wald tests show zero effects on all combined coefficients, including for the middle output producers.

The results from sorting by farm size are reported in Table 7.6. The profit response from the use of bST is statistically zero for all farm size groups. However, when the bST coefficient is added to the various use history coefficients, one combined effect is statistically different from zero with a Wald test value of 5.59. The largest farms that continuously used bST earned US$229 more per cow than the largest farms that did not use bST. Although detailed cow management practices are not collected on farms, larger farms may have management practices or facilities where they can profitably benefit from using bST. This might include more effective cow monitoring and feeding programmes.

Table 7.4. Impact of bST on production change per cow; sample sorted by production per cow.

Independent variables (standard error of each estimate is in parentheses below each coefficient estimate)	Production per cow (kg)		
	Low	Middle	High
bST use (δ)	419	575	−179
	(129)	(162)	(236)
Continuous bST(ϕ_{11})	745	−126	1249
	(236)	(242)	(356)
bST use/drop(ϕ_{10})	1061	−913	111
	(195)	(305)	(411)
bST wait/use(ϕ_{01})	470	−224	119
	(232)	(303)	(496)
AGE	−172	13	53
	(56)	(59)	(64)
AGE^2	1.6	−0.1	−0.6
	(0.55)	(0.48)	(0.72)
EDUC	−33	160	157
	(127)	(133)	(185)
BUS_ORG	196	65	140
	(138)	(133)	(177)
MILK_SYS	−233	72	−8
	(147)	(136)	(200)
MILKPR	−119	−77	−51
	(32)	(34)	(44)
Constant	5922	600	−1153
	(145)	(1429)	(1647)
Wald tests			
Ho: $\delta + \phi_{11} = 0$ chi-square (prob)	34.28 (0.00)	7.10 (0.01)	15.91 (0.00)
Ho: $\delta + \phi_{10} = 0$ chi-square (prob)	56.38 (0.00)	1.67 (0.20)	0.03 (0.86)
Ho: $\delta + \phi_{01} = 0$ chi-square (prob)	14.84 (0.00)	1.05 (0.30)	0.01 (0.90)

Table 7.5. Impact of bST on short-run profit per cow; sample sorted by production per cow.

Independent variables (standard error of each estimate is in parentheses below each coefficient estimate)	Profit per cow (US$)		
	Low	Middle	High
bST use (δ)	–91.05	176.10	–17.78
	(67)	(77)	(118)
Continuous bST(ϕ_{11})	–38.35	–261.93	24.64
	(128)	(110)	(164)
bST use/drop(ϕ_{10})	–33.05	–115.86	–132.54
	(103)	(121)	(172)
bST wait/use(ϕ_{01})	–35.12	–154.39	–182.12
	(130)	(144)	(194)
AGE	14.48	–46.17	10.20
	(30)	(27)	(28)
AGE^2	–0.19	0.416	–0.083
	(0.29)	(0.28)	(0.30)
EDUC	23.77	–35.16	157.10
	(66)	(69)	(75)
BUS_ORG	146.98	25.93	261.25
	(72)	(67)	(72)
MILK_SYS	–136.21	–86.22	–345.35
	(79)	(62)	(70)
MILKPR	80.90	59.15	66.52
	(17)	(19)	(23)
Constant	–836.0	1047.0	–519.0
	(760)	(680)	(741)
Wald tests			
Ho: $\delta + \phi_{11} = 0$ chi-square (prob)	1.45 (0.23)	0.23 (0.63)	1.01 (0.31)
Ho: $\delta + \phi_{10} = 0$ chi-square (prob)	1.25 (0.26)	1.35 (0.25)	0.01 (0.95)
Ho: $\delta + \phi_{01} = 0$ chi-square (prob)	1.07 (0.31)	0.02 (0.88)	0.91 (0.34)

Conclusions

Data from the same 138 dairy farms for the years 1994 to 1997 were used to determine if bST generates profits for adopters. bST has been commercially available since 1994 and slightly over half of these farms used bST during that time, with a number of them stopping or starting bST use during the 4-year period. Data from these same farms from 1993 were used to sort farms into groups by production per cow, profit per cow and farm size.

On average, farms that are using bST are experiencing an output response per cow, but are not profiting from using bST. If the bST impact coefficient is combined with bST history coefficients, assuming no selection bias, then only well managed (high profit) farms and the largest farms make money from bST. Assuming self-selection bias and measuring impact by the bST use coefficient only, shows that middle production per cow farms make money from bST.

The average profit response from the use of bST was statistically zero, but larger and well managed farms may be making a profit from bST. That implies that some farmers may be losing money using bST, although no statistical negative profit response from bST use was measured for any subgroup.

Table 7.6. Impact of bST on short-run profit per cow; sample sorted by farm size (number of cows).

Independent variables (standard error of each estimate is in parentheses below each coefficient estimate)	Farm size (cows)		
	Small	Medium	Large
bST use (δ)	−4.88	55.59	−77.64
	(98)	(65)	(118)
Continuous bST (ϕ_{11})	77.93	−43.83	306.59
	(150)	(115)	(152)
bST use/drop (ϕ_{10})	5.87	−101.21	387.29
	(147)	(110)	(230)
bST wait/use (ϕ_{01})	61.82	−130.06	169.25
	(182)	(143)	(150)
AGE	3.09	−24.03	20.98
	(39)	(30)	(26)
AGE^2	−0.09	0.22	−0.20
	(0.38)	(0.30)	(0.27)
EDUC	14.99	59.34	−3.24
	(83)	(71)	(65)
BUS_ORG	179.83	90.40	154.83
	(89)	(73)	(60)
MILK_SYS	−223.66	−181.48	−23.50
	(116)	(70)	(107)
MILKPR	97.60	67.66	37.82
	(21)	(20)	(20)
Constant	−711.0	279.0	−746.0
	(936)	(820)	(678)
Wald tests			
Ho: $\delta + \phi_{11} = 0$ chi-square (prob)	0.00 (0.99)	0.02 (0.90)	5.59 (0.02)
Ho: $\delta + \phi_{10} = 0$ chi-square (prob)	0.41 (0.52)	0.02 (0.90)	2.00 (0.16)
Ho: $\delta + \phi_{01} = 0$ chi-square (prob)	0.11 (0.74)	0.27 (0.60)	0.30 (0.58)

References

Butler, L.J. (1992) Economic evaluation of bST for onfarm use. In: Hallberg, M.C. (ed.) *Bovine Somatotropin and Emerging Issues: an Assessment.* Westview Press, Boulder, Colorado, pp. 145–171.

Card, D. (1996) The effect of unions on the structure of wages: a longitudinal analysis. *Econometrica* 64, 957–979.

Davidson, J. and MacKinnon, J. (1993) *Estimation and Inference in Econometrics.* Oxford University Press, New York.

Greene, W. (1997) *Econometric Analysis.* Prentice Hall, Englewood Cliffs, New Jersey.

Knoblauch, W.A. and Putnam, L.D. (1998) *Dairy Farm Management Business Summary, New York State, 1997.* R.B. 98–06. Department of Agricultural, Managerial, and Resource Economics, Cornell University, Ithaca, New York.

Marion, B.W. and Wills, R.L. (1990) A prospective assessment of the impacts of bovine somatotropin: a case study of Wisconsin. *American Journal of Agricultural Economics* 72, 326–336.

Patton, R.A. and Heald, C.W. (1992) Management of bST-supplemented cows. In: Hallberg, M.C. (ed.) *Bovine Somatotropin and Emerging Issues: an Assessment.* Westview Press, Boulder, Colorado, pp. 73–98.

Stefanides, Z. and Tauer, L.W. (1999) The empirical impact of bovine somatotropin on a group of New York dairy farms. *American Journal of Agricultural Economics* 81, 95–102.

Tauer, L.W. (1992) Impacts of bST on small versus large dairy farms. In: Hallberg, M.C. (ed.) *Bovine Somatotropin and Emerging Issues: an Assessment.* Westview Press, Boulder, Colorado, pp. 207–217.

Tauer, L.W. and Knoblauch, W.A. (1997) The empirical impact of bST on New York dairy farms. *Journal of Dairy Science* 80, 1092–1097.

8 The Importance of Feed Management Technologies in the Decision to Adopt Bovine Somatotropin: an Application to California Dairy Producers

Irene Henriques[1] and Leslie J. Butler[2]

[1]*Schulich School of Business, York University, Toronto, Ontario, Canada;* [2]*Department of Agricultural and Resource Economics, Social Science and Humanities, University of California, Davis, California, USA*

Introduction

Bovine somatotropin (bST) is a large, naturally occurring bovine peptide hormone involved in growth regulation and mobilization of body energy stores. Although the exact mechanism by which bST increases milk yield is not fully understood, its impact on yields may be dramatic. A large, rapidly growing body of experimental results suggests immediate responses of 10–20% increases in yield in dairy cows receiving bST (Cassell, 1992).

Prior to the introduction of bST in 1994, milk production in California had more than doubled from 1971 to 1991. According to Zepeda *et al.* (1991), much of the increase in production was due to rising productivity per cow as a result of rapid adoption of improved management and production technologies. Since the introduction of bST, animal science studies have shown that the consistency of feeding and management practices has become even more important (Coppock, 1992; Shaver, 1998). These studies suggest that the use of bST requires more comprehensive diet formulation based on detailed feed testing. According to Coppock (1992), those dairy producers who have a nutrition programme based on continuous and comprehensive feed testing, with the intended nutrient profile confirmed in the final mixture, plus feeding systems, such as total mixed rations – which allow cows continuous access to feed – have a much greater probability of success when they adopt bST.

Using 1997 data for California dairy producers, the purpose of this study is to address the following questions:

1. Is the adoption of certain feed management technologies an important explanatory variable influencing the adoption of bST?
2. Of the dairy producers who no longer use bST but had adopted it in the past, were feed management technologies (or the lack thereof) an important explanatory variable?
3. Are the adoption of bST and feed management technologies interrelated and, if so, how?

The chapter is divided as follows. The next section surveys the more recent literature on bST adoption. Then the main feed management technologies used in the dairy are briefly discussed. The models we use to examine the bST adoption decision (both current and past) and the role of feed management practices in this decision are the subjects of a later section. The data set is then described and in the final section the results are presented.

Previous Research

Research on bST adoption can be divided into two groups:[1] *ex ante* and *ex post*. The *ex ante* research has focused on whether dairy producers will or will not adopt bST prior to its commercialization while *ex post* research focuses on the decision to adopt once the technology has become commercially available. Stefanides and Tauer (1999) and Barham (1996) are two studies that have gathered *ex ante* and *ex post* survey data on bST (from New York dairy farmers and Wisconsin dairy farmers, respectively). In Wisconsin and many parts of the USA and the rest of the world,[2] bST has been the focus of extensive public debate prior to its commercialization. Barham (1996) finds that although adoption levels of bST in Wisconsin are strongly related to factors suggested by conventional models, his survey data and regression results show that bST adoption in Wisconsin is strongly consistent with a politicized, constrained path.[3]

In the case of New York dairy farms where the controversy over bST was not as great as in Wisconsin, Stefanides and Tauer (1999) estimated bST adoption functions and measured the impact of bST on milk output and profitability per cow. They found that their results were consistent with previous *ex ante* bST adoption studies and that the use of bST significantly increases the milk output per cow net of other explanatory variables but that the impact on profits was not significantly different from zero.

No study, with the exception of Barham (1996), has controlled for the use of dairy feed management practices. The reason this may be important is because animal studies have noted that cows on bST require more energy as their milk production increases (Coppock, 1992; Grant and Keown, 1996). Barham (1996) includes the use of total mixed rations (TMR) as an independent variable and finds that TMR use increased the likelihood of being a bST adopter but had no impact on the likelihood of being a non-adopter or taking on a wait and see stance. While Barham (1996) does not specify the reason for including TMR in his analysis, his results support our premise that feed management practices should be included in the analysis of bST adoption.

Administering bST to dairy cows has two distinct effects on the lactation curve. First, there is an immediate increase in milk production causing the lactation curve to shift upwards a few days following administration, and second, use of bST increases the persistency of lactation causing higher levels of milk production to be maintained for a longer period (Grant and Keown, 1996). The consequence of these effects is that the producer's ability to keep sufficient amounts of a well-balanced ration available at all times is critical to the success or failure of bST use. In other words, feed management practices may be an important determinant of continued bST adoption.

Feed and Management Practices/Technologies

The consistency of feeding and management practices is extremely important in dairy production. Before the commercialization of bST, much of the increase in dairy productivity was due to rising productivity per cow as a result of rapid adoption of improved feeding and management technologies (Zepeda *et al.*, 1991). Examples of important feed and management practices include total mixed rations, feed buffers and the use of computers in the operation of the dairy farm.

TMR blend all feedstuffs (forage, grain, protein and minerals) into one complete ration. The advantages of feeding TMR (Shaver, 1998) include increased milk production, use of low cost alternative feeds, control of forage, lower incidence of metabolic and digestive disorders and reduced labour inputs for feeding. Some disadvantages of feeding TMR include: (i) exclusion or difficulty with baled hay since this must be chopped before being incorporated into a TMR; (ii) fixed equipment cost (mixer, silage conveyors), some modification of existing facilities may be necessary; and (iii) ration formulation is more demanding since errors in formulation will affect many cows.

Feed or dietary buffers are compounds that were originally added to the diets of dairy cows to improve milk fat content (West, 1998). The need for feed buffers developed because of changes in the way dairy cows are fed. Buffers are used largely to

[1] See Stefanides and Tauer (1999) for an excellent discussion of *ex ante* research that has been undertaken to date.

[2] In fact, as of 15 January 1999, the Canadian government has banned the use of bST.

[3] According to Barham (1996), marketing and regulatory actions were taken soon after bST's commercial release, which inevitably constrained farmer adoption choices.

offset acidic conditions produced by the relatively high grain rations fed today. In the past when cows were fed large quantities of forage and relatively small amounts of grains, buffers were not needed. However, genetically superior dairy cows capable of high milk yield (West, 1998) and cows on bST (Grant and Keown, 1996) require increasing nutrient density (especially energy) in the diet which is supplied to a great extent by larger quantities of grains.

Computers are becoming an increasingly useful tool in the effective management of dairy operations. Computers can be used to store production records, store financial records, collect data in the milking parlour, computerize feeding and aid in making herd management decisions.

Model

Independently defined univariate logit or probit models are usually used to examine the adoption decision. In the case where technologies may be interrelated, estimating the impact of adoption for each sub-group of adopters and non-adopters separately could lead to a 'self-selection' bias. This leads to biased and inconsistent parameters of a single equation model because the same omitted variables may influence the decisions for all interrelated components, leading to correlation in the error terms of equations explaining adoption decisions. Multiple technology choices have also been analysed using multinomial logit models (Caswell and Zilberman, 1985; Zepeda, 1990), but these models require restrictive assumptions such as the assumption of the independence of irrelevant alternatives (Cramer, 1991).

In this chapter, we model the bST adoption decision in two ways. First, we estimate the adoption decision (current and past) using a univariate probit model and include feed management technologies as independent variables. The objective here is to see whether feed management practices are important determinants in the decision to both adopt and continue to adopt bST.

The proposed model is a binary choice probit model. Using the latent regression framework, we specify the following model.[4]

$$R_i^* = Z_i \gamma + u_i \qquad (8.1)$$

where R_i^* is an unobserved index variable, Z_i represents explanatory variables, and u_i is an error term. The observed dummy variable is the farmer's decision to adopt ($R_i = 1$) or not ($R_i = 0$), where $R_i = 1$ if $R_i^* > 0$ and $R_i = 0$ if R_i^* ″ 0. The error term is assumed to be normally distributed with zero mean and variance equal to one.

The second approach asks whether the three technologies are interrelated and, if so, how. A multivariate probit model is estimated. Three probit equations reflecting the bST adoption decision, the feed buffer adoption decision and the total mixed rations adoption decision are estimated simultaneously using full information maximum likelihood (Greene, 1997). The multivariate probit model is a direct extension to M equations of the bivariate probit model, that is,

$$R_{im}^* = Z_{im}^* \gamma_m + u_{im}, \quad m = 1, \ldots, M \qquad (8.2)$$

where R_{im}^* is an unobserved index variable for equation m, Z_{im} represents explanatory variables for equation m, and u_{im} is an error term for equation m. The observed dummy variable is the farmer's decision to adopt technology m ($R_{im} = 1$) or not ($R_{im} = 0$), where $R_{im} = 1$ if $R_{im}^* > 0$ and $R_{im} = 0$ if R_{im}^* ″ 0. The error terms are distributed as multivariate normal with mean vector **0** and covariance matrix **V** with diagonal elements equal to 1.0. Each individual equation is a standard probit model.

Data

The data for this study were collected through a mail and telephone survey of 1000 randomly selected California dairy producers in 1997, which represented 50% of the California dairy industry. There were 626 responses (107 via the telephone survey and 519 via the mail survey).[5] Survey participants

[4] This assumes that we are not dealing with a sequential adoption model (Byerlee and Polanco, 1986; Khanna, 1999). In this case, dairy producers need not adopt any of these feed management practices before using bST and vice versa. Unlike site-specific crop management, which is a bundle or package of component technologies where the two key components are diagnostic techniques (such as soil tests) and application techniques (such as variable rate technology) (Khanna, 1999), feed management techniques are *not* a prerequisite for adopting bST. In other words, bST adoption does not occur in a sequential or step-wise manner with respect to the adoption of total mixed rations, feed buffers or use of computers.

[5] Note that due to missing values, there were 388 usable responses for the bST adoption equation and 306 usable responses for the producers who had adopted bST in the past.

Table 8.1. Variable description and statistics.

Variable description

Variable name	Explanation
CURBST	Dummy variable, equals 1 if producer adopted bST, zero otherwise
EVERBST	Dummy variable, equals 1 if producer has adopted bST in the past, zero otherwise
AGE	Equals 1 if producer is under 35 years of age, 2 if producer is between 35 and 44, 3 if producer is between 45 and 54, 4 if producer is between 55 and 64, 5 if the producer is between 65 and 74 and 6 if the producer is 75 years or older
EDUC	Equals 1 if producer has some high school education, 2 if producer has a high school diploma/GED, 3 if producer has some college education, 4 if the producer has an associate degree, 5 if the producer has a bachelors degree and 6 if the producer has a graduate degree
NUMCOWS	Number of cows in producer's herd. The natural log of this variable is used, LNCOWS
MILKPROD	Average milk production year^{-1} rolling herd average lb^{-1}
MILKCOW	Average milk production per cow year^{-1}. As a herd average, it also includes milk from cows not treated with bST
XMILK	Number of times a day a producer's herd is milked
AFFNOBST	Dummy variable, equals 1 if cooperative or creamery asked producer to sign an affidavit stating that he/she would *not* use bST, zero otherwise
AD_TMR	Dummy variable, equals 1 if producer adopted total mixed rations (TMR), zero otherwise
AD_FB	Dummy variable, equals 1 if producer adopted feed buffers (FB), zero otherwise
PCRECORD	Dummy variable, equals 1 if producer uses a personal computer (PC) for record keeping, zero otherwise

Descriptive statistics, all results based on non-missing observations

Variable	Mean	SD	Minimum	Maximum	Cases
CURBST	0.290843806	0.454559838	0	1	557
EVERBST	0.276255708	0.447656137	0	1	438
AGE	2.90538336	1.29888856	1	6	613
EDUC	2.93311037	1.43381302	1	6	598
NUMCOWS	860.914754	933.466435	12	9000	610
MILKPROD	19839.7342	5395.74689	7	40000	538
MILKCOW	45.6125074	43.4429070	0.0111	346.154	538
XMILK	2.11812298	0.337732506	1	4	618
AFFNOBST	0.354779412	0.478886780	0	1	544
AD_TMR	0.701013514	0.458201058	0	1	592
AD_FB	0.675126904	0.468724135	0	1	591
PCRECORD	0.705387205	0.456252720	0	1	594

were asked a number of questions including: (i) their bST adoption decision in 1997; (ii) past bST use; (iii) adoption of feed management technologies such as total mixed rations and feed buffers; (iv) adoption of personal computers to manage their dairy operations; and (v) dairy farm production activity and socio-economic information.

The variables used in this study to explain adoption (current, CURBST and past, EVERBST) include proxies for scale economies, human capital, innovativeness and technical ability. The effect of scale of operation is proxied by the natural log of the number of cows in a producer's herd (LNCOWS), the number of times the herd is milked (XMILK)[6] and average milk production per cow (MILKCOW). Note that since MILKCOW is a milk average, it also includes milk production from cows not treated with bST. The human capital component is proxied using the producer's education level (EDUC) and age (AGE). Innovativeness and technical ability are proxied by the use of personal computers (PCRECORD), the adoption of TMR (AD_TMR) and the adoption of feed buffers (AD_FB). Moreover, in California, some coopera-

[6] This variable is included in the multivariate probit feed management equations.

tives or creameries have asked dairy producers to sign an affidavit stating that they would *not* use bST. To take this factor into account, a dummy variable (AFFNOBST) equal to one if a producer was asked to sign such a document is included in our analysis. Table 8.1 provides a summary of the variable descriptions and statistics.

Results

The individual binary choice probit model results for dairy producers who adopted bST in 1997 are presented in Table 8.2. More educated dairy producers were more likely to adopt bST while a producer's age had no significant impact on bST adoption. With regard to the scale effect, producers with larger herds were significantly more likely to adopt bST. Average milk production per cow (MILKCOW), however, had no significant impact on bST adoption. Not surprisingly, signing an affidavit stating that the producer would not use bST significantly reduces the likelihood that a producer adopted bST. A Wald test to determine whether the coefficient on AFFNOBST is equal to –1 (i.e. all individuals who sign such an affidavit did not use bST) was not rejected (χ^2 = 1.45 with a *P*-value of 0.228). In so far as innovativeness and technical ability is concerned, the adoption of feed buffers and the use of personal computers each had a positive and significant impact on a producer's likelihood to adopt bST. The adoption of TMR had no

Table 8.2. Binary probit model estimates for bST adoption.

	bST adoption in 1997		Ever used bST	
Variable	Probit estimates	Marginal effects	Probit estimates	Marginal effects
Constant	–5.371 (1.199)***		–4.480 (1.399)***	
AGE	–0.070 (0.067)	–0.020 (0.019)	–0.172 (0.072)**	–0.051 (0.021)**
EDUC	0.181 (0.056)***	0.052 (0.016)***	–0.069 (0.066)	–0.020 (0.019)
LNCOWS	0.467 (0.159)***	0.134 (0.047)***	0.567 (0.193)***	0.168 (0.058)***
MILKCOW	0.005 (0.0037)	0.0013 (0.0011)	–0.0034 (0.0043)	0.0010 (0.058)
AFFNOBST	–0.784 (0.179)***	–0.225 (0.050)***	–0.494 (0.188)***	–0.146 (0.056)***
AD_FB	0.616 (0.209)***	0.177 (0.059)***	–0.033 (0.207)	–0.0099 (0.062)
AD_TMR	0.260 (0.194)	0.075 (0.056)	0.548 (0.214)**	0.162 (0.063)***
PCRECORD	0.856 (0.264)***	0.246 (0.072)***	0.781 (0.244)***	0.232 (0.070)***
Number of observations	388		306	
Log likelihood	–176.428		–141.842	

Standard errors are in parentheses; *** denotes that the estimated coefficient is statistically significant at the 1% level; ** statistically significant at the 5% level; and * statistically significant at the 10% level.

significant impact on the likelihood of adopting bST.

Table 8.2 also presents the probit and marginal effect estimates of producers who had adopted bST in the past (and who are not using it in 1997). Older dairy producers were less likely to have ever adopted bST while a producer's education had no significant impact. With regard to the scale effect, producers with relatively larger herds were more likely to ever have adopted bST. Not surprisingly, signing an affidavit stating that the producer will not use bST significantly reduces the likelihood that a producer ever adopts bST. Again, a Wald test to determine whether the coefficient on AFFNOBST is equal to –1 was performed. In this case, however, the hypothesis was rejected ($\chi^2 = 7.21$ with a *P*-value of 0.007) suggesting that although some producers had signed the affidavit, they continued using bST. In so far as innovativeness and technical ability is concerned, TMR and the use of personal computers each had a positive and significant impact on a producer's likelihood to ever adopt bST. The adoption of feed buffers, however, had no significant impact on a producer's likelihood to ever adopt bST.

Since the second and third columns of Table 8.2 estimate the likelihood of adoption in 1997 and the following two columns estimate the likelihood of previous adoption decisions (producers who are no longer using bST in 1997), we can compare these columns to give us some idea of the reasons

Table 8.3. Multivariate probit model estimates of bST adoption, feed buffer adoption and TMR adoption.

Variable	Coefficient	Standard error	$P[\|Z\| > z]$
Index function for CURBST (01)			
Constant	–5.316	1.188	0.0000
AGE	–0.060	0.071	0.3974
EDUC	0.199	0.057	0.0005
LNCOWS	0.543	0.161	0.0008
MILKCOW	0.004	0.004	0.3067
AFFNOBST	–0.768	0.183	0.0000
PCRECORD	0.918	0.289	0.0015
Index function for AD_FB (02)			
Constant	–5.077	1.379	0.0002
AGE	0.072	0.056	0.1964
EDUC	0.115	0.056	0.0382
LNCOWS	0.475	0.126	0.0002
MILKCOW	0.003	0.002	0.1896
XMILK	0.767	0.406	0.0588
PCRECORD	0.453	0.185	0.0145
Index function for AD_TMR (03)			
Constant	–3.364	1.315	0.0105
AGE	–0.016	0.060	0.7873
EDUC	0.056	0.054	0.3030
LNCOWS	0.309	0.149	0.0388
MILKCOW	–0.004	0.003	0.1741
XMILK	1.026	0.356	0.0040
PCRECORD	–0.119	0.186	0.5224
Correlation coefficients			
ρ (01,02)	0.321	0.112	0.0043
ρ (01,03)	0.146	0.113	0.1973
ρ (02,03)	0.376	0.092	0.0000
Number of observations	388		
Log likelihood function	–579.8899		

why previous adopters discontinued using bST. To do so, we look at the factors that increase the likelihood of current adoption and ask whether these hold for those producers who used bST in the past. Whereas higher education and the adoption of feed buffers increased the likelihood of using bST in 1997, such was not the case for past users of bST. The latter suggests that the lack of such important determinants such as a higher education and the use of feed buffers may have led producers to discontinue using bST.

The second approach we take asks whether the three technologies, bST adoption, feed buffer adoption and TMR adoption, are interrelated and, if so, how. A multivariate probit model is estimated in which three probit equations reflecting the bST adoption decision, the feed buffer adoption decision and the total mixed ration adoption decision are estimated simultaneously using full information maximum likelihood (Greene, 1997). These results are presented in Table 8.3.

The estimated parameters show that the null hypothesis that the covariance parameter ρ is zero is rejected at the 1% level for two out of the three covariance relationships: the relationship between bST adoption (CURBST) and feed buffer adoption (AD_FB) and the relationship between feed buffer adoption (AD_FB) and TMR adoption (AD_TMR). A positive value for ρ indicates that unobserved factors that influenced the adoption of bST also increased the likelihood of adopting feed buffers (ρ(01,02)) and the unobserved factors that influenced the adoption of TMR increased the likelihood of adopting feed buffers (ρ(02,03)). The covariance parameter between the adoption of TMR and bST was not significantly different from zero.

The bST equation results are similar to those found in our binary probit equation model. A dairy producer's education, herd size and use of a personal computer each had positive and significant impact on the likelihood of adopting bST. The signing of a no bST affidavit had a negative influence. Turning to the feed buffer equation, the adoption of feed buffers is positively influenced by herd size, education, the number of times the herd is milked and the use of a personal computer. With respect to the TMR equation, the relative size of a producer's herd and the number of times the herd is milked each

Table 8.4. Bivariate probit model estimates of bST adoption and feed buffer adoption.

Variable	Coefficient	Standard error	$P[\|Z\| > z]$
Index function for CURBST (01)			
Constant	–5.477	1.201	0.0000
AGE	–0.058	0.072	0.4215
EDUC	0.196	0.056	0.0005
LNCOWS	0.530	0.161	0.0010
MILKCOW	0.0051	0.004	0.2960
AD_TMR	0.284	0.201	0.1566
AFFNOBST	–0.763	0.181	0.0000
PCRECORD	0.926	0.281	0.0010
Index function for AD_FB (02)			
Constant	–5.138	1.370	0.0002
AGE	0.068	0.056	0.2227
EDUC	0.119	0.055	0.0312
LNCOWS	0.474	0.125	0.0002
MILKCOW	0.003	0.003	0.1866
XMILK	0.803	0.405	0.0475
PCRECORD	0.447	0.183	0.0148
Correlation coefficient ρ (01,02)	0.2687	0.117	0.0219
Number of observations	388		
Log likelihood function	–381.5746		

increased the likelihood that the producer adopts TMR.

The fact that the covariance parameter ρ between bST adoption and TMR adoption was not significantly different from zero suggests that it may be more appropriate to estimate the bST equation and the feed buffer equation jointly (and include the TMR variable as an independent variable in the bST equation). These results are presented in Table 8.4. The estimated parameters show that the null hypothesis that the covariance parameter ρ is zero is rejected at the 1% level for two out of the three covariance relationships, indicating the validity of estimating the two equations jointly. A positive ρ indicates that unobserved factors that influenced the adoption of bST also increased the likelihood of adopting feed buffers.

A dairy producer's education, herd size and use of a personal computer each have a positive and significant impact on the likelihood of adopting bST. The adoption of TMR had no significant impact on the likelihood of adopting bST. Signing a no bST affidavit reduces the likelihood of adopting bST. The adoption of feed buffers is positively influenced by a dairy producer's education, a producer's herd size, the number of times the herd is milked and the use of a personal computer.

Conclusions and Implications

This chapter shows that feed management technologies are important determinants of current and continuing adoption of bST. More specifically we found that feed buffers had a significantly larger impact in current bST adoption decisions than did TMR, which had no significant impact. Furthermore, our results suggest that the non-adoption of feed buffers may have contributed to a dairy producer's decision to discontinue using bST.

We also asked whether bST adoption and the adoption of feed management technologies were interrelated. To address this question, we estimated a three equation multivariate probit model. The first equation modelled the bST adoption decision, the second modelled the feed buffer adoption decision and the third modelled the TMR adoption decision. We found that the bST adoption decision was interrelated with the feed buffer adoption decision but not interrelated with the TMR adoption decision. Given the latter, a bivariate probit model of the bST adoption decision and the feed buffer adoption decision was estimated. The results obtained support our original hypothesis that the adoption of feed buffers can be viewed as an interrelated technology.

In general, our results show that, indeed, feed management practices are important indicators of whether dairy producers will use and continue to use bST. This suggests that efforts should be made to communicate the importance of these feed management technologies in the successful adoption of bST.

References

Barham, B.L. (1996) Adoption of a politicized technology bST and Wisconsin dairy farmers. *American Journal of Agricultural Economics* 78, 224–234.

Byerlee, D. and Polanco, E.H. (1986) Farmers stepwise adoption of technological packages: evidence from Mexican Altiplano. *American Journal of Agricultural Economics* 68, 519–527.

Cassell, B.G. (1992) Impact of bovine somatotropin on DH1 and genetic evaluations. National Agricultural Library, bovine somatotropin related documents from National Dairy Base. www.nalusda.gov/bic/bst/ndd/the_importance_of_nutrition.html

Caswell, M.F. and Zilberman, D. (1985) The choices of irrigation technologies in California. *American Journal of Agricultural Economics* 67, 224–234.

Coppock, C.E. (1992) The importance of nutrition. In: National Agricultural Library, bovine somatotropin: related documents from National Dairy Base. www.nalusda.gov/bic/bst/ndd/the_importance_of_nutrition.html

Cramer, J.S. (1991) *The Logit Model.* Edward Arnold, London.

Grant, R. and Keown, J. (1996) Feeding the bovine somatotropin (bST) treated dairy cow. In: *NebGuide*, GP91-1041-A. Cooperative Extension, Institute of Agriculture and Natural Resources, University of Nebraska, Lincoln. www.ianr.unl.edu/pubs/dairy/g1041.htm

Greene, W.H. (1997) *LIMDEP: User's Manual and Reference Guide Version 7.0.* Econometric Software Inc., Bellport, New York.

Khanna, M. (1999) Sequential adoption of site-specific technologies and its implications for nitrogen productivity: a double selectivity model. Paper presented at the Annual Meeting of the American Agricultural Economics Association, Nashville, Tennessee, 8–11 August 1999.

Shaver, R. (1998) Keynote address: the fundamentals of profitable milk production. 1998 Western Canadian Dairy Seminar. www.afns.ualberta.ca/wcds/wcd98/ch01.htm

Stefanides, Z. and Tauer, L.W. (1999) The empirical impact of bovine somatotropin on a group of New York dairy farms. *American Journal of Agricultural Economics* 81, 95–102.

West, J.W. (1998) Buffers – what and when to use. 1998 Western Dairy Seminar. www.afns.ualberta.ca/wcds/wcd98/ch05.htm

Zepeda, L. (1990) Adoption of capital versus management intensive technologies. *Canadian Journal of Agricultural Economics* 38, 457–469.

Zepeda, L., Butler, L.J. and Carter, H.O. (1991) Simulating BST introduction in California for dairy policy analysis. *Western Journal of Agricultural Economics* 16(2), 228–237.

9 The Potential Effect of Recombinant Bovine Somatotropin on World Dairying

Lovell S. Jarvis[1]
Department of Agricultural and Resource Economics, University of California, Davis, California, USA, and The Giannini Foundation

Introduction

A large number of dairies in the USA, as well as dairies in 16 other countries, currently inject their cows with recombinant bovine somatotropin (rbST) to produce milk more efficiently. The use of rbST was controversial when it was first proposed and its use remains controversial even in the USA, where it has been approved for use since February 1994 (e.g. McGuirk and Kaiser, 1991; Roush, 1991; Kronfeld, 1993). Its use is more controversial in a number of other countries, including Canada and the European Union, where rbST use is still not permitted. Use of rbST was perhaps more controversial than other biotechnologies because it was one of the first major technologies available for use to produce human food. Moreover, milk – some say the most perfect food – has a special emotional and diet role for many consumers. rbST use has been opposed on the basis that it will: (i) endanger humans that consume the milk produced; (ii) harm cows which are treated with it; and (iii) destabilize dairy industries by dramatically increasing milk supply, decreasing the milk price and causing special harm to poorer farmers who may not find it profitable to adopt rbST. There is no scientific evidence that any of these fears is justified, though use of rbST is intended to reduce the cost of milk production and thus ought to lower milk prices.

This chapter reviews the economics of rbST in developed and developing countries, at the farm and industry-wide levels. It also discusses the importance of attitudinal as opposed to economic factors regarding the adoption of rbST, the effect of rbST on international comparative advantage in milk production, and the lessons that the rbST experience may offer regarding the general use of agricultural biotechnology. I conclude that rbST appears to be a fairly 'mundane' technology that does not appear to introduce specific health risks to humans or animals or economic risks to dairy sectors. When managed properly, rbST provides a moderate once-and-for-all cost reduction on farms where it is utilized. It thus offers a substantial aggregate long-run benefit to consumers. rbST also rewards management skills and higher quality feed, whether concentrate or pasture, and thus will probably shift comparative advantage slightly towards countries that possess these in relative abundance.

Background

bST, a naturally produced hormone in cows, is a

[1] Lovell S. Jarvis is Associate Dean for Human Sciences, College of Agricultural and Environmental Sciences, and Professor, Department of Agricultural and Resource Economics, University of California, Davis. This chapter draws on and extends information contained in Jarvis (1996).

homeorhetic control that coordinates the metabolism of many tissues in dairy cattle. Among other effects, it triggers milk production in cows' mammary glands. It is scientifically known as bovine somatotropin (bST) and as bovine growth hormone (bGH). Although bST was identified over a century ago, the high cost of its production restricted research and practical applications until recently.

rbST is a genetically engineered synthetic analogue of the natural hormone. Scientists identified the gene responsible for production of the natural bST and, using standard genetic-engineering techniques, duplicated the gene and spliced it into the DNA of *Escherichia coli* (*E. coli*) K-12 bacterium. These bacteria can be grown in fermentation tanks, and rbST is produced with the organism. The *E. coli* bacteria are killed and ruptured and large amounts of rbST can then be inexpensively extracted for injection into cows (Bauman *et al.*, 1985; Roush, 1991).

The injection of rbST produces a biological reaction that is essentially the same as that which occurs in dairy cows that naturally produce higher levels of bST. If appropriately managed and nourished, cows injected with rbST produce more milk. As milk production rises, feed consumption also usually rises. Because the absolute nutrient requirements for bodily maintenance do not change, a higher proportion of nutrient intake is channelled to milk synthesis. rbST also partitions calories to milk production at the expense of body fat. Since cows that are injected with rbST generally produce more milk, and more milk per unit of feed consumed, use of rbST reduces milk costs, provided that the cost of rbST is sufficiently low.

Four pharmaceutical firms applied to the Federal Drug Administration (FDA) for permission to market rbST in the USA. After lengthy delays to allow for a comprehensive review of rbST's efficacy and animal and human safety, the FDA determined that rbST is safe and, in November 1993, approved the variety produced by Monsanto (recombinant methionyl bovine somatotropin, Sometribove, marketed as POSILAC) for on-farm use. rbST became available for sale in the USA in February 1994.[2] Monsanto is currently the only producer of rbST. According to Monsanto, the health organizations of 53 countries have determined that dairy and meat products from dairy cows receiving rbST are safe for human consumption. However, only 17 countries have approved rbST use and, of these, only the USA is a developed country.[3]

Not surprisingly, rbST is most profitable when it produces a large absolute increment in milk production. The increment obtained correlates positively with dairy farm characteristics found predominantly in developed countries: cow genetic potential to produce milk, skilled management, adequate high quality feed and favourable production environment. Thus, the inability of Monsanto to market rbST in the European Union, Japan and Canada sharply restricts its broader use.

The Committee for Veterinary Medicinal Products of the European Economic Community concluded that rbST was safe for use (Commission of the European Communities, 1993). The Economic Community (EC) nonetheless imposed a moratorium on the commercial use of rbST, primarily out of concern that its adoption would exacerbate EC milk surpluses. In late 1994, the Economic Commission extended the initial moratorium until the end of 1999. In March 1999, the European Union (EU) extended the moratorium again, citing concern regarding the tests previously used to determine rbST safety, specifically regarding the dose–effect relationship between rbST, insulin-like growth factor-1 (IGF-1), and human health. rbST acts on lactating cells in the mammary gland through a messenger substance called IGF-1. Milk from rbST-treated animals contains higher-than-normal levels of IGF-1. Because IGF-1 acts as a messenger for human growth hormone in humans, if IGF-1 survives digestion and enters the human bloodstream, it could cause health problems. Canada also imposed a moratorium on the use of rbST. Though Health Canada found rbST posed no threat to human safety, it expressed concern for animal safety, principally a higher incidence of udder

[2] Production of POSILAC began in 1989, at a time when it was illegal for a US firm to export a drug or chemical that had not been approved for use in the USA. Thus, Monsanto contracted with a chemical firm located in Austria to produce POSILAC. This is currently the only production site, though Monsanto is building a plant to manufacture POSILAC in Augusta, Georgia. POSILAC has been marketed in Brazil and Mexico since 1989.

[3] These are: Brazil, Colombia, Costa Rica, Egypt, Honduras, Israel, Jordan, Korea, Mexico, Namibia, Panama, Peru, South Africa, Turkey, UAE, US and Zimbabwe. Outside the USA, POSILAC is marketed by Lilly under contract from Monsanto.

infections, a shorter productive-life (burnout) and worsening reproductive parameters. Canada extended its moratorium in January 1999, referencing the human health issue raised by the EU and continued animal health concerns.

The FDA recently examined and scientifically refuted the new arguments of the EU and Canada, indicating that the tests used include standard hazard assessment procedures that have been applied to determine 'the safety of vitamins, food additives and drugs, including hormones, for over twenty-five years'. The FDA asserts that the amount of IGF-1 and truncated forms excreted in milk following rbST treatment is 'safe for all consumers, including infants'.[4] The FDA has also rejected the claim that rbST causes significant animal health problems.

Western Europe and Canada continue to oppose use of rbST on what appear to be mainly economic considerations,[5] though a general fear of biotechnology in Western Europe remains a major consideration as well. Although US farmers have been adopting rbST, many surveys indicate that a considerable number of farmers express unwillingness to use rbST. These farmers cite concerns similar to those expressed officially by the EU and Canada, that is, concern that their cows will be harmed and/or that consumers will be scared by rbST and that milk consumption will drop (e.g. rbST surveys of California dairy industry carried out by L.J. (Bees) Butler, 1995). Some of these farmers are doubtlessly influenced by the popular and scientific debate, others by their perceptions following observed efforts to utilize rbST, and still others by a fear of what appears to be 'unnatural'. Despite these concerns, longer-term studies have found that cows receiving rbST evidence no particular stress, lack of heat tolerance, health effects nor any significant gap in calving interval (e.g. Bauman, 1987; Chalupa *et al.*, 1996; Bauman *et al.*, 1999). Cows treated with rbST do evidence a higher level of udder infections, but no higher than untreated cows that yield equal amounts of milk, and all of these infections are curable with antibiotics. There is no evidence in the USA that consumer demand has been affected by the presence of rbST.[6]

Microeconomics of rbST Use and its Adoption

Initial yield trials showed considerable variability of milk production response to rbST treatment. For example, at the 1989 meetings of the American Dairy Science Association and the American Society of Animal Science, Bauman *et al.* (1989) reported trials in which rbST-related production increases ranged from 2.5 to 30%, or about 1.3 lb to 15.4 lb of milk day^{-1} of treatment. Herd management was a critical factor in determining response, but it was then believed that the absolute level of response was positively correlated with initial animal milk yields. Surprisingly, subsequent analysis of on-farm rbST use suggests that the absolute magnitude of the potential response to rbST in the US herd is largely independent of the pre-rbST level of production. That is, when properly managed, response is fairly constant at about 10–12 lb day^{-1} of treatment (W. Weiland, 1999, personal communication).[7] Although genetic potential is a factor in rbST response, animals within the US dairy herd are sufficiently similar that their rbST response is potentially similar. Thus, most of the observed variation in rbST response in US herds can be attributed to differences in herd management. Poorly managed herds achieve a lower response than well-managed herds, with nutrition being the main limiting factor.

A recent study by Bauman *et al.* (1999) provides further support for this view. The study utilizes panel data for 1990–1998 from several continuing surveys of dairy herd productivity in Pennsylvania and New England. Using a list of rbST users and non-users provided by Monsanto, the authors identified 164 dairies that began using rbST between February and June 1994, continued using rbST through March 1998, had treated at least 50% of their cows, and used only Holstein cows. The

4 FDA's determination is supported by numerous scientific and regulatory bodies, including the Joint Food and Agricultural Organization/World Health Organization Expert Committee on Food Additives (JECFA).

5 Canada's primary concern is that use of rbST will increase Canadian milk production and thereby jeopardize Canada's milk quota system.

6 Currently it is impossible to distinguish between bST and rbST in milk. Thus, promises or assertions are the only means that brand-name products have to ensure consumers that their milk is rbST-free.

7 Since the response to rbST is small during the early stages of lactation, it is not profitable to administer rbST until about the 90th day of lactation. Thus, rbST has a smaller proportional effect on total output per lactation than on daily output when rbST is being administered.

authors identified 170 control herds that also used only Holstein cows and that did not use rbST at any time. The 340 herds contained more than 27,000 cows in milk, generating over 2 million cow test day records. Control herds were slightly smaller than rbST herds and had lower total and per cow production both before and after rbST became available. Adjusting for cow age, stage of pregnancy, stage of lactation, and months fresh, the authors determined the annual average change in milk yields for rbST and control herds. Assuming that parameters other than rbST (e.g. weather, feed supply, farmer education, milk price) affect the two sets of herds in a similar way, the authors attribute changes in the difference in yields between rbST and control herd yields after 1994 to rbST. The data show that milk yields for rbST and control herds fell slightly in 1990 and 1991, were constant in 1992 and 1993, and rose in all years 1994–1998. However, yields for rbST-using herds rise sharply in 1994 relative to yields of control herds. The yield gap widens slightly thereafter and then remains roughly constant with both sets of herds manifesting continuing yield growth (presumably from management improvements and adoption of duplication technologies other than rbST). The data indicate that rbST use increased milk yields by about 6 lb per cow milked on test day. However, only about 65% of a herd's cows are eligible for treatment with rbST at any point in time;[8] the data indicate that rbST increased the yields of treated cows by about 9.2 lb. Moreover, most farmers do not treat all their cattle.[9] Assuming that sampled farmers treated 80% of their cows, rbST treatment increased cow yields by 11.5 lb day^{-1} of treatment.[10]

If these results are fairly representative of dairies that effectively manage rbST, how profitable is rbST? And how does profitability vary as the response varies? Calculations can be made easily. The change in the daily profit per cow from use of rbST equals:

$$\Delta\pi = P_M \frac{dM}{dR} - \sum V_i \frac{dX_i}{dM}\frac{dM}{dR} - C_R \quad (9.1)$$

where π = daily profit per cow, P_M = price of milk, M = milk output, V_i = the cost of input i (other than rbST) and X_i = the quantity of input i. For simplicity, let R = the quantity of rbST used and C_R = its total cost. There is no assumption that inputs are chosen optimally as these are likely to vary across farms.

Rewriting, we get:

$$\Delta\pi = \left(P_M - \sum V_i \frac{dX_i}{dM}\right)\frac{dM}{dR} - C_R \quad (9.2)$$

Monsanto sells POSILAC in a standard 14-day prolonged-release dose for subcutaneous injection. Experimental trials indicated that milk production responds strongly to rising amounts of rbST over a range and then plateaus, making it relatively easy to choose the profit-maximizing dose. However, some farmers, feeling that the response is stronger with a somewhat larger dose, inject all cows every 10 or 11 days. In general, farmers must provide a continuous high quality feed ration and other complementary inputs as needed. A farmer's ability to manage nutrition and other inputs, maintaining cow health and comfort, strongly affects the milk increment achieved.

Monsanto currently sells POSILAC for US\$5.80 per 14-day dose, or US\$0.414 day^{-1}.[11] The price is increasingly discounted as a higher

[8] Cows are treated for the last 215 days of their 305 day lactation, and a few cease treatment because of poor condition or ill health.

[9] Monsanto believes that the sample of effective rbST users in the Bauman *et al.* (1999) study treat about 80–85% of their cows. Though most rbST users decide not to treat some of their animals in the belief that they will not respond well to rbST, animal scientists recommend that farmers treat all cows. Everett (1999, personal communication) believes that the lack of response of individual animals is likely to be a result of management and says that it is difficult to statistically separate management factors from genetic factors in the determination of milk yields for individual animals with the data available.

[10] The distribution of average herd response in this sample is symmetrical and the SE is small (R. Everett, 1999, personal communication). The distribution of milk response to rbST within herds is fairly uniform and appears to be uncorrelated with observable cow characteristics. The survey data show no significant change between rbST-treated and control herds in cow age, cow health, length of lactation or cow reproduction rates, and thus show no sign of health or reproductive problems (W. Weiland, personal communication).

[11] Monsanto sells POSILAC at a price that appears broadly consistent with an assumption of profit-maximizing behaviour. The price is substantially above the cost of production, which is estimated at US\$0.05 or less (Marion and Wills, 1990). However, Monsanto markets POSILAC directly to farmers and provided considerable free technical assistance to adopting farmers during the first 3 years that POSILAC was marketed. It wanted to ensure that

proportion of a farmer's herd is treated; the lowest discounted price is US$5.25 per 14-day dose. Thus, depending on whether cows are treated every 10 or every 14 days, and on the discount obtained, the daily cost of rbST ranges between US$0.38 and 0.58. Marion and Wills (1990) estimated that marginal feed costs are about US$0.0289 lb^{-1} of milk produced for Wisconsin dairy herds and that other costs like farm labour, power, veterinary services and milk transport increased at US$0.0087 lb^{-1} of milk produced. Adopting their estimates of variable costs assuming that the farm gate price of milk (which varies regionally) is US$11.50 per hundredweight (US$0.115 lb^{-1}), each incremental lb of milk produced yields the farmer US$0.0774, gross of the cost of rbST itself.[12] See Table 9.1. Since each cow is treated for about 215 days during each lactation, the increase in annual profit per cow rises rapidly.[13] Although rbST appears to be a scale-neutral technology in terms of its specific application, the absolute incentive for adoption varies proportionately with herd size, assuming that management is scale neutral. In fact, US dairy herds vary greatly in size both within and across regions. For example, while Wisconsin herds average about 50 cows, California herds average more than 500 cows (Reed, 1994). Similarly, within California, herds range from 50 cows to 10,000 cows.

Farms having only 50 cows that achieve an increment of about 11–12 lb per cow would appear to gain US$4000–5000 annually, a significant profit increment given the dairy's size. However, if the same farm achieves an increment of only 8 lb, and then treats only 60% of its cows, the benefit is about US$1000. Assuming rbST is sold at US$0.414 per daily dose, a dairy farmer must achieve an average increment of 5.34 lb to break even. It appears that most producers should at least achieve this increment, making adoption of rbST relatively low risk. Most farms must introduce changes in management to manage rbST effectively. Some farmers may lose money in the first several months following adoption. Subsequently, however, most adopters appear to achieve a significant economic benefit from adoption. When managed well, milk production increases in a predictable, consistent, significant manner. The data are increasingly clear that treating cows with rbST creates no unusual animal health or reproductive problems. None the less, the likely gains to smaller dairies from adoption of rbST are relatively small and not likely to be a panacea. In contrast, a dairy with 2000 cows that treats 80% of its cows and achieves an average increase of 12.5 lb per cow gains an estimated annual profit increase of about US$190,000.

It was widely expected that rbST would have a differential effect on the profitability of different types of dairy farms within the US (Boehlje *et al.*, 1987; Fallert *et al.*, 1987; Butler and Carter, 1988; Kaiser and Tauer, 1989). Considerable empirical evidence on the adoption and use of new technologies suggests a positive correlation between farm size and farmer education on the one hand, and the rate and extent of adoption of (profitable) new

Table 9.1. Profit effect of treating cows with rbST, assuming different milk increments.

Milk increment lb day^{-1} (dM/dR)	Daily profit increment per cow ($\Delta\pi$) (US$)	Annual profit increment per cow (215 $\Delta\pi$) (US$)	Annual profit increment per herd (assuming 100 cows treated) (US$)
5	(0.03)	(6.45)	645
7.5	0.17	36.55	3,655
10	0.36	77.40	7,740
12.5	0.55	118.25	11,825
15	0.75	160.13	16,013

Assumptions: P_M = US$0.115 lb^{-1}, C_R = US$0.414, and V_iX_i = US$0.0376 lb^{-1} (the sum of input cost other than rbST).

adopting farmers got good results and continued use, and Monsanto also wanted to better understand on-farm use and results. Monsanto still provides technical assistance, but at a lower level.

12 Marion and Wills (1990) first worked out a variant of this simple model.

13 As the response to rbST is not strong during the first 60 days of the lactation cycle, Monsanto recommends that dairy farmers treat essentially all cows after the 60th day of lactation.

technologies on the other, and it seems likely that the same will hold for adoption of rbST. Indeed, Saha *et al.* (1994) found that, *ex ante*, the willingness of Texas dairy producers to adopt rbST is positively related to herd size and education. Klotz *et al.* (1995) found the same is true of dairy farmers in California. Given regional differences in farm size and farm productivity, rbST would seem likely to differentially affect regional milk production as well.

Monsanto does not report sales figures for any of its products. However, Monsanto reported in May 1999, that 13,000 US dairy producers were using rbST and that adopting farms have approximately 30% of the 9 million dairy cows in the US herd. Although the data provided do not seem wholly internally consistent, Monsanto claims that the number of adopters was growing at about 3600 year^{-1} in 1996, 1997 and 1998, which suggests that the combined number of new adopters in the first 2 years of diffusion, 1994 and 1995, was about 2000–3000. There are 120,000 farms with ten or more cows each in the USA, and about 45,000 farms with 50 or more cows. Monsanto provides no information about the farm size distribution of adopting farmers but it is a reasonable guess that most very small producers are not potential users simply because the potential gains are not worth the effort. If valid, then approximately 25–30% of the farms with more than 50 cows have adopted. Monsanto claims that sales of rbST have been rising at about 30% year^{-1} in the last 2 years and continued to do so in early 1999, a figure approximately consistent with the rate of increase of new adopters. Thus, there is no evidence of a changing average herd size for adopters, nor that adopters have significantly changed the proportion of the herd treated over time.

Assuming that treated cows increase milk production by 10 lb day^{-1} and that each cow is treated for 215 treatment days per lactation (with roughly one lactation year^{-1}), rbST is increasing the yields of treated cows by about 2150 lb year^{-1}, or about 14% of the 15,000 lb US average yield. Assuming that 70% of animals in adopting herds are currently treated, rbST use is increasing the output of treated herds by about 10%. If these herds contain 30% of the US dairy herd, rbST has increased US milk production by 3%, assuming no indirect effects on herd size or the adoption of other technologies. Although this is an important contribution to US milk production, the increase achieved thus far by use of rbST is dwarfed by the impact of other technologies that have affected the US dairy sector during the same period. Output per cow has been rising at a rate of about 2% year^{-1} over the last several decades and the data in Bauman *et al.* (1999) suggest that this rate of technological progress has been maintained in the last 5 years on farms that have not adopted rbST. If so, rbST accounts for a much smaller part of the increment in cow milk yields than have other technologies, in the aggregate. This fact bears emphasis. Numerous farmers and policy analysts have expressed opposition to rbST on the grounds that its use will sharply increase milk output, reduce milk prices and create economic havoc in the dairy industry. However, these voices do not suggest that other types of technical change should be banned even though their aggregate effect is substantially larger.

If rbST has contributed to an increase, even if modest, the milk price has probably been affected. What is likely to happen in the longer run? How much will the price of milk fall in equilibrium if all dairies eventually adopt rbST? Following considerable debate and a number of studies that predicted a large price decline, Perrin (1991) used data from Fallert *et al.* (1987) and Marion and Wills (1990) to show that the equilibrium price decline associated with rbST use should be modest. Although animal milk yields increase by about 15%, costs also increase so that the decline in the unit cost of production is fairly small. Using the available data, Perrin estimated unit cost savings of 0.5% and 4.4%.[14] Such small unit cost savings suggest little potential for rbST adoption to cause large-scale disruption of the US milk market. A study by Stennes (1989), using a linear programming model, derives comparable unit cost savings of about 6% from rbST use in Canada.[15]

Adoption of rbST in other developed countries such as Canada, the EU, other Western European

[14] Assuming application of rbST to herds averaging 20,000 lb of milk year^{-1}, a 15% production response and a price of US$12.50 per cwt. (cwt. is hundredweight), the reduction in unit costs achieved by using rbST rises to 6.9%.

[15] Even a significant milk price decline should not reverse rbST adoption as its use yields US$37 in net incremental revenue per cow year^{-1} (versus no adoption) when the milk price has fallen to $9 cwt.$^{-1}$, assuming an 11 lb response. Given the low cost of manufacturing rbST, a decline in the price of milk would surely lead manufacturers to reduce the price of rbST if needed.

countries and Japan depends principally on whether the use of rbST is permitted. If it is permitted, it seems difficult to imagine that it will not be used. However, dairies in these countries are generally smaller than dairies in the US and, since the incentive to adopt is importantly related to the number of cows milked, a slower rate of adoption can be predicted in these countries. These countries will be able to observe the experience of the US with rbST and to implement additional regulations regarding rbST use if such regulations are deemed useful. To the extent that milk prices fall significantly in response to rbST use, there will be pressures in other countries to restrict the expansion of adopting farms, for example using farm quotas. Such regulations would further reduce the incentives for adoption.

rbST Use in Less Developed Countries

Although it appears that rbST will be adopted by most larger dairy farmers in the US during the next decade and would also be adopted by many farmers in other developed countries if its use were approved, significant adoption is not likely to occur in developing countries even if the price of rbST is significantly discounted there.

Dairy farmers in many developing countries face a milk price similar to that found in developed countries (Jarvis, 1996). Although the international price of milk was sometimes as low as \$0.055 lb^{-1} during the 1980s, domestic prices in developing countries usually exceeded the international price by a significant margin due to domestic protection (Sere *et al.*, 1990). However, most developing countries are located in tropical regions with a relatively unfavourable milk production environment including high temperature and humidity, low quality pastures and a high incidence of disease and parasites. The cattle populations in these regions are composed predominantly of Zebu (*Bos indicus*) breeds and their crosses. Such cattle display considerable tolerance of the climatic, nutritional and disease challenges present in these areas, but their genetic capacity for milk production is limited even under more favourable conditions. Not surprisingly, milk yields in less developed countries are usually significantly lower than those in the US, Canada and Western Europe. The average annual yields in Africa and the Middle East are about 1100 lb, less than one-tenth those achieved in the US and Canada. Yields in Latin America and Asia are about 2200 lb and 5500 lb, respectively. Although milk yields vary considerably across regions, the average yields are representative of most herds in these areas.

High quality management, improved nutrition and timely veterinary care have been shown to be crucial to profitable application of rbST in the US. These inputs will be at least as important in developing countries where the environmental challenge is greater, yet these inputs are scarce in developing countries. Milk marketing and processing networks are underdeveloped and often unreliable in terms of offering a secure outlet for production. Markets are segmented by unreliable and expensive transport. Production is highly seasonal, and additional output of the sort that would be stimulated by rbST during the period of high production is often of much less value.

Bos taurus dairy cattle in developed countries have milk yields ranging from 2000 to 10,000 kg $year^{-1}$ while *B. indicus* dairy cattle in developing countries have milk yields ranging from about 500 to 2500 kg $year^{-1}$. Substitution of *B. taurus* for *B. indicus* cattle has been perceived as a means of increasing milk yields in developing countries. Many purebred dairy cattle were imported into developing countries after the Second World War (Vaccaro, 1979; Jasiorowski *et al.*, 1988). However, the efforts to introduce purebred *B. taurus* animals into most developing country regions proved disastrous. *B. taurus* dairy cattle perform poorly in most developing country regions because there is an inherent incompatibility between these animals and the developing country context into which they are placed. Animals with the genetic potential to produce high milk yields are genetically unsuited to the harsh environment, disease/parasite challenge and poor nutrition available in developing countries (Vercoe and Frisch, 1980). Purebred *B. taurus* cattle suffer high morbidity and mortality, have longer calving intervals, require more feed (and of higher cost), and produce less milk than do local varieties.

Purebred *B. taurus* cows have been introduced successfully in some, usually highland, regions where the environmental context is more similar to that faced in developed countries. *B. taurus* cattle in less developed countries have also shown a significant response to rbST when the production conditions have been adequate, for example in Brazil (Fontes *et al.*, 1997) and in Zimbabwe (Phipps *et al.*, 1991). However, *B. taurus* herds account for a steadily declining share of milk production in the developing countries because they have not been competitive with milk-producing herds of local and, especially, crossbred animals (Jarvis, 1986; Sere and Rivas, 1987; Nicholson, 1990; Vaccaro, 1994). *B. indicus* purebred and crossbred cattle types now produce most milk in developing countries. Crossbreeds are the cattle whose response will determine whether rbST is applied in most developing countries (Cunningham and Syrstad, 1987).

Although few rbST trials have been undertaken in less developed countries, the studies published suggest that rbST achieves a significant percentage yield increase in buffalo (Ferrara *et al.*, 1989; Ludri *et al.*, 1989), *B. indicus* and *B. indicus–B. taurus* crossbred cattle. However, the absolute increase achieved in *B. indicus* cattle is small. For example, Phipps *et al.* (1991) found that rbST increased milk yields by about 2.8 lb day^{-1} for well managed *B. indicus* cattle in Zimbabwe. This increment is far too low to allow profitable use. However, Phipps *et al.* found that treating crossbred cattle with rbST achieved an increase of 5.3 lb day^{-1}, a yield increment that is at the margin of profitability. Subsequent research indicates that a 250-mg dose achieves the best result (Phipps *et al.*, 1996). Fontes *et al.* (1997) obtained very similar results in studies with Brazilian crossbred cows. As 250 mg is a smaller dose than that used on *B. taurus* cattle, Monsanto might use this smaller size dose as justification to sell rbST at a lower price in many developing countries. Other inputs are also somewhat cheaper in developing countries. Still, it appears that rbST will be profitable for only a small proportion of developing country dairy herds during the next decade or two, mainly those comprising *B. taurus* cattle in favourable production environments.

Conclusion

Though the controversy surrounding the introduction of rbST has often given it the aura of something complex and mysterious, it increasingly appears to be a fairly mundane technology. When administered to *B. taurus* cattle in favourable production conditions and managed effectively, rbST consistently offers a moderate, but limited reduction in milk production costs. These conditions prevail in nearly all developed countries. rbST use appears to cause no health threat to animals treated, nor to humans who consume meat and milk products from treated animals.

The use of rbST will not be profitable in most parts of most developing countries in the foreseeable future. The yield increases obtained from administering rbST to *B. indicus–B. taurus* crossbred animals, which produce the bulk of milk in developing countries, are too small. Milk production in most developing countries is carried out within a much less favourable environment than exists in developed countries. Climate is often more harsh. In tropical areas, temperatures and humidity are higher. The disease–parasite challenge is greater, subjecting animals to stress and reducing their ability to produce high milk yields. Producers have fewer management skills, particularly as regards the use of technologies like rbST, which require improved animal husbandry and precise adjustment of other inputs, especially nutrition, for profitability. Some farmers also face lower producer prices, caused partly by a lack of transport infrastructure and inefficient milk collection, processing and distribution systems. The use of rbST will increase production risks along nearly every dimension of the production system. This higher risk will also discourage producers from adopting rbST, particularly since credit markets function less well.

Looking at the complementarities between rbST, as a new technology, and existing factors of production, it seems clear that the use of rbST will increase the return to skilled management working in favourable production conditions. Thus, rbST is a technology that is more complementary with productive factors in developed countries than in developing countries. rbST will be adopted earlier and in greater degree in developed countries than in developing countries. In the case of those developing countries with the least favourable production environments, the adoption lag may be measured in decades. To the extent that the introduction of rbST reduces the unit cost of milk production in developed countries more than in developing countries, this new biotechnology will improve the competitive position of the developed countries.

Adoption of rbST will result in higher milk production and in lower milk prices. Although rbST will contribute more to declining prices in developed countries, to the extent that the international milk market is integrated across countries, consumers throughout the world will gain. Dairy farms with cows with relatively better management and/or larger herds should profit most from rbST and, therefore, are likely to be the first adopters. Using Perrin's analytical framework, but with calculations using parameters from the relatively more efficient dairy firms who will increasingly dominate the dairy industry, the unit cost of milk is estimated to decline by a maximum of about 5%. Given that the price elasticity of demand in the US is low, milk consumption will increase by only about 2%, which suggests that unless sizeable quantities of milk can be exported, production will increase little. Although adoption of rbST will improve the competitive advantage of developed countries, the effect on international markets is likely to be relatively small. Developed countries will not rely on the large milk export subsidies to sell their surplus milk; given the moderate levels of protection that are still permitted in developing countries and consumers' preference for fluid milk, developing countries are unlikely to greatly increase milk imports.

If rbST is adopted, increasing milk yields, the number of dairy cows will decrease further, as well as the number of dairy farms. Taken on that criterion alone, the use of rbST may not appear particularly socially attractive. None the less, the total benefits of adoption could amply exceed total costs, even if use of rbST is constrained to the USA. Some benefits will occur to producers who can profitably adopt, particularly early in the diffusion process, but US consumers will capture most benefits in the long run. Assuming milk consumption of approximately 70 Mt in the US, an eventual 5% reduction in the farm level cost of production and constant processing and distribution margins, would translate into a decline of about \$0.004 lb^{-1} at the consumer level. Consumer surplus, a crude approximation of the change in consumer benefits, would increase by about US\$560 million annually. Monsanto will also profit so long as its patents protect production of rbST. Assuming that 5,000,000 cows, producing on average 28,000 lb $year^{-1}$, are each treated with rbST for 215 days, and that Monsanto's net operating profit is US\$0.25 per dose, Monsanto would earn US\$216 million annually. If the use of rbST is eventually adopted in other major developed country milk producing regions, the total annual benefits to consumers (and Monsanto) might be roughly five times as great as the estimate above. Although terribly crude, these estimates provide some magnitude of the potential benefits from the rbST technology.

Canada and the countries making up the Economic Union have delayed rbST use out of concern that it might produce such an increase in milk production as to sharply reduce market price, requiring painful adjustment among dairy farmers, and/or that its use, directly or indirectly, poses a threat to human health. However, if the US market absorbs rbST without a politically unacceptable disruption of the milk market and if use of rbST results in no major human or animal health problems, it seems likely that most other developed nations will eventually approve adoption. Their delay in adoption offers some potential for the USA to temporarily increase its comparative advantage in milk production. Some countries may continue to resist the introduction of rbST given their deeper aversion to the concept of biotechnology, but if their competitive position in the world milk market erodes there will be pressure for approval of rbST use. Ironically, countries like Canada and those in the EU probably face greater risk of economic destabilization to their dairy sectors the longer they postpone rbST. The longer rbST is used in the USA, the greater the amount of information available regarding its profitability, safety and appropriate management. This valuable information should increase the profitability of adoption by producers and, thus, if and when use of rbST is approved, it should be adopted more rapidly, with consequent potential for greater economic disruption.

In the longer run, rbST use may be feasible in a wider range of developing countries. The extent depends mainly on whether ongoing, broad-scale rural development leads to improved management skills, nutrition, herd genetics, herd health and a marketing infrastructure. Improvements of this sort are precisely the developments that are needed to achieve significant gains in milk production even in the absence of rbST. Moreover, milk production has been rising at a faster rate in developing countries than in developed countries during the last two decades (Jarvis, 1996). This rapid rate of increase has been achieved by the adoption of 'traditional' technologies borrowed largely from developed countries. Continued adoption of such technologies is highly suitable to these countries' current resource endowments and exploitation of this stock of 'traditional' technologies offers great potential for these countries to achieve higher milk output at declining cost. The existence of rbST ought not to distract developing countries from this path. Indeed, since incorporation of these technologies is basic to being able to make profitable use of rbST in the future, it should enhance their importance and the emphasis on bringing them about. Further, they may be able to incorporate these traditional technologies sufficiently rapidly to reduce milk costs faster than developed countries can do so via the adoption of new technologies like rbST.

References

Bauman, D.E. (1987) Bovine somatotropin: the Cornell experience. In: *Proceedings of National Invitational Workshop on Bovine Somatotropin, 21–23 September 1987.* Westport Plaza, St Louis, Missouri, pp. 46–56.

Bauman, D.E., Eppard, J.E., DeGeeter, M.J. and Lanza, G.M. (1985) Responses of high producing dairy cows to long term treatment with pituitary somatotropin and recombined somatotropin. *Journal of Dairy Science* 68, 1352–1362.

Bauman, D.E., Hard, D.L., Crooker, B.A., Partridge, M.S., Garrick, K., Sandles, L.D., Erb, H.N., Franson, S.E., Hartness, G.F. and Hintz, R.L. (1989) Long term evaluation of a prolonged-release formulation N-methionyl bovine somatotropin in lactating dairy cows. *Journal of Dairy Science* 72, 642–651.

Bauman, D.E., Everett, R.W., Weiland, W. and Collier, R.J. (1999) Production responses to bovine somatotropin in Northeast dairy herds. *Journal of Dairy Science* 82, 2564–2573.

Boehlje, M., Cole, G. and English, B. (1987) Economic impact of bovine somatotropin on the U.S. dairy industry. In: *Proceedings of National Invitational Workshop of Bovine Somatotropin, 21–23 September 1987.* Westport Plaza, St Louis, Missouri, pp. 167–176.

Butler, L.J. (1995) rBST and Industrial Dairying in the Sunbelt. Paper presented at 1995 Annual Meeting and Science Innovation Exposition of the American Association for the Advancement of Science, February 1995. American Association for the Advancement of Science, Washington, DC.

Butler, L.J. and Carter, H.O. (1988) *Potential impacts of bovine somatotropin on the U.S. dairy industry: a West Coast perspective,* paper No. 88–4. Agricultural Issues Center, University of California.

Chalupa, W., Vecchiarelli, B., Galligan, D.T., Ferguson, J.D., *et al.* (1996) Responses of dairy cows supplemented with somatotropin during weeks 5 through 43 of lactation. *Journal of Dairy Science* 79, 800–812.

Commission of the European Communities (1993) *Final Scientific Report of the Committee for Veterinary Medicinal Products on the Application for Marketing Authorization submitted by the Monsanto Company for Somatech, 500 mg Sometribove (Recombinant Bovine Somatotropin) for Subcutaneous Injection.* Commission of the European Communities, Brussels, 27 January.

Cunningham, E.P. and Syrstad, O. (1987) *Crossbreeding* Bos indicus *and* Bos taurus *for Milk Production in the Tropics.* FAO Animal Production and Health Paper 68. FAO, Rome.

de Vaccaro, L. (1979) The performance of dairy cattle breeds in tropical Latin America and programmes for their improvement. In: Balaine, D.S. (ed.) *Dairy Cattle Breeding in the Humid Tropics.* Haryana Agricultural University, Hissar, India.

de Vaccaro, L. (1994) Technology for development in the area of animal breeding. In: Anderson, J.R. (ed.) *Agricultural Technology: Policy Issues for the International Community.* CAB International, Wallingford, UK.

Fallert, R., McGuckin, T., Betts, C. and Bruner, B. (1987) *bST and the Dairy Industry: A National, Regional and Farm-level Analysis,* Agricultural Economics Report No. 579. US Department of Agriculture, Washington, DC.

Ferrara, L., Di Liccia, A., Manniti, F., Piva, G.G., Masaero, F., Fiorentini, L. and Litta, G. (1989) The effect of somidobove (biosynthetic bovine somatotropin) on the production and quality of milk of buffaloes (*Bubalus bubalis*) raised in Italy. *Asian-Australasian Journal of Animal Sciences* 2, 2–7.

Fontes, C., Jr, Meserole, V.K., Mattos, W., Barrow, R.P., Wu, Z. and Huber, J.T. (1997) Response of Brazilian crossbred cows to varying doses of bovine somatropin. *Journal of Dairy Science* 80, 3234–3240.

Jarvis, L.S. (1986) *Livestock Development in Latin America.* World Bank, Washington, DC.

Jarvis, L.S. (1996) *The Potential Effect of Two New Biotechnologies on the World Dairy Industry.* Westview Press, Boulder, Colorado.

Jasiorowski, H.A., Stolman, M. and Reklewski, Z. (1988) *The International Friesian Strain Comparison Trial: A World Perspective.* FAO, Rome.

Kaiser, H.M. and Tauer, L.W. (1989) Impact of bovine somatotropin on U.S. dairy markets under alternative policy options. *North Central Journal of Agricultural Economics* 11, 59–73.

Klotz, C., Saha, A. and Butler, R.J. (1995) The role of information in technology adoption: the case of rbST in the California dairy industry. *Review of Agricultural Economics* 17, 287–298.

Kronfeld, D.S. (1993) Recombinant bovine growth hormone: its efficacy, safety and impacts on herd health management, economics, and public health. In: Liebhardt, W.C. (ed.) *The Dairy Debate: Consequences of Bovine Growth Hormone and Rotational Grazing Technologies.* University of California Sustainable Agriculture Research and Education Program, Davis, California.

Ludri, R.S., Upadhay, R.C., Singh, M., Guneratne, J.R.M. and Basson, R.P. (1989) Milk production in lactating buffalo receiving recombinantly produced bovine somatotropin. *Journal of Dairy Science* 72, 2283–2287.

McGuirk, A.M. and Kaiser, H.M. (1991) bST and milk: benefit or bane? *Choices* First Quarter, 20–26.

Marion, B.W. and Wills, R.L. (1990) A prospective assessment of the impacts of bovine somatotropin: a case study of Wisconsin. *American Journal of Agricultural Economics* 63, 685–699.

Nicholson, C.F. (1990) An optimization model of dual purpose cattle production in the humid lowlands of Venezuela. MS thesis, Cornell University, Ithaca, New York.

Perrin, R.K. (1991) *Ex-ante* evaluation of experimental technology. Working paper. Department of Agricultural and Resource Economics, North Carolina State University, October.

Perrin, R. (1997) The impact of technological change as a competitive industry. *Journal of Agricultural and Resource Economics* 22(2), 345–355.

Phipps, R.H., Weller, R.F., Craven, N. and Peel, C.J. (1990) Use of prolonged-release bovine somatotropin for milk production in British friesian dairy cows. *Journal of Agricultural Science, Cambridge* 115, 95–104.

Phipps, R.H., Madakadze, C., Mutsvangwa, T., Hard, D.L. and de Kerchove, G. (1991) Use of bovine somatotropin in the tropics: the effect of sometribove on milk production of *Box indicus* dairy crossbred and *Bos taurus* cows in Zimbabwe. *Journal of Agricultural Science, Cambridge* 117, 257–263.

Phipps, R.H., Hard, D.L. and Adriaens, F. (1996) Use of bovine somatotropin in the tropics: the effect of sometribove on milk production in Western, Eastern and Southern Africa. *Journal of Dairy Science* 80, 504–510.

Reed, B. (1994) For wages and benefits, bigger dairies may be better. *California Agriculture* 48, 9–13.

Roush, W. (1991) Who decides about biotech: the clash over bovine growth hormone. *Technology Review* 14, 28–34.

Saha, A., Love, H.A. and Schwarb, R.B. (1994) Adoption of emerging technologies under output uncertainty. *American Journal of Agricultural Economics* 76, 836–846.

Sere, C. and Rivas, L. (1987) Dairy production in dual purpose herds: evidence from tropical Latin America. In: Pinstrup-Andersen, P. (ed.) *Economics of Dairy Development in Selected Countries and Policy Implications.* Unpublished manuscript, proceedings of International Food Policy Research Institute conference by the same name, Copenhagen.

Sere, C., Li Pun, H.H. and Estrada, R.D. (1990) The rationale of developing dairy production in the third world. *Proceedings of the XXIII International Dairy Congress, Montreal, 8–12 October 1990.* Vol. 1, pp. 146–157.

Stennes, B.K. (1989) Bovine somatotropin and the Canadian dairy industry: an economic analysis. MS thesis, University of British Columbia, Vancouver, British Columbia.

Vercoe, J.E. and Frisch, J.E. (1980) Animal breeding and genetics with particular reference to beef cattle in the tropics. In: Verde, L.S. and Fernandez, A. (eds) *Proceedings of the IV World Conference on Animal Production.* Argentine Association of Animal Production, Buenos Aires, pp. 452–464.

10 Gains to Yield-increasing Research in the Evolving Canadian Canola Research Industry

Richard S. Gray, Stavroula Malla and Peter W.B. Phillips
Department of Agricultural Economics, University of Saskatchewan, Saskatoon, Saskatchewan, Canada

Introduction

Canola is a good example of how research and development can improve the comparative advantage of an industry. In the late 1950s rapeseed was a very minor crop in Canada that was used to make lubricant. By the 1980s, genetic, agronomic and processing research had transformed rapeseed into canola, a new crop that contained premium edible oil and a protein meal used for livestock feed. In the 1990s canola became the dominant crop in the black soil zone of the prairies, and a large processing industry had been built around it. In the past few years biotechnology has been used to develop transgenic herbicide tolerant varieties and hybrid varieties that have been widely adopted by Canadian producers. Throughout this 40-year period of development, research has been an important catalyst for growth in the industry.

The funding of canola research in Canada has undergone many changes since its inception in the mid-1950s, when Agriculture Canada began a programme to improve rapeseed processing methods. Over time research has shifted from a modest public research programme to a large research industry dominated by private sector participation. In 1970, 83% of research spending was public investment. Ten years later, the percentage of public versus private research investment was 63% vs. 38% respectively. By 1997 the private sector's share had grown to 80% of the total (Canola Research Survey, 1997). This funding shift is evident in the registration of new varieties. Prior to 1973 all varieties were public, while in the 1990–1998 period 86% of the varieties were private (Nagy and Furtan, 1977; Prairie Pools Inc., 1977–1992; Canola Council of Canada, 1998). This large shift in emphasis from public to private research is due to the large increase in private sector investment rather than a reduction in public research.

The change in the private funding of research has coincided with a change in the ownership of the property rights for the research and, implicitly, who benefits from the resulting returns to the investment. In 1987, virtually all of the canola varieties were open-pollinated and non-transgenic, and were not protected by plant breeders' rights until 1990. This meant that virtually all of the acreage was grown without a production agreement, giving producers the right to retain production for future seed use and to sell non-registered seed to their neighbours. In contrast, by 1998, over half of the acreage was planted to herbicide-tolerant (HT) varieties; producers were required to sign a technology agreement or to purchase a specific herbicide. An estimated 30% of the acreage was seeded to hybrid varieties (Canola Council of Canada, 1998) and much of the remaining acreage was seeded to varieties with plant breeders' rights. These changes have put plant breeders in a far better position to capture value from genetic innovation.

Previous research has shown very high rates of

return for canola research. The evaluation of public investment in canola research and development (R&D) was first published in 1978 by Nagy and Furtan. For the period 1960–1974 they calculated the internal rate of return (IRR) from improved yield research to be 101%. Ulrich *et al.* (1987) updated the estimates of IRR in canola research for the period 1951–1982 and calculated the IRR from improved yield research to be 51%. Ulrich and Furtan (1985) incorporated trade effects and found the estimated Canadian IRR from higher yielding varieties to be 50%. Despite the dramatic changes in the industry since 1982, we could not find a more recent comprehensive analysis.

There have also been some recent advances in the estimation of the returns to research that have yet to be applied to canola. Many studies have used econometrics to examine the effect of R&D investment on agricultural productivity (e.g. Thirtle and Bottomley, 1988; Huffman and Evenson, 1989, 1992, 1993; Pardey and Craig, 1989; Leiby and Adams, 1991; Chavas and Cox, 1992; Alston and Carter, 1994; Evenson, 1996). Some of the more recent econometric studies did not impose the shape and length of adoption lag but applied statistically based transformation of the data and generally have found lower rates of return (e.g. Akgüngör *et al.*, 1996; Makki *et al.*, 1996; Myers and Jayne, 1997). Alston *et al.* (1998) also explicitly dealt with the concept of knowledge depreciation, which is not common in the agricultural R&D literature. These new approaches will have relevance for estimating the IRR for canola research.

Given the dramatic changes that have recently occurred in the canola industry, there is a need to re-examine the returns to research in the sector. In particular, has the entrance of private industry, the change in the property rights, the introduction of biotechnologies, and the changing role of the public institutions resulted in a change in the benefits created? A contemporary evaluation of the situation will be useful as a guide for further investment decisions in the canola industry. Furthermore, this evaluation methodology may provide insights or raise important questions for emerging biotechnology research in other sectors.

The objective of this chapter is to estimate the returns to yield-increasing Canadian canola research over time as a means of examining whether the changes in the canola research industry have affected the returns to research. The chapter begins by presenting a simple economic model to show how changing property rights and government involvement can affect both the level and the return to research. This is followed by a description of the framework used to estimate the returns to research. The econometric model and the data used to estimate the relationship between research expenditure and yield over time are then presented. This is followed by a description of how the estimated parameters are then applied historically to the canola market to estimate the return to research under different scenarios. The chapter concludes with a discussion of the results, the implications for policy and the need for further research.

An Economic Model to Examine the Effect on Property Rights and the Return to Research

In the absence of enforceable property rights, many of the products of research can be copied or reproduced. This creates a 'public good' market failure resulting in underinvestment in research activities. As shown in Fig. 10.1, in the absence of complete property rights the private marginal benefit (*MBp*) that can be captured from the marketplace is less than the public or social benefits (*MBs*) of the research. A private research firm will equate the marginal cost (*MC*) of doing research with the private demand (or private marginal benefit) for the research and produce a quantity of research *Qp*. At this amount of research the social marginal benefit of research is far greater than the marginal cost of doing research. In this case, the marketplace fails to produce the socially optimal amount of research *Qs*, where the marginal cost of research is equal to marginal social benefit. If the government provides a quantity of research *Qg* – *Qp*, this research creates a social benefit equal to the additional area under the social benefit curve while incurring costs equal to the much smaller area under the marginal cost curve. In this instance there is a high rate of return to public research, which has been found in many empirical studies. This illustration may characterize the situation in canola research until the mid-1980s before the private sector played a major role.

Government has also addressed the incomplete set of property rights for research goods by providing assistance to private firms doing research. This assistance has come in many forms. Research tax credits have been used in many countries and in many sectors to stimulate research. Recently, grants

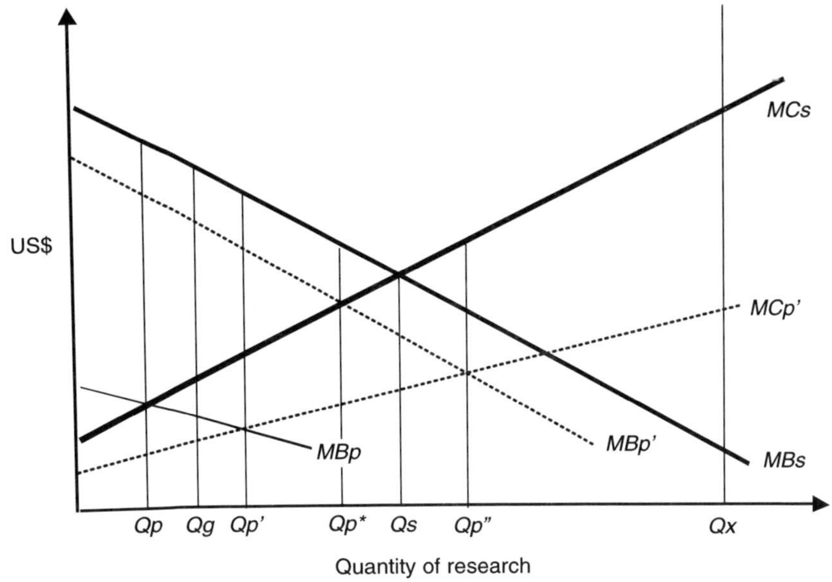

Fig. 10.1. The marginal private and social costs and benefits of research.

have provided money to match private expenditures on research. Infrastructure has also been provided at a reduced cost in many jurisdictions. Indirect support from the public sector has also come in the form of the public education of research scientists and the free provision of the output of public research. This public assistance to private research has lowered the private cost of doing research, allowing the private sector to do more research. In Canada the public sector has always provided some direct support for private sector research on canola. Investment tax credits and infrastructure grants have existed for some time (Gray *et al.*, 1999). Since 1995 the government has also offered matching investment initiatives (MII), which match private research expenditure in approved projects. This has allowed the private sector to play a greater role in research provision. In Fig. 10.1 this is equivalent to lowering the marginal cost of private research to *MCp'*, which, in the absence of complete property rights, moves the private investment towards the socially optimal level *Qs* to a level of *Qp'*.

Recently, governments, and to some extent the private sector, have addressed the 'public good' market failure in research by establishing effective property rights over the products of research. As outlined by Malla *et al.* (1998) the assignment of intellectual property rights provided some of the added ability to capture value from research. The adoption of plant breeders' rights in Canada can forbid the sale of registered varieties without royalty payments. This assignment followed a number of milestones, including the US Patent Office decision of 1985 to grant patents for whole plants. Many of the seeds produced during the 1990s had very specific attributes. Herbicide-tolerant canola requires the use of a specific herbicide in order to be useful. Similarly, canola with particular oil characteristics needed specialized processing and marketing in order to be viable. The development of hybrid varieties has given private firms a greater ability to capture value from their genetic material. The first hybrid variety was introduced in 1989. These varieties, although often protected with plant breeders' rights and production contracts, do not require the enforcement of contracts to maintain control over the use of the genetics.

The establishment of enforceable property rights has the effect of moving the private marginal benefit (market demand) curve towards the social demand curve. As was discussed earlier, this has had the effect of increasing the demand for private research and the amount of private research provided by the private sector, thus partially addressing the market failure. In the absence of government support for research this moves the private investment from *Qp* to *Qp** in Fig. 10.1. The establishment of property rights changes the optimal role for government. If government provides support for private research, once property rights are established this further increases the private incentive to do research. If the property rights are nearly complete and research support is significant the private sector can provide more research than is socially desirable, as represented by point *Qp"* in Fig. 10.1. Thus, both correcting the public good failure and subsidizing research can result in excessive research.

A few final points apparent in Fig. 10.1 are worthy of note. The highest benefit-to-cost ratio, which will generate the highest IRR, will be at some level of research less than the social optimum. The socially optimal quantity of research occurs where the marginal social benefit is equal to the marginal social cost. At this point total net benefits of research are maximized. Additional research beyond this point is socially wasteful, costing more on the margin than what is produced. At these excessive levels of research (anything less than the *Qx* in Fig. 10.1) the total benefits can still be greater than total costs, and the IRR can still be above market rates. Importantly, a positive overall return to research, or an IRR greater than market rates, does not imply that more research is socially desirable, rather, it suggests that the research programme taken as whole has produced net benefits.

This simple economic model presented in Fig. 10.1 illustrates several important concepts for research policy. The first is that in the absence of enforceable property rights, the private sector will underinvest in research, creating a role for government to address the research shortage. Second, the assignment of property rights to research products can increase the amount of private investment towards the socially optimal amount. Third, if enforceable property rights have been established the subsidization of private research could lead to socially wasteful overinvestment in research. Finally, an assessment of total research benefits or the rate of return on total investment are not good indicators that on the margin more research is socially desirable.

Estimating the Returns to Agricultural Research

This section contains a brief description of the conceptual framework used to estimate the returns to canola research.[1] The process of creating new crop varieties can be described in four phases, as shown in Fig. 10.2. The first phase is the research phase, which involves effort over a number of years to create varieties with commercially desirable genetic traits. The second phase is the gestation phase, which is the period when potential varieties undergo private and public testing and multiplication, preparing the variety for potential registration and commercial sale. In the adoption phase, after commercial release of the variety, the varieties are adopted and grown by producers, contributing to increased productivity. In the fourth phase, these new varieties become part of the germplasm and knowledge stock from which newer varieties are created. This fourth phase continues even after the particular variety is no longer grown. Over time the contribution to the knowledge stock depreciates as pests adapt themselves to the germplasm and new techniques replace older ones (Alston *et al.*, 1998). These four phases of crop variety development, and the long lags between investment and output, have made the estimation of returns to research difficult and a subject of considerable debate.

Estimating the Relationship between Expenditure and Yield Increase

The empirical procedure began by constructing a yield index of different canola varieties to the same base variety (Torch = 100). The annual yield index was created from an average of the yield index for varieties grown each year, weighted by the seeded acreage. The data on the relative yield of different varieties, were obtained from various issues of Saskatchewan Agriculture and Food, *Varieties of Grain Crops in Saskatchewan.* These data are based on the research station yield trials at a number of locations across Saskatchewan, which were designed to measure varietal performance due to genetic causes. The data on the percentage acreage of each canola variety were obtained from three sources. The first source was a 1977 study by Nagy and Furtan,

[1] A conceptual model for measuring these returns and many of the empirical issues that have to be dealt with are explained in some detail in the book entitled *Science Under Scarcity* by Alston *et al.*, 1997.

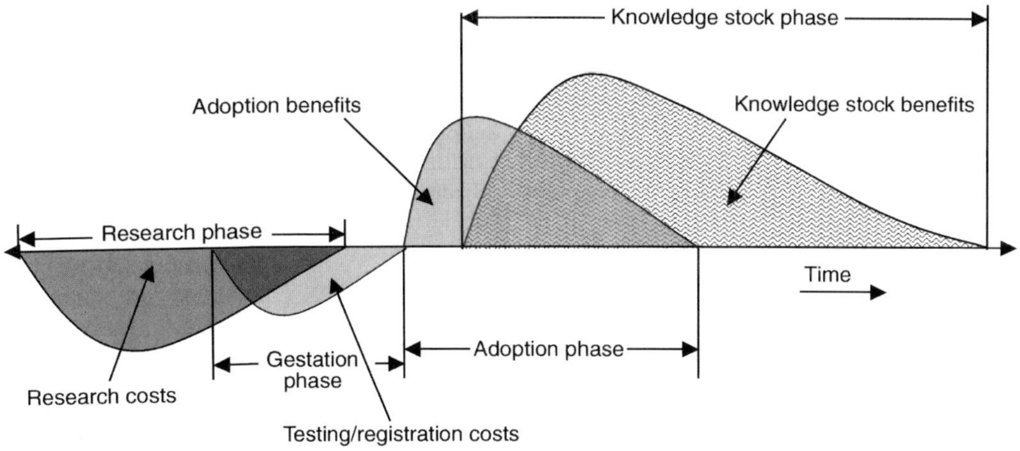

Fig. 10.2. Four phases of crop development and the path for R&D costs and benefits.

which covered the period 1960–1976. The second source was collected from various issues of Prairie Pools Inc., *Prairies Grain Variety Survey* (1977–1992). The final source was the authors' estimates based on the Manitoba Crop Insurance Corporation *Variety Summary* (1998). The annual yield index since 1970 is shown in Fig. 10.3.

In addition to the genetic stock there are two other factors that will influence the yield index of the varieties grown. First of all there are two types of rapeseed/canola grown that are grown in Canada: Argentine species (*Brassica napus* L.) and Polish species (*Brassica rapa* L.). Argentine varieties are higher yielding than Polish varieties (15–20%) while Polish varieties mature faster. The proportion of area grown to each will vary from year to year, with more Polish varieties seeded when spring is late. In order to capture the effect of planting Argentine versus Polish varieties a variable indicating the proportion of area seeded to Argentine

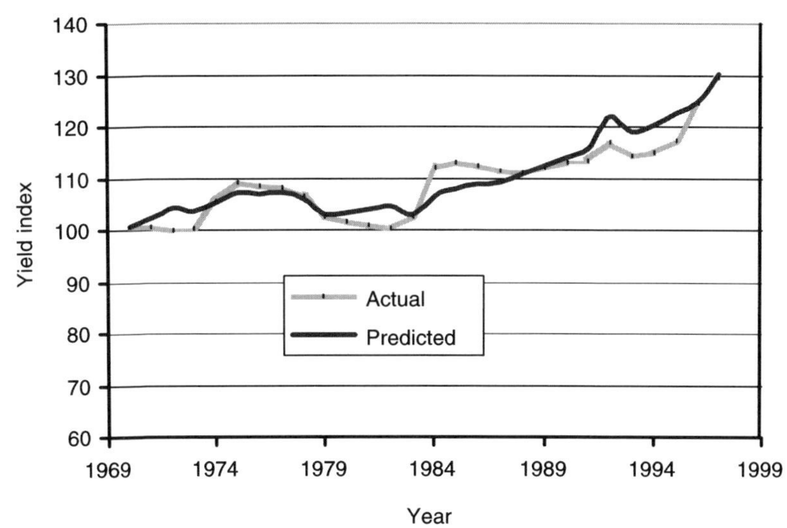

Fig. 10.3. Actual vs. predicted yield index change.

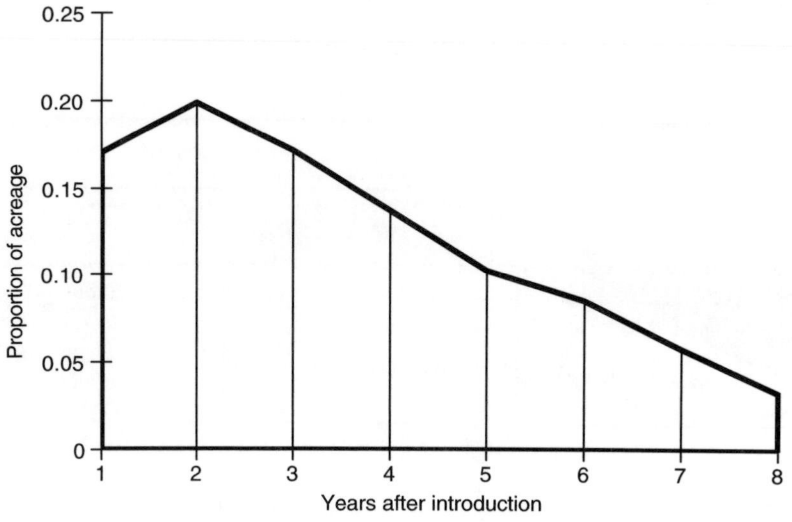

Fig. 10.4. Average adoption curve for new canola varieties.

varieties is included in the regression. The other yield factor that needs to be accounted for is the switch from rapeseed to canola varieties. The selection for low erucic acid and glucosinolate in canola quality has been attained at the expense of seed yield. Figure 10.3 reveals the reduction in the annual weighed average yield index from 1978 to 1984 when the changeover was made. This result is similar to other findings, which have shown that the combined yield of Argentine and Polish type canola varieties decreased from the middle 1970s to the beginning of the 1980s (e.g. Forhan, 1993; Malla, 1996). To account for the yield effect of the conversion of rapeseed to canola, a variable was created that represents the percentage of total rapeseed/canola varieties seeded that were canola varieties. This variable is also included in the regression.

The total research expenditure per year was calculated by multiplying the total person-years invested in the research by the total research cost, fixed and variable (Canola Research Survey, 1997), for each year. A person-year is used to define either a professional person-year or a technical person-year, where a professional person-year corresponds to full-time annual work dedicated to professional research and a technical person-year corresponds to full-time annual work on technical research (as reported in the Inventory of Canadian Agri-Food Research). The data on canola research professional and technical person-years were obtained from five sources: Canola Research Survey 1997–1998; Nagy and Furtan (1977); ISI (1997); ICAR (1998); and Phillips (1997). Where there were discrepancies in the overlapping periods from the data source, the earlier estimates were indexed upwards to reflect later estimates.[2]

An average adoption curve for canola varieties was estimated rather than assuming a specific adoption lag structure. The individual adoption rate of each rapeseed/canola variety was calculated by dividing the acreage sown of each variety by the maximum acreage sown of that variety for each year after the year of introduction. These adoption rates were then averaged for all varieties and weighted to sum to one. The average adoption curve is reported in Fig. 10.4.

The weighted average adoption rate of rapeseed/canola varieties was applied to the cost data to create a variable of the weighted lag research expenditure. The adoption curve means that on average the acreage planted today to specific varieties is a function of when in the past the varieties were introduced. The annual weighted yield index is therefore an average of the yield of varieties previously introduced weighted by the respective coefficients on the adoption curve. Mathematically,

[2] For more details on the data source and the calculation see Gray *et al.* (forthcoming).

$$Y_t = \sum_{i=1}^{n} \gamma_i y_{t-i} \qquad (10.1)$$

where Y_t is the yield index in year t, γ_i is the weights on the adoption curve, and y_{t-i} is the average reported yields of the varieties introduced in year $t-i$. Given this relationship it follows that the annual change in the weighted yield index is a weighted average of the changes in the yield of the varieties introduced, or:

$$\mathrm{d}Y_t = \sum_{i=1}^{n} \gamma_i \mathrm{d}y_{t-i} \qquad (10.2)$$

where dY and dy represent the change in the weighted index and the yield of new varieties. If change in new variety yield is proportional to the lagged expenditures on research, or $\mathrm{d}y_t = \rho X_{t-g}$, where X_{t-g} is the expenditure in year $t-g$, and g is the gestation lag, then:

$$\mathrm{d}Y_t = \rho \sum_{i=1}^{n} \gamma_i X_{t-i-g} \qquad (10.3)$$

Making these assumptions, the change in the annual weighted yield index is proportional to the adoption curve weighted yield expenditures.

To estimate the effect of research expenditure on the average yield index we specified and estimated the following regression.

$$\mathrm{d}Y_t = b_R \log X_{t-g} + b_C\, \mathrm{d}DCAN_t + b_A\, \mathrm{d}DARG_t + \varepsilon_t \qquad (10.4)$$

where $\mathrm{d}Y_t$ is a change in the annual weighted average yield index for a year t; X_{t-g} is the annual adoption lag weighted research expenditure for a year t minus the gestation lag (a number of years between making an investment and beginning to have an effect on yield index); $DCAN_t$ is a change in a percentage of the total canola/rapeseed varieties that are canola varieties, which takes a value between 0 and 1; and $DARG_t$ is a change in a percentage of the total canola/rapeseed varieties that are Argentine (*B. napus*) varieties, which takes a value between 0 and 1.

In order to determine the appropriate specification of the model, the time series properties of the variables in the model were examined. In the research expenditure series the Dickey-Fuller test and Phillips-Perron test revealed that a unit root could not be rejected in favour of a stationary when it is measured 'in level'. However, when the research expenditure is measured in logarithm, the unit root hypothesis is rejected in favour of stationarity. Hence, the logarithms of research expenditure were used in this analysis. For the yield index series, the unit root hypothesis could not be rejected in either the 'in level' or the logarithmic form. By taking the first difference of the yield index, the variable is judged to be stationary about a linear trend. Thus, the analysis uses the first difference form. The specification error test was used to determine the adequacy and final specification of the models. The error test was performed using the Ramsey's Regression Specification Test (RESET).[3]

The regression results are reported in Tables 10.1 and 10.2. Table 10.1 shows the regression results using a 4-year gestation lag between research expenditure and the release of a new variety. This 4-year gestation lag was inferior only to a 1-year lag using the Akaike information criterion. However, the 1-year gestation lag regression results reported in Table 10.2 are very close to the 4-year gestation lag results.[4]

The results appear to be robust, having passed the specification tests and given that the explanatory variables are statistically significant at the 5% level, and that the coefficients have the expected signs.[5] The R^2 showed that the explanatory variables explain just less than half of the variation of the year-to-year change in yield. The predicted line in Fig. 10.3 represents the fitted values from the regression estimate.

The coefficient on the Argentine variable in Table 10.1 indicates that a complete switch from Polish to Argentine varieties would increase yield by 17.86 index points. Similarly, the complete switch

[3] The Ramsey's RESET test adds extra regressors to the original regression and examines the hypothesis that the coefficients on the forecast vectors are all zero. The null hypothesis is rejected whenever the associated probabilities of the output from the test (F-statistic and likelihood ratio test) were less than 0.05, which indicate evidence of specification error.

[4] We use the results in Table 10.1 for the calculation of the rate of return because of prior information that suggests that it takes several years of research and testing to develop and introduce a new variety.

[5] A squared log expenditure term was added to the regression to create a more flexible fit but was rejected as the adjusted R-squared decreased.

Table 10.1. Regression results for gestation lag of 4 years.

Dependent variable: d*Y* Independent variables		Coefficient	*T*-statistic[a]
$\log X_{t-4}$		0.425	2.760
DCAN		−17.095	−2.351
DARG		17.868	3.830
$R^2 = 0.466$			
Akaike info criterion: 1.771			
$(n = 27)$[b]			
Ramsey's Reset test			
F-statistic	0.186		Probability-values: 0.671
Likelihood ratio test	0.217		Probability-values: 0.641

[a] The estimated coefficients are significant within a 95% confidence interval.
[b] R^2 is the coefficient of determination; *n* is the number of observations.

Table 10.2. Regression results for gestation lag of 1 year.

Dependent variable: d*Y* Independent variables		Coefficient	*T*-statistic[a]
$\log X_{t-1}$		0.410	3.096
DDCAN		−17.407	−2.537
DDARG		17.638	4.046
$R^2 = 0.464$			
Akaike info criterion: 1.655			
$(n = 30)$[b]			
Ramsey's Reset test			
F-statistic	0.267		Probability-values: 0.610
Likelihood ratio test	0.307		Probability-values: 0.580

[a] The estimated coefficients are significant within a 95% confidence interval.
[b] R^2 is the coefficient of determination; *n* is the number of observations.

from rapeseed to canola varieties reduced the average yield by 17.09 index points. These are large effects and may have implications for the value of non-yield traits in canola.

The coefficient of the research expenditure is 0.425 which means, holding all other variables constant, that a 1% increase in the annual lag weighted research expenditure in year $(t - 4)$ increases the yield index level by 0.00425 index points (Table 10.1). Given that the yield index was 127 in 1997, a 1% increase in the annual lag weighted research expenditure in year $(t - 4)$ increases, on average, the yield index by approximately 0.0033% at 1997 yields.

Net Returns to Canola Research

To calculate the social return from the yield-increasing research, the econometric estimates of the yield increase due to research expenditure are applied to the historical production of canola. The regression results reported in Table 10.1 are used to predict the amount of yield increase due to research expenditure in each year. As an approximation, it is assumed that the additional yield due to genetic improvement came at no resource cost and thus benefits are in direct proportion to revenue each year.[6] Benefits in 1997 dollars are estimated by multiplying the quantity of canola seed (production) by the price of

[6] This is consistent with the treatment in the research plots where additional yield is measured with no additional use of crop inputs. There would be additional costs for harvesting and transport to the elevator which are small per tonne and not accounted for in our analysis.

Table 10.3. Net present value (US$ millions, 1997) under different scenarios.

Scenario	A	B	C	D	E	F
Depreciation	0.01	0.05	0.01	0.05	0.01	0.05
Real discount rate	0.03	0.03	0.06	0.06	0.09	0.09
Year of yield change						
1971	66.8	44.8	63.2	41.8	59.4	38.6
1972	71.2	48.2	67.1	44.8	62.9	41.2
1973	76.7	52.5	72.2	48.7	67.4	44.6
1974	80.4	55.1	75.3	50.7	69.9	46.0
1975	85.7	58.1	79.9	53.1	73.8	47.7
1976	92.3	61.6	85.9	56.1	79.0	50.0
1977	101.2	68.0	94.2	62.0	86.8	55.4
1978	104.4	69.7	97.1	63.3	89.2	56.3
1979	103.0	66.3	95.3	59.6	86.9	52.1
1980	295.4	107.8	198.0	86.8	153.7	71.0
1981	294.4	107.5	194.6	85.3	149.2	68.6
1982	297.4	110.2	194.6	86.6	147.7	69.0
1983	300.5	113.0	194.4	87.8	146.0	69.1
1984	302.3	112.8	191.8	85.8	141.4	65.8
1985	302.7	111.7	187.7	82.5	135.2	61.0
1986	304.5	112.7	185.2	81.3	130.5	58.2
1987	306.4	115.0	183.4	81.5	126.7	56.9
1988	305.1	115.7	179.1	80.3	121.0	54.2
1989	302.1	114.3	172.8	76.7	113.0	49.0
1990	303.6	116.4	169.9	76.0	107.8	46.4
1991	305.0	118.9	167.0	75.8	102.7	44.2
1992	305.5	120.6	162.8	74.4	96.1	40.6
1993	305.9	122.6	158.5	73.3	89.5	37.2
1994	300.0	118.0	147.5	65.1	75.8	26.6
1995	288.7	107.6	130.4	50.7	55.5	9.3
1996	277.6	97.5	113.2	36.2	35.0	−8.6
1997	268.9	90.8	99.1	25.1	17.8	−22.9

canola seed, and deflating by the consumer price index.[7] The present values of research benefits are estimated by first calculating all future yield increases due to the yield increases in a particular year. This uses the notion that there is a stock of knowledge that is subsequently built on. This calculation was made using various rates of depreciation. These future yield increases are then applied to the revenue in each future year to calculate the future benefits. For 1997 and beyond it was assumed that 1997 revenue would continue indefinitely. Once the future stream of benefits is calculated this is then brought back to the present value in the year of introduction using a discount rate. The present value of costs for varieties grown in year t is calculated from the present value in year t of the weighted expenditures lagged by the gestation period and the adoption curve. The net present value (NPV) of research is calculated from the difference between the present value of the benefits and the cost of the research using a number of depreciation and real discount rates.

The NPV results are reported in Table 10.3. What is most striking is that for each of the scenarios the NPV peaks in the 1980s and declines thereafter. For instance, with 5% depreciation and a 6% real discount rate, the increase in the research expenditures resulted in an increase in net present

7 The price of canola seed in Canada was obtained from Saskatchewan Agriculture and Food, *Agricultural Statistics 1997*. The price of canola seed was the Saskatchewan farm price in Canadian dollars per metric tonne. The data for the consumer price index (CPI) was obtained from Statistics Canada (1998a). Finally, the quantity of canola production was obtained from Statistics Canada (1998b).

Table 10.4. Estimated internal rate of return for yield increase (1967–1993).

Lag Depreciation Year of yield change	4 year 0%	4 year 1%	4 year 5%	4 year 10%	1 year 5%
1971	0.398	0.382	0.324	0.260	0.471
1972	0.384	0.369	0.313	0.252	0.442
1973	0.371	0.356	0.303	0.245	0.425
1974	0.350	0.336	0.285	0.228	0.404
1975	0.336	0.322	0.271	0.215	0.389
1976	0.328	0.313	0.261	0.205	0.375
1977	0.327	0.312	0.261	0.208	0.380
1978	0.318	0.304	0.253	0.201	0.375
1979	0.304	0.289	0.236	0.180	0.355
1980	0.366	0.343	0.262	0.182	0.397
1981	0.353	0.331	0.253	0.174	0.380
1982	0.344	0.323	0.248	0.172	0.369
1983	0.333	0.314	0.243	0.170	0.359
1984	0.319	0.301	0.232	0.159	0.338
1985	0.304	0.286	0.218	0.145	0.315
1986	0.290	0.272	0.208	0.137	0.299
1987	0.276	0.260	0.200	0.134	0.289
1988	0.263	0.248	0.192	0.130	0.275
1989	0.247	0.233	0.181	0.121	0.256
1990	0.235	0.222	0.173	0.118	0.244
1991	0.223	0.211	0.167	0.117	0.231
1992	0.210	0.199	0.159	0.113	0.213
1993	0.198	0.188	0.153	0.112	0.197
1994	0.179	0.170	0.137	0.100	0.171
1995	0.152	0.145	0.114	0.079	0.136
1996	0.126	0.119	0.092	0.060	0.102
1997	0.105	0.099	0.075	0.047	0.076

value from US$41 million in 1971 to a peak of US$88 million in 1983 which then began to decline as the increase in expenditures exceeded the growth in benefits. By 1997 the net present value of yield increases had declined to US$25 million.

Table 10.4 shows the estimated IRR for the change in yield increases for the years 1970–1997, under different assumptions about deprecation rates and lag structures. The first four columns show the IRR from the regression results from Table 10.1, with a 4-year gestation lag and 0%, 1%, 5% and 10% depreciation rates, respectively. Not surprisingly, the IRR declines as the deprecation rate increases because the investment is not a durable. The last column of Table 10.4 shows the IRR with a 1-year gestation period. As expected, this increases the IRR and illustrates once again that the IRR can be sensitive to assumptions about the gestation lag. What is most striking in this table is the general decline in the IRR as the level of investment has increased. While the IRR was clearly excessive in the early 1970s it had declined to market levels by the mid-1990s.

Given that many biotechnologies became predominant during the 1980s the declining NPV and IRR provides little support to the notion that biotech has led to significant increases in the returns to research. However, in 1997 about 35% of area was sown to herbicide tolerant varieties and thus the research may have produced other benefits not measured as a yield increase. Yet, given the very recent introduction of herbicide tolerant varieties, this phenomenon does not explain the decline in the NPV prior to 1996.

The increase in the level of expenditures and the declining IRR approaching market rates suggests that the assignment of property rights and matching grants has corrected the public good market failure. Further extrapolation of these rates of return would suggest an over-investment in the sector resulting in

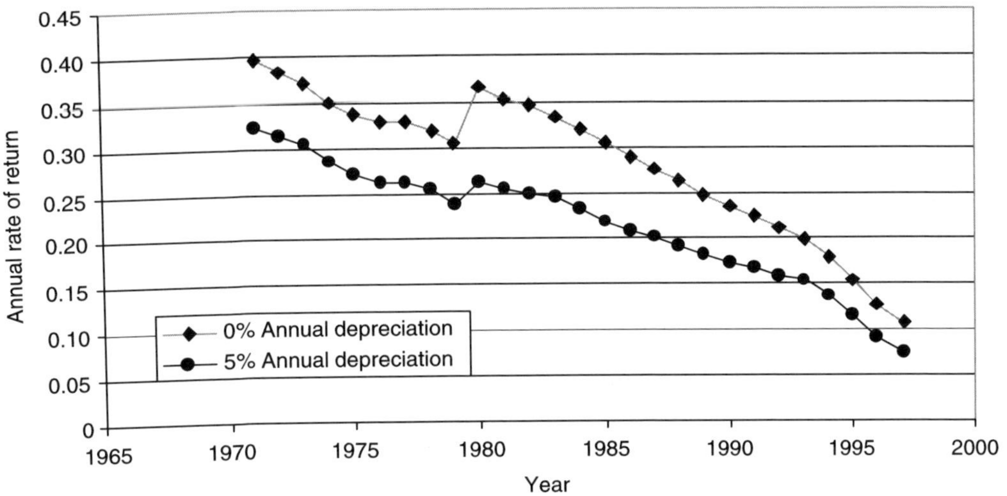

Fig. 10.5. Estimated IRR for canola yield-increasing research (1971–1997).

low private and social IRR for investment. The decline in the NPV of research as expenditure increases is consistent with moving beyond the optimal amount of research.

Figure 10.5 shows the IRR with 0% and 5% depreciation rates. One interesting feature of these series is the increase in the rate of return in the late 1970s when canola acreage surged in response to growing on-farm wheat stocks brought about by grain transport constraints. Note, however, revenue during the mid-1990s was near record levels and, despite this, the rate of return is low.

Conclusions

The objective of this study was to examine the returns to research for investment in the canola sector in Canada for the period 1960–1997. Many changes took place in the industry during this period. A small but very successful public research programme eventually became dominated by a large influx of private research investment induced by property rights and technologies that provided a greater opportunity to capture the benefits from research. During this private growth period the technologies used for genetic improvement shifted from traditional breeding to the use of many biotechnologies. This study focused on the net social benefit from yield-increasing canola research.

The rate of return from canola research has been on the decline throughout the study period. Specifically, the IRR from the high rate era of the 1960s and 1970s declined and became more realistic in the 1990s. Moreover, the total net present value of yield-increasing research peaked some time during the early 1980s and has subsequently declined, suggesting an overinvestment in research. This result indicates that the increase in private research and development efforts did not actually yield as much net benefit as one would expect when witnessing a large amount of private investment flowing into an otherwise publicly funded research area. Therefore, further investment in canola research and development may not be as profitable a venture as the investment stampede would lead us to believe.

This study challenges the current government policy in canola research. The canola research industry is heavily subsidized and property rights for canola seed are well established. Given that property rights allow private firms to capture the full social benefit of investment this will attract capital and drive the rate of return towards normal levels. If, on top of that, government subsidizes the costs of private research, then it is certainly possible to create overinvestment in an industry. The analysis presented shows declining net present value of investment. Consequently the industry might already be operating beyond the point where the marginal benefit is equal to the marginal cost. Hence, this study indicates a need for a much closer examination of policy in this industry.

The general result that biotechnology has yet to produce measurable high social returns in the canola sector raises some very important questions. Clearly, genetic traits other than yield have economic value; would the incorporation of these effects change the general conclusion? If the net present value has fallen, does the vintage of a crop, to a large extent, dictate the rate of return to research? Is there a natural cycle to crop development, which has an increasing and then a decreasing return to research investment? If this is true, then should public investment be targeted to crops on the basis of vintage rather than historic rates of return? A related question is: what has been happening to the rate of the return in other crops and in other sectors? Of particular importance is whether the falling rate of return in canola is the result of the assignment of IPRs that have become general to all crops, or do market failures continue to exist in other crops where hybrids and other physical reproductive barriers do not exist? Answering these will provide some important insights into the best policies to govern the rapidly expanding biotechnology industry.

References

Akgüngör, S., Makanda, D.W., Oehmke, J.F., Myers, R.J. and Choe, Y.C. (1996) Dynamic analysis of Kenyan wheat research and rate of return. Contributed paper *Proceedings from the Conference on Global Agricultural Policy for the Twenty-First Century,* Melbourne, Australia, 26–28 August 1996, pp. 333–366.

Alston, J.M. and Carter, H.O. (1994) *Valuing California's Agricultural Research and Extension,* Publication No. VR-1. University of California Agricultural Issues Center, Davis, California.

Alston, J.M., Norton, G.W. and Pardey, P.G. (1997) *Science Under Scarcity: Principles and Practice for Agricultural Research Evaluation and Priority Setting.* Cornell University Press, Ithaca, New York.

Alston, J.M., Craig, B.J. and Pardey, P.G. (1998) *Dynamics in the Creation and Depreciation of Knowledge, and the Returns to Research,* GPTD Discussion Paper No. 35 (August). International Food Policy Research Institute, Washington, DC.

Canadian Food Inspection Agency (1998) *Canadian Varieties. January 1, 1923 to June 24, 1998.* Special Tabulation from the Plant Health and Production Division. www.inspection.gc.ca

Canola Council of Canada (1998) Canola growers' manual. www.canola-council.org/manual/canolafr.htm

Canola Research Survey (1997) College of Agriculture. University of Saskatchewan.

Chavas, J.-P. and Cox, T.-L. (1992) A nonparametric analysis of the effects of research on agricultural productivity. *American Journal of Agricultural Economics* 74, 583–591.

Evenson, R.E. (1996) Two blades of grass: research for U.S. agriculture. In: Antle, J.M. and Sumner, D.A. (eds) *The Economics of Agriculture,* Vol. 2, *Papers in Honor of D. Gale Johnson.* University of Chicago Press, Chicago, Illinois, pp. 171–203.

Forhan, M. (1993) Canola breeding goes private. *Canola Guide,* 28–31.

Gray, R., Malla, S. and Phillips, P. (1999) The public and not-for-profit sectors in a biotechnology-based, privatizing world: the canola case. In: Lesser, W.H. (ed.) *Proceedings of NE-165 Conference, Transitions in Agbiotech: Economics of Strategy and Policy.*

Huffman, W.E. and Evenson, R.E. (1989) Supply and demand functions for multiproduct U.S. cash grain farms: biases caused by research and other policies. *American Journal of Agricultural Economics* 71, 761–773.

Huffman, W.E. and Evenson, R.E. (1992) Contributions of public and private science and technology to U.S. agricultural productivity. *American Journal of Agricultural Economics* 74, 752–756.

Huffman, W.E. and Evenson, R.E. (1993) *Science for Agriculture: a Long-Term Perspective.* Iowa State University Press, Ames, Iowa.

ICAR (Inventory of Canadian Agri-Food Research) (1998) Special tabulation upon request.

ISI (Institute for Scientific Investigation) (1997) Citations database, special tabulation of academic publications based on key word search for canola. November 1997.

Leiby, J.D. and Adams, G.D. (1991) The returns to agricultural research in Maine: the case of a small northeastern experiment station. *Northeastern Journal of Agricultural and Resource Economics* 20, 1–14.

Makki, S.S., Tweeten, L.G. and Thraeu, C.S. (1996) Returns to agricultural research: are we assessing right? Contributed paper *Proceedings from the Conference on Global Agricultural Science Policy for the Twenty-First Century.* Melbourne, Australia, 26–28 August 1996, pp. 89–114.

Malla, S. (1996) The distribution of the economic and health benefits from canola research. MSc thesis. University of Saskatchewan, Saskatoon, Saskatchewan.

Malla, S., Gray, R. and Phillips, P.W.B. (1998) *The Evolution of Canola Research: an Example of Success.* An interim report prepared for Agriculture Development Fund, Saskatchewan Agriculture and Food, Department of Agricultural Economics, University of Saskatchewan, Saskatoon, Saskatchewan.

Manitoba Crop Insurance Corporation (1998) *Variety Summary, 1993–1997.*

Myers, R.J. and Jayne, T. (1997) Regime shifts and technology diffusion in crop yield growth paths with an application to maize yields in Zimbabwe. *Australian Journal of Agricultural and Resource-Economics* 41, 285–303.

Nagy, J.G. and Furtan, W.H. (1977) *The Socio-Economic Costs and Returns from Rapeseed Breeding in Canada.* Department of Agricultural Economics, University of Saskatchewan, Saskatoon, Saskatchewan.

Nagy, J.G. and Furtan, W.H. (1978) Economic costs and returns from crop development research: the case of rapeseed breeding in Canada. *Canadian Journal of Agricultural Economics* 26, 1–14.

Pardey, P.G. and Craig, B. (1989) Causal relationships between public sector agricultural research expenditures and output. *American Journal of Agricultural Economics* 71, 9–19.

Phillips, P. (1997) Manual search of ISI Citations Index. July.

Prairie Pools Inc. (1977–1992) *Prairie Grain Variety Survey.* Various issues.

Ravenscraft, D. and Scherer, F.M. (1982) The lag structure of returns to research and development. *Applied Economics* 14, 603–620.

Saskatchewan Agriculture and Food (1997) *Agricultural Statistics.* Saskatchewan Agriculture and Food, Regina, Saskatchewan.

Saskatchewan Agriculture and Food (Various issues) *Varieties of Grains Crops in Saskatchewan.* Saskatchewan Agriculture and Food, Regina, Saskatchewan.

Statistics Canada (1998a) Direct cansim time series: CPI and all goods for Canada. www.datacenter.chass.utoronto.ca:5680/cansim/cansim.html

Statistics Canada (1998b) Direct cansim time series: canola – rapeseed/production, metric tonnes, canola – Canada. www.datacenter.chass.utoronto.ca:5680/cansim/cansim.html

Thirtle, C.G. and Bottomley, P. (1988) Is publicly funded agricultural research excessive? *Journal of Agricultural Economics* 31, 99–111.

Ulrich, A.W. and Furtan, W.H. (1985) *An investigation in the rates of return from the Canadian Crop Breeding Program.* Report by Department of Agricultural Economics, University of Saskatchewan, Saskatoon, Saskatchewan.

Ulrich, A., Furtan, H. and Downey, K. (1987) *Biotechnology and Rapeseed Breeding: Some Economic Considerations.* Science Council of Canada, Ottawa, Ontario.

11 Determinants of GMO Use: a Survey of Iowa Maize–Soybean Farmers' Acreage Allocation[1]

Corinne Alexander,[1] Jorge Fernandez-Cornejo[2] and Rachael E. Goodhue[1]

[1] Department of Agricultural and Resource Economics, University of California, Davis, California, USA; [2] Economic Research Service, USDA, Washington DC, USA

American maize and soybean farmers have rapidly adopted genetically modified (GM) seed with traits such as herbicide tolerance and pest resistance since their introduction in 1996. By 1999, GM varieties accounted for 33% of commercial maize acreage in major states and 57% of commercial soybean acreage (USDA, 2000). However, the future use of GM crops is uncertain given the current controversy in Europe and Japan. Fernandez-Cornejo *et al.* (2000) suggest that the traditional methods to predict adoption are insufficient given this market uncertainty. In this chapter, we use results from focus groups and a pre-planting survey of Iowa maize–soybean farmers to examine individual farmers' decisions to adopt GM crops.

GM crops carrying herbicide tolerant genes were developed to survive specific broad spectrum herbicides, increasing the number of effective post-emergent herbicides available to farmers. The most common herbicide-tolerant crops are Roundup Ready (RR) crops which are resistant to glyphosate. Roundup Ready traits are commercially available for soybeans, maize, canola and cotton. Liberty Link (LL) maize is another GM herbicide tolerant crop resistant to glufosinate-ammonium. The only insect resistant crops commercially available are Bt crops containing a gene from a soil bacterium *Bacillus thuringiensis*. Bt maize produces a protein that is toxic to the European corn borer (ECB), a major maize pest.

The decision to adopt a new technology depends on the expected profitability of the innovation which is determined by the characteristics of the farmer, the farm, the innovation, the institutional setting of the farm, and the farmers' beliefs about the innovation (Kinnucan *et al.*, 1990). There are several well-established links between the adoption decision and characteristics of the farmer and farm. First, farmers who have more allocative ability, which can be measured in part by human capital measures such as schooling and experience, are more likely to profit from an innovation (Welch, 1970; Kislev and Shchori-Bachrach, 1973; Huffman, 1974; Feder *et al.*, 1985). Second, farmers' risk preferences will affect both the adoption decision and the extent of adoption. Farmers who are more risk averse will be less likely to adopt a new, uncertain innovation (Feder *et al.*, 1985; Chavas and Holt, 1990; Pope and Just, 1991). Last, farm size has been shown to play a significant role in adoption (Lindner, 1983; Feder *et al.*, 1985; Feder and Slade, 1986). Larger farms are more likely to adopt an innovation

[1] The views expressed are those of the authors and do not necessarily correspond to the views or policies of the US Department of Agriculture.

for many reasons. For example, larger farms are able to spread fixed costs over more units of output and, consequently, the potential profits from 'lumpy' innovations increase with farm size.

We focus on three aspects of the adoption process. First, we evaluate whether producer and farm characteristics are related to farmers' response to the current genetically modified organism (GMO) controversy, as expressed in their 2000 planting intentions, in any systematic way. Second, we examine the early adopters of GM maize and show that the early adopters of GM crops match the stylized profile of early adopters of other innovations. We focus on farmer characteristics, including age, experience, schooling and primary occupation, and farm characteristics, including number of planted acres, share of land owned by the farmer and the gross farm income. We do not include measures of farmers' risk preferences; however, in a related paper, we explain farmers' acreage allocation intentions for 2000 using their risk preferences (Alexander *et al.*, 2000). Third, we examine the relationship between adoption and farmers' concerns regarding ECB and weed infestations, both of which may affect the perceived efficacy of the innovation.

Data

The University of California, Davis (UC Davis) collected these data with the cooperation of the Iowa Farm Bureau Federation. Mail surveys were sent to 1000 Iowa Farm Bureau members who grow maize and plant at least 100 acres of row crop. We restricted the sample to farms of 100 acres of row crop or more in order to focus on the farms that produce the majority of the maize in Iowa. According to the 1997 Census of Agriculture, farms of 100 acres or more account for 58.6% of maize-producing farms, and produced 90.2% of the Iowa maize crop in 1997 (USDA, 1999). The first wave of the survey was mailed on 9 February 2000 and a second copy was sent to non-respondents on 1 March 2000. After excluding the undeliverable surveys,[2] we obtained 389 usable responses at a response rate of 38.9%.

Prior to the survey, we conducted three focus groups in cooperation with the Iowa Farm Bureau Federation: two in Mason City in north-central Iowa on 14 and 15 December 1999, and one in Albia in south-central Iowa on 5 January 2000. We discussed how farmers choose which maize seed to plant and we pre-tested the survey. The Iowa Farm Bureau provided us with membership lists and allowed us to conduct the meetings at their branch offices.

Operator Demographics and Farm Characteristics

In this section, we report the demographic and farm characteristics of the survey respondents. In the following section, we compare the respondent group with data from the 1997 Census of Agriculture (USDA, 1999). Ninety-four per cent of the farmers grow both maize and soybeans. Ninety-nine per cent of the farmers grow maize.[3] Lucerne is the next most common crop, grown by half of the farmers. Thirty per cent of the farmers grow oats and just a few farmers grow other grains or speciality crops (Table 11.1).

Table 11.1. Percentage of farmers who grow each crop.

Crop	Percentage
Maize	99.9
Soybeans	95.4
Lucerne	49.6
Oats	30.1
Other grains	1.8
Speciality crops	3.1

Sixty-one per cent of the farmers in our sample raised livestock. Of the farmers who raised livestock, over half had a cow/calf operation. Over a third of the farmers had feedlots and hog finishing operations. About one in five had a hog farrowing operation and one in eight had a dairy operation. The least common livestock operations include sheep, poultry and other livestock (Table 11.2).

The survey asked questions about operator demographics, which include operator's age, number of years actively farming, highest level of formal education and principal occupation. The questions about farm characteristics include county and zip code, planted acres owned by the farmer, planted acres rented by the farmer, family members who work on-farm, people employed by the farm and the

[2] Undeliverable means that the farmers were either retired, renting out their land or deceased.

[3] One sample selection criterion was farms that grow maize. The Farm Bureau membership information is based on past activities so the sample included farms that have grown maize in the past.

Table 11.2. Percentage of livestock operators who engaged in each activity.

Livestock operation	Percentage
Hog finishing	35.1
Hog farrowing	18.6
Cow/calf	57.4
Feedlot	37.6
Dairy	12.0
Sheep	5.4
Poultry	2.9
Other livestock	2.9

Table 11.3. Operator demographics and characteristics of their farms.

	All farmers
Sample size	389
Average age	54
Average years' experience	32
Share farming primary occupation	84%
Average acres farmed	560
Average share land owned	52%

total gross farm income (GFI) in 1999. Table 11.3 presents selected descriptive statistics and Tables 11.4 to 11.8 present more detailed information on operator demographics and farm characteristics.

The average operator in the sample is 54-years-old and has been actively farming for 32 years. Eighty-four per cent of the operators in the sample say that farming is their primary occupation. The average farm size is 560 acres and the average operator owns 52% of the land he or she farms. The median farmer holds a high school diploma.

GFI is used as a measure of farm size, in addition to total acres, because it provides information about activities such as hog finishing, which do not require many acres, but do require substantial management effort and other inputs. The median farmer in our sample has a GFI of US$100,000–249,999.

There are several significant correlations between the operator demographics and farm characteristics.[4] The age of the operator and years of farming experience are significantly and positively correlated. The age of the operator and years of formal education are significantly and negatively correlated, indicating that

Table 11.4. Percentage of UC Davis respondents and their highest level of formal education attained.

Education category	Percentage
Grade school	3.4
Some high school	2.4
High school diploma	45.4
Some college work	14.8
Some vocational technical work	8.2
2-year community college degree	8.2
4-year college degree	12.4
Some postgraduate work	2.9
Postgraduate degree	2.4

Table 11.5. Percentage of UC Davis farms in each total gross farm income category in 1999.

Income category	Percentage
Less than US$10,000	2.2
US$10,000–19,999	6.0
US$20,000–24,999	3.3
US$25,000–39,999	5.7
US$40,000–49,999	5.4
US$50,000–99,999	15.4
US$100,000–249,999	35.0
US$250,000–499,999	19.8
US$500,000 or more	7.3

Table 11.6. Percentage of farms in each size category.

Acres	UC Davis 1999	NASS 1997
1–49	0	18.3
50–179	15.4	27
180–499	35.9	31.9
500–999	35.4	16.3
1000 or more	13.0	7.5

Table 11.7. Percentage of farms in each total gross farm income category.

Income category	UC Davis 1999	NASS 1997
Less than $10,000	2.2	26
US$10,000–49,000	20.3	24.2
US$50,000–99,999	15.4	15.1
US$100,000–249,999	35.0	21.1
US$250,000 or more	27.1	13.6

Table 11.8. Operators by principal occupation.

	UC Davis 1999	NASS 1997
Farming	84%	62%
Other	16%	38%

4 We calculate the Pearson's product-moment correlation and test the null hypothesis that the correlation is significantly different from zero with a *t*-test with $n-2$ degrees of freedom.

younger farmers tend to have more schooling. Older farmers own a larger share of the land they farm than younger operators, but they also have smaller farms. Consistent with the negative correlation between age and schooling, farmers with more education are significantly more likely to own a smaller share of the land they farm and to have larger farms.

Full-time farmers have significantly higher GFIs than part-time farmers. Part-time farmers have significantly more formal education than full-time farmers. There is no correlation between age and whether the respondent is a full-time or part-time farmer.

Younger, less experienced farmers are significantly more likely to have higher GFIs and to operate larger farms. Farmers with a higher level of formal education have significantly higher GFIs. Again, these relationships are consistent with the significant negative correlation between age and schooling.

Comparison with NASS Data

The differences between the UC Davis survey respondents and Iowa farmers as a whole can be explained by two sampling decisions: the UC Davis sample was based on Iowa Farm Bureau members, and farms that had at least 100 acres of row crop. For this reason, the farmers in the UC Davis survey tend to operate larger farms than Iowa farmers as a whole. According to the 1997 Census of Agriculture, 41.4% of Iowa maize producers farm less than 100 acres. These farmers were not sampled in the UC Davis survey due to the minimum farm size requirement. However, producers that farm 100 acres or more grew 90.2% of the 1997 Iowa maize crop (USDA, 1999). Hence, farmers in the UC Davis survey represent the majority of Iowa maize production.

Based on GFI, the average UC Davis survey farm is larger than the average Iowa farm, as expected due to the minimum acreage requirement. Twenty-six per cent of Iowa farmers have a GFI of less than US$10,000 compared with about 2% of UC Davis respondents. Even more striking, 62% of UC Davis farmers had a GFI of US$100,000 or more compared with 34% of Iowa farmers.

A larger share of the farmers in the UC Davis sample are full-time farmers than Iowa farmers as a group. Eighty-four per cent of UC Davis survey respondents describe farming as their primary occupation compared with 62% of Iowa farmers. Since full-time farmers have larger farms than part-time farmers, this difference is consistent with the larger farms in the UC Davis sample.

Planting Intentions

There is a great deal of uncertainty about the future use of bio-engineered seed. Prior to planting in 1999, European governments explicitly banned the import of specific maize hybrids with stacked traits. GMO use in the food chain has become even more contentious over the past year. There is a great deal of uncertainty regarding the European response, as well as the response of consumers and regulators in Japan and the USA. To date, American consumers' reaction has been limited. However, several large US based corporations including Frito Lay, Gerber and Heinz have committed to delivering non-GM food products to their customers. Farmers are aware of the potential marketing difficulties that may emerge as a result of this uncertain demand. In interviews and focus groups with farmers in December 1999 and January 2000, Iowa grain farmers expressed concern about planting GM crops due to marketing considerations. Producers cited three marketing concerns: first, that their local grain elevator would refuse all GM crops;[5] second, that GM crops would face a price penalty; and third, that grain elevators would delay accepting GM grain for some indeterminate period of time after harvest.

Based on farmers' planting intentions for 2000, we identify four groups of farmers: (i) a *GM users* group which grew GM crops in 1999 and plans to continue growing GM crops in 2000; (ii) a *conventional only* group which did not grow GM crops in 1999 and plans to grow only conventional crops in 2000; (iii) a *disadopters* group which plans to disadopt GM crops in 2000; and (iv) a *new adopters* group which plans to adopt GM crops for the first time in 2000.[6] We apply these categories to producers' acreage allocation responses to GM maize

[5] Apparently in 1999 the neighbour of a focus group participant had to transport his grain an additional 25 miles because the nearest elevator refused all GM grain.

[6] For GM maize, we define new adopters as farmers who did not plant GM maize in 1997, 1998 or 1999, but intend to plant GM maize in 2000. For soybeans, we only have data on acres planted in 1999 and intentions for 2000. Hence, we define adopters as farmers who did not plant GM soybeans in 1999, though they may have planted them in previous years.

and GM soybeans separately because their reactions differ consistently for maize and soybeans. Over half of the producers have the same planting intentions for both maize and soybeans, that is, they fall into the same category for both crops. However, 15% of producers are GM users for soybeans and conventional only for maize. Another 8% who are GM users for soybeans plan to disadopt GM maize in 2000. Less than 5% of the producers are GM users for maize and disadopting GM soybeans. Overall for the 2000 crop year, farmers are more likely to reduce acreage or disadopt GM maize than GM soybeans. In fact, many producers are planning to increase their GM soybean acreage.

Comparing the maize and soybean categories

We find that the profile of farmers in each group is consistent across maize and soybeans. The two largest groups are GM users and conventional only. The GM users group is younger, less experienced, has more schooling, has a larger share of full-time farmers and operates larger farms than the conventional only group.

The disadopters are demographically similar to the GM users, but they tend to operate farms that are smaller than those of the GM users. Disadopters may be more sensitive to the current market conditions than the GM users because of their smaller farm size.

The new adopters are the smallest group. There are at least two potential explanations that are consistent with this behaviour. First, the current market uncertainty may cause potential adopters to delay adoption. Indeed, one wonders why there are any adopters, given the market uncertainty. One possible explanation is that GM crops may be a substitute for human capital.[7] The new adopters have the least amount of schooling of the four groups. Second, there may be few adopters because the diffusion of GM crops may be approaching the ceiling. The new adopters fit the profile of followers that adopt innovations towards the end of the diffusion process; they have the least amount of schooling, are older than GM users and disadopters, and operate relatively small farms.

GM maize

With respect to maize, over half of the farmers are GM users and 27% are conventional only. More farmers decided to disadopt than adopt GM maize; 15% are disadopters while only 3% are new adopters. The large number of disadopters is consistent with the increase in market uncertainty. Another factor affecting disadoption decisions may be the lack of ECB infestation in Iowa in the 1998 and 1999 seasons. In focus groups, many producers indicated that they were considering reducing their use of Bt maize due to the lack of realized returns in previous seasons. In a later section, we discuss the relationship between adoption and concern regarding ECB infestation in more detail.

Conventional-only farmers were the oldest, with an average age of 58. They have the smallest average farm size of 380 acres. Consistent with the smaller farm size, these farmers had generally lower GFIs. Only 36% had a GFI of US$100,000 or more. They are also more likely to be part-time farmers (Table 11.9).

Table 11.9. Operator demographics and characteristics of their farms for each group.

	GM maize in 1999 and 2000	No GM maize in 1999 and 2000	Disadopted GM maize in 2000	Adopted GM maize in 2000
Sample size	185	94	51	14
Average age	52	58	51	55
Average years' experience	29	35	29	35
Share farming primary occupation	87%	73%	94%	93%
Average acres farmed	717	380	619	583
Average share land owned	44%	62%	40%	58%

[7] In the focus groups and supplementary interviews, several farmers and one grain elevator manager said that herbicide tolerant crops and pest resistant crops are easier to manage. According to the grain elevator manager, Roundup Ready soybeans 'make everyone good bean farmers'.

Table 11.10. Highest level of formal education attained for farmers in each group (%).

Education category	GM maize in 1999 and 2000	No GM maize in 1999 and 2000	Disadopted GM maize in 2000	Adopted GM maize in 2000
Grade school	1.7	6.5	4.0	7.7
Some high school	2.2	2.2	2.0	0.0
High school diploma	38.1	46.2	56.0	53.9
Some college work	19.9	12.9	4.0	15.4
Some vocational technical work	8.3	5.4	14.0	15.4
2-year community college degree	9.4	9.7	4.0	7.7
4-year college degree	15.5	10.8	14.0	0.0
Some postgraduate work	2.2	4.3	0.0	0.0
Postgraduate degree	2.8	2.2	2.0	0.0

GM users are younger and less experienced than conventional-only farmers. They operate larger farms, with an average size of 717 acres. GM users have much higher GFIs than conventional only farmers; 75% have a GFI of US$100,000 or more.

Disadopters are demographically similar to GM users, with an average age of 51 and 52, respectively. Both groups have an average of 29 years' experience. Disadopters have less schooling than GM users. Only 38% of disadopters had formal schooling beyond high school compared with 58% of GM users (Table 11.10). Even though disadopters tend to have smaller farms than GM users, they have relatively large farms with an average farm size of 619 acres and 73% have a GFI of US$100,000 or more.

New adopters are older than GM users and disadopters, with an average age of 54. They have a bit more farming experience than GM users and disadopters with an average of 32 years. They have the least amount of formal schooling compared with the other three groups. New adopters have larger farms than conventional-only farmers but smaller than disadopters and GM users. The average farm size is 570 acres and 57% have GFIs of US$100,000 or more (Table 11.11).

GM soybeans

Over 70% of the farmers are GM users with respect to Roundup Ready soybeans and only 15% are conventional only. Twice as many farmers decided to disadopt GM soybeans as adopt; 9% are disadopters and 4% are adopters.

Like GM maize users, GM soybean users are younger and less experienced than conventional-only soybean farmers. The average age of GM soybean users is 53 with 31 years' farming experience and the average age of conventional-only soybean farmers is 58 with 36 years' experience (Table 11.12). There is no correlation between use of GM soybeans and full-time versus part-time farming. GM users have more formal schooling than conventional-only farmers. Fifty-three per cent of GM users have formal schooling beyond high school compared with 37% of conventional-only farmers (Table 11.13). Like GM maize users, GM soybean users operate larger farms. The average farm size for GM users is 656 acres compared with 448 acres for conventional-only. They have a much higher GFI; 69% have GFI of US$100,000 or more compared with only 42% of conventional-only farms (Table 11.14).

As with maize, disadopters are demographically most similar to GM users. The average age of disadopters is 52 and they have an average of 28 years'

Table 11.11. Total gross farm income for each group (%).

Income category	GM maize in 1999 and 2000	No GM maize in 1999 and 2000	Disadopted GM maize in 2000	Adopted GM maize in 2000
Less than $10,000	0.6	5.6	0.0	0.0
US$10,000–49,999	12.8	33.33	16.3	9.1
US$50,000–99,999	11.1	24.4	10.2	27.3
US$100,000–249,999	38.3	22.2	49.0	36.4
US$250,000 or more	37.2	14.4	24.5	27.3

Table 11.12. Operator demographics and farm characteristics for each group.

	Some RR soybeans in 1999 and 2000	No RR soybeans in 1999 and 2000	Disadopted RR soybeans in 2000	Adopted RR soybeans in 2000
Sample size	255	54	31	16
Average age	53	58	52	56
Average years' experience	31	36	28	35
Share farming primary occupation	85%	89%	81%	81%
Average acres farmed	656	448	522	425
Average share land owned	50%	53%	36%	49%

Table 11.13. Highest level of formal education attained for each group (%).

Education category	Some RR soybeans in 1999 and 2000	No RR soybeans in 1999 and 2000	Disadopted RR soybeans in 2000	Adopted RR soybeans in 2000
Grade school	2.4	3.7	0.0	18.8
Some high school	1.2	5.6	0.0	3.2
High school diploma	43.3	53.7	38.7	56.3
Some college work	16.6	13.0	9.7	6.3
Some vocational technical work	9.3	3.7	12.9	0.0
2-year community college degree	9.3	9.3	0.0	12.5
4-year college degree	11.7	11.1	25.8	6.3
Some postgraduate work	2.4	0.0	9.7	0.0
Postgraduate degree	3.6	0.0	0.0	0.0

Table 11.14. Total gross farm income for each group (%).

Income category	Some RR soybeans in 1999 and 2000	No RR soybeans in 1999 and 2000	Disadopted RR soybeans in 2000	Adopted RR soybeans in 2000
Less than US$10,000	1.2	4.0	3.3	6.7
US$10,000–49,999	15.9	36.0	20.0	20.0
US$50,000–99,999	14.2	18.0	16.7	13.3
US$100,000–249,999	37.8	22.0	40.0	40.0
US$250,000 or more	30.9	20.0	20.0	20.0

experience. Notably, disadopters have the most schooling of any group with 58% attending school beyond high school. They also tend to have smaller farms than GM users but their farms are larger than conventional-only with an average of 522 acres; 60% have a GFI of US$100,000 or more.

Again, with an average age of 56, adopters are older than GM users and disadopters. They have almost as much farming experience, 35 years, as the conventional-only farmers. Adopters have the least amount of education; only 25% completed work beyond high school. Based on the demographic characteristics of the respondents as a whole, we would expect that older farmers would have fewer years of formal education. However, conventional-only farmers are older than the adopters, and have more schooling. The adopters have the smallest average farm size at 425 acres. While the adopters have the smallest average farm size, 60% have a GFI of US$100,000 or more.

Adoption intensity and GM soybeans

We found that GM soybean users differ in their adoption intensity. Sixty-nine per cent of GM users, or 39% of the respondents, planted 100% of their soybean acreage to Roundup Ready soybeans. We refer to these farmers as complete adopters. The others partially adopted Roundup Ready soybeans.

The complete adopters and partial adopters have about the same average age and experience. Partial adopters are slightly more likely to be full-time farmers. The partial adopters have substantially more schooling than the complete adopters; 57% of partial adopters have progressed beyond high school compared with 49% of complete adopters. The partial adopters also operate larger farms than the complete adopters; they farm an average of 740 acres, and 78% have a GFI of US$100,000 or more compared with complete adopters with an average of 580 acres and 62% have a GFI of US$100,000 or more (Tables 11.15–11.17).

Early Adopters of GM Maize

The adoption literature has identified some stylized facts about early adopters. First, larger farms tend to adopt new innovations earlier (Lindner, 1983; Feder *et al.*, 1985). One explanation for this empirical regularity is that fixed costs of adoption, such as the cost of acquiring information about the innovation, are relatively smaller for larger farms (Feder and Slade, 1986). Second, farms with more human capital, including formal schooling, experience and farmer health, tend to adopt new innovations earlier (Kislev and Shchori-Bachrach, 1973; Huffman,

Table 11.15. Operator demographics, farm characteristics and adoption intensity of GM soybeans.

	100% RR soybeans in 1999 and 2000	Some RR soybeans in 1999 and 2000
Sample size	150	67
Average age	53	54
Average years' experience	32	31
Share farming primary occupation	83%	87%
Average acres farmed	580	740
Average share land owned	55%	47%

Table 11.16. Highest level of formal education attained and adoption intensity of GM soybeans (%).

Education category	100% RR soybeans in 1999 and 2000	Some RR soybeans in 1999 and 2000
Grade school	3.5	1.5
Some high school	0.7	3.0
High school diploma	47.6	38.8
Some college work	16.8	14.9
Some vocational technical work	8.4	7.5
2-year community college degree	9.8	7.5
4-year college degree	8.4	17.9
Some postgraduate work	2.1	1.5
Postgraduate degree	2.8	7.5

Table 11.17. Total gross farm income and adoption intensity of GM soybeans (%).

Income category	100% RR soybeans in 1999 and 2000	Some RR soybeans in 1999 and 2000
Less than US$10,000	0.7	1.5
US$10,000–49,999	21.0	7.7
US$50,000–99,999	16.0	12.3
US$100,000–249,999	32.9	43.1
US$250,000 or more	29.4	35.4

1974; Feder *et al.*, 1985). Consistent with the literature, we find that larger farms and farmers with more human capital were more likely to adopt GM maize in 1997.

The survey asked farmers to report how many acres they planted in conventional maize and maize with specialized traits from 1997 to 1999 and how many acres they intend to plant in 2000.[8] In 1997 22% of the respondents planted GM maize. These farmers were younger than the rest of the sample, with an average age of 48 years compared with the sample average of 54 (Table 11.18). The farmers who planted GM maize in 1997 also had less experience, but they had a higher level of schooling (Table 11.19). Eighty-nine per cent of the farmers who adopted GM maize in 1997 said that farming is their primary occupation compared with 84% of the sample as a whole.

According to both measures of farm size, larger farms adopted GM maize in 1997. The average

Table 11.18. Operator and farm characteristics of those who grew GM maize in 1997.

	Farmers who grew GM maize in 1997[a]
Sample size	85
Average age	48
Average years' experience	26
Share farming primary occupation	89%
Average acres farmed	729
Average share land owned	41%

[a] We do not have information on GM maize planted in 1996. These farmers may have grown GM maize in 1996 or they may have adopted GM maize in 1997.

Table 11.19. Highest level of formal education attained by those who grew GM maize in 1997.

Education category	1997 Adopter (%)	Overall sample (%)
Grade school	1.2	3.4
Some high school	2.5	2.4
High school diploma	34.5	45.4
Some college work	16.0	14.8
Some vocational technical work	13.6	8.2
2-year community college degree	12.4	8.2
4-year college degree	17.3	12.4
Some postgraduate work	1.2	2.9
Postgraduate degree	1.2	2.4

Table 11.20. Total gross farm income of those who grew GM maize in 1997.

Income category	1997 Adopters (%)	Overall sample (%)
Less than $10,000	0	2.2
US$10,000–49,999	5	20.3
US$50,000–99,999	10	15.4
US$100,000–249,999	38	35.0
US$250,000 or more	48	27.1

8 In the focus groups we tested questions that asked farmers how many acres they planted in GM maize in 1996. Of the 20 farmers, none had planted GM maize in 1996, so we dropped the question to lessen the response burden.

number of acres farmed was 729, well above the sample average of 560 acres. Further, farmers who adopted GM maize in 1997 had a larger GFI (Table 11.20). On average these farmers owned 41% of the land they farmed compared with the sample average of 52%, which reflects that younger farmers tend to own a smaller share of the land they operate.

About 30% of the farmers in our sample adopted GM maize for the first time in 1998. These farmers are demographically similar to the farmers in the whole sample, but they tend to operate larger farms. For instance, the average age is 54 and the farmers have an average of 33 years' experience. The level of formal schooling is comparable to the rest of the sample and 85% of the operators list farming as their primary occupation.

By both measures of farm size, farmers who adopted GM maize in 1998 tend to have larger farms. They farmed an average of 702 acres compared with 560 for the sample average. The GFI for GM maize adopters was relatively higher; they were more likely to have a GFI above US$100,000 than the overall sample. However, on average, they owned a smaller share of the land they farm (Tables 11.21–11.23).

Table 11.21. Operator and farm characteristics of those who adopted GM maize in 1998.

	Farmers who adopted GM maize in 1998[a]
Sample size	115
Average age	54
Average years' experience	33
Share farming primary occupation	85%
Average acres farmed	702
Average share land owned	46%

[a] We know these farmers did not grow GM maize in 1997 but they may have planted GM maize in 1996.

Table 11.22. Highest level of formal education attained by those who adopted GM maize in 1998.

Education category	1998 Adopter (%)	Overall sample (%)
Grade school	2.6	3.4
Some high school	2.6	2.4
High school diploma	42.6	45.4
Some college work	19.1	14.8
Some vocational technical work	7.0	8.2
2-year community college degree	7.0	8.2
4-year college degree	15.7	12.4
Some postgraduate work	0.9	2.9
Postgraduate degree	2.6	2.4

Table 11.23. Total gross farm income of those who adopted GM maize in 1998.

Income category	1998 Adopters (%)	Overall sample (%)
Less than US$10,000	1	2.2
US$10,000–49,999	18	20.3
US$50,000–99,999	13	15.4
US$100,000–249,999	42	35.0
US$250,000 or more	26	27.1

Concerns that Affect Adoption of GM Maize

A producer's decision to adopt a new innovation depends on his or her subjective beliefs about its net benefit, which depends on his or her beliefs about the total potential benefits and about the efficacy of the new technology in capturing these potential benefits. For example, in order for farmers to perceive benefits from planting Bt maize, they must first believe that the ECB can cause economically significant yield loss, so there is substantial potential benefits from more effective ECB control. Secondly, the farmers must believe that Bt maize is a more economically effective method of controlling ECB than the alternatives. For the case of herbicide tolerant crops, farmers must first believe that weed damage can cause economically significant yield loss and second that it is more cost effective to control weeds with the herbicide tolerant crops. In both cases, adoption is more likely for producers who view the problem addressed by the GM variety as more significant. We would expect that farmers who are very concerned about yield loss from weeds or ECB would be more likely to adopt herbicide tolerant crops or Bt crops, respectively.

Relatively high producer concern about weeds or ECB can be ascribed to the following: (i) high levels of expected pest pressure on the farm; (ii) high levels of producer awareness of the pest pressure; or (iii) the producer is more risk averse and therefore more likely to try to reduce the income variance associated with pest pressure, relative to others in the sample. For the purposes of this analysis, we do not attempt to identify the sources of producers' concerns about pest pressure. Rather we focus on whether their stated level of concern is correlated with their decision to plant GM crops.

In the survey we asked farmers to rate weed pressure in maize production and damage by the ECB as a major concern, minor concern or not a concern. We also asked farmers about the practices they use to control weeds on their maize fields and the practices they use to estimate ECB populations on their maize fields. We find that producer concern about damage from ECB is significantly and positively correlated with adoption of Bt maize. However, producer concern about damage from weed pressure on their farms is, at best, weakly correlated with adoption of herbicide tolerant maize and uncorrelated with herbicide tolerant soybeans.

Herbicide tolerant maize and weed pressure

When we pre-tested the survey in the focus groups, farmers had an interesting reaction to the question on concern about weed pressure. They said their answer depended on whether or not they were using herbicides; if they do not use herbicides, then weed pressure is a major concern but if they use herbicides then weed pressure is not a concern. Farmers also commented that they know which fields will have severe weed pressure when they make their planting decisions: 'If I use my herbicides and keep my field clean, I don't have any yield loss from weeds. If I didn't use [herbicides], then I got problems.'

In the survey, producers who grow Liberty Link maize are significantly more likely to cite weed pressure as a major concern on their farm (Table 11.24). However, there is no statistical correlation between producers who grow Roundup Ready maize and their concern about weed pressure indicating that other considerations affect the decision to adopt Roundup Ready maize.

Table 11.24. Percentage of farmers who say weed pressure is a major concern on their farm.

Crop	Percentage
Roundup Ready maize	69
Liberty Link maize	84
No herbicide tolerant maize	68

Farmers who say weed pressure is a major concern on their farm are more likely to employ certain weed control practices. In particular, they are more likely to use mechanical cultivation, crop rotation, plant seed with fast seedling development, plant herbicide resistant seed and they are less likely to delay planting date (Table 11.25). About 70% of farmers use herbicides, regardless of their concern about weed pressure.

Insect resistant maize and ECB pressure

In the focus group discussions, farmers commented that Bt maize is significantly more effective at controlling ECB than the alternative of applying chemical insecticides. 'The only reason they sell Bt maize is because it's so hard to get a treatment and if it's only 80% accurate to start with, then you lessen that [because the plane doesn't arrive before the ECB have burrowed in the stalk].' However, the farmers

Table 11.25. Percentage of farmers who say weed pressure is a major concern and weed control practices.

	Uses this weed control practice	Does not use this practice
Mechanical cultivation	71	64
Rotary hoeing	72	68
Crop rotation	71	64
Herbicides	69	71
Seed with fast seedling development	82	67
Herbicide resistant seed	79	66
Delay planting date	62	70

also said Bt maize was not profitable in 1998 and 1999 because ECB pressure was very low. As a consequence, some of the focus group participants plan to disadopt Bt maize because of the low payoff. Others will continue to plant Bt maize because they believe it is good insurance against the possibility of high ECB pressure. 'Of course, the Bt takes care of the corn borers and there haven't been any for 2 years. However, one of these days…'

In the survey, producers who say that damage by the ECB is a major concern for maize production on their farm are significantly more likely to plant Bt maize (Table 11.26). Forty-eight per cent of producers who plant Bt maize cite damage by ECB as a major concern compared with only 17% of producers who do not plant Bt maize.

Table 11.26. Percentage of farmers who cite damage by the ECB as a major concern.

Crop	Percentage
Bt maize	48
No Bt maize	17

These same producers are significantly more likely to measure ECB pressure on their farms. Sixty per cent of producers who hire someone to scout their fields for ECB and 42% of those who scout their own fields cite damage by ECB as a major concern compared with 27% of those who do not hire scouts and 16% of those who do not scout their own fields. Only 5% of producers who do not measure ECB pressure cite ECB damage as a major concern compared with 46% of those who do measure ECB pressure (Table 11.27).

Overall, farmers are better able to cope with weed pressure than ECB pressure. Farmers know which fields will have severe weed pressure and they cannot predict ECB pressure. One farmer said, 'Mostly, you raise your own [weed] pressure the year before. You know where it is. . . The insects jump the fence.' In addition, farmers said that alternative herbicides were effective at controlling weed pressure but it is difficult to effectively control ECB with insecticides. This suggests that beliefs regarding the net benefit of adoption encourages the adoption of Bt maize relative to the adoption of herbicide-tolerant maize.

Conclusion

This chapter examined three aspects of adoption of GM maize and soybeans by Iowa farmers. First, we examined Iowa farmers' planting intentions for 2000 and identified four distinct groups. Second, we showed that the early adopters of GM maize fit the stylized facts of early adopters. Third, we showed that farmers who are more concerned about yield damage to their maize crops from either weeds or insects are more likely to adopt LL maize and Bt maize, respectively. These findings are based on a

Table 11.27. Percentage of farmers who scout fields and measure ECB pressure.

	Uses this practice	Does not use this practice
Hire someone to scout fields	60	27
Scout own fields	42	16
Do not measure ECB pressure	5	46

survey of Iowa Farm Bureau members who have at least 100 acres and they may not reflect the representative maize–soybean farmer in the US. However, Iowa is an important state for maize and soybean production. In 1998, Iowa was the number one producer of both maize for grain and soybeans for beans, producing 18% of the total US maize crop and 18% of the total US soybean crop (Sands and Holden, 1999).

References

Alexander, C., Fernandez-Cornejo, J. and Goodhue, R.E. (2000) Effects of the GMO controversy on corn-soybean farmers' acreage allocation decisions. Department of Agricultural and Resource Economics Working Paper, University of California, Davis.

Chavas, J.-P. and Holt, M.T. (1990) Acreage decisions under risk: the case of corn and soybeans. *American Journal of Agricultural Economics* 72(3), 529–538.

Feder, G. and Slade, R. (1986) The acquisition of information and the adoption of new technology. *American Journal of Agricultural Economics* 66, 312–320.

Feder, G., Just, R.E. and Zilberman, D. (1985) Adoption of agricultural innovations in developing countries: a survey. *Economic Development and Cultural Change,* 33, 255–298.

Fernandez-Cornejo, J., Alexander, C. and Goodhue, R.E. (2000) Dynamic technology diffusion with disadoption: the case of crop biotechnology. Department of Agricultural and Resource Economics Working Paper, University of California, Davis.

Huffman, W.E. (1974) Allocative efficiency: the role of human capital. *Quarterly Journal of Economics* 91, 59–80.

Kinnucan, H., Hatch, U., Molnar, J.J. and Pendergrass, R. (1990) Adoption and diffusion potentials for bovine somatotropin in the southeast dairy industry. *Auburn University, Agricultural Experiment Station Bulletin* 605, February.

Kislev, Y. and Shchori-Bachrach, N. (1973) The process of an innovation cycle. *American Journal of Agricultural Economics* 55, 28–37.

Lindner, R.K. (1983) Farm size and time lag to adoption of a scale neutral innovation. Mimeograph. University of Adelaide, Adelaide.

Pope, R.D. and Just, R.E. (1991) On testing the structure of risk preferences in agricultural supply analysis. *American Journal of Agricultural Economics* 73, 743–748.

Sands, J.K. and Holden, H.R. (1999) *Iowa Agricultural Statistics 1999.* United States Department of Agriculture and Iowa Farm Bureau, Des Moines, Iowa.

United States Department of Agriculture (1999) 1997 Census of Agriculture. Technical Report 15, National Agricultural Statistical Service.

US Department of Agriculture (2000) Prospective plantings, technical report, National Agricultural Statistical Service. www.usda.mannlib.cornell.edu/reports/nassr/field/pcp-bbp/psp10300.txt

Welch, F. (1970) Education in production. *Journal of Political Economy* 78, 35–59.

12 Estimating Adoption of GMO Soybeans and Maize: a Case Study of Ohio, USA

David A. Darr and Wen S. Chern
Department of Agricultural, Environmental and Development Economics, The Ohio State University, Columbus, Ohio, USA

> Biotechnology's been around almost since the beginning of time. It's cavemen saving seeds of a high-yielding plant. It's Gregor Mendel, the father of genetics, cross-pollinating his garden peas. It's a diabetic's insulin, and the enzymes in your yogurt... Without exception, the biotech products on our shelves have proven safe.
>
> (US Agriculture Secretary Dan Glickman; 13 March 1997)

Introduction

This statement simply states the US opinion of biotechnology and genetically modified (GM) plants as of 1997. While other statements have been made concerning voluntary labelling of non-GM food, the USA still strongly believes in the advantages that can be gained through adoption of genetic modification. Glickman's agency, the United States Department of Agriculture, has tracked adoption of GM varieties since their inception.

This chapter takes a specific look at just one state's adoption of genetically modified organisms (GMOs). We will look at Ohio, a state on the eastern edge of the Corn Belt that is a substantial producer of maize and soybeans in the US. In previous reports, Ohio's data have been included in Corn Belt statistics. We are interested in looking at Ohio alone. Our objectives for this study include conducting a survey of Ohio farmers regarding GMO use, collecting data on attitudes and beliefs concerning GMO adoption, estimating GMO acreage in Ohio, and investigating factors affecting Ohio farmers' adoption of GMO seed varieties.

Background

For the past few years, biotechnology has been the primary source of debate in the agriculture industry (Northern Light: www.special.northernlight.com/gmfoods). Since 1996, when Roundup Ready soybeans were introduced to the public, farmers have been adopting GM seed varieties at rates faster than ever previously witnessed in agriculture. With the introduction of the Roundup Ready soybean, farmers could now apply a single broad-spectrum herbicide over the top of the soybean crop and kill a majority of weeds without harming the beans (Carpenter and Gianessi, 1999). This breakthrough promised farmers lower herbicide use, costs and crop damage by minimizing trips across the field. While this was not the first product resulting from modern gene-splicing techniques, it was a product that saw high levels of consumer (farmer) acceptance and adoption. Using particle gun bombardment, researchers at Monsanto were able to insert a gene for glyphosate tolerance into the soybean seed (Monsanto: www.biotechbasics.com). Table 12.1 illustrates adoption rates for major crop producing states as estimated by the United States Department of Agriculture (USDA: www.aphis.usda.gov/biotechnology) for 1996, 1997 and 1998. The adoption rate of GM soybeans in the USA increased from 7.4% in 1996 to 44.2% in 1998.

Maize has seen similar adoption trends following the introduction of Bt maize and herbicide resistant maize in 1996 (Biotechnology Industry Organization: www.bio.org). Bt maize contains a protein that is deadly to the European corn borer (ECB), a major cause of damage in US maize crops. The ECB feeds on the leaves and stalks of maize plants, thus weakening the stalk and making harvest difficult. Bt maize contains its own insecticide to prevent damage from this pest (Hyde *et al.*, 1999). Herbicide-resistant maize has traits similar to soybeans in that a broad-spectrum herbicide can be applied over the maize crop to kill weeds that can rob valuable nutrition from the ground. Again, USDA tracks adoption of GM maize varieties and has published the results shown in Table 12.1.

One common trait of all GMOs produced so far is that there is a technology fee associated with purchase of the seed. The purpose of this is for the life-science companies to recoup their investment of research into the product, given that they have a patent on the technology. The technology fees for soybeans and maize can increase the cost of production by up to US$10 per acre (McBride and Brooks, 1999). The fee is currently charged on a per unit basis on all product sold in the US. At the international level, the technology fee is assessed depending on patent law in the country of sale.

Ohio

Ohio is a very important state for US agriculture. In 1997, Ohio ranked fifth in the country in soybean production and sixth in production of maize for grain. This is despite the fact that Ohio has the second fewest acres of farmland in the important North Central region. This region is defined by the National Agricultural Statistics Service (NASS) and includes Illinois, Indiana, Iowa, Kansas, Michigan, Minnesota, Missouri, Nebraska, North Dakota, Ohio, South Dakota and Wisconsin. Ohio is distinguishable by its dependence on small family farms. This is characterized by the fact that Ohio has the lowest average farm size in the North Central region and is actually more comparable in average farm size to states in the north-east US. It is estimated that 21.5% (3,200,000 acres) of Ohio farmland was used for maize production in 1999, while 30.2% (4,500,000 acres) of farmland went towards production of soybeans (Ohio Agricultural Statistics Service: www.nass.usda.gov/oh). These are by far the largest crop volumes produced in Ohio.

Previous Work

When one considers the genetic modification of farm crops to be a new technology available to farmers, it seems reasonable to compare this to the adoption of hybrid maize that occurred from the 1930s to the 1950s. Similar to GM, hybrid seeds were put under much scrutiny and concerns about safety were voiced. In his seminal paper, Griliches (1957) investigated factors responsible for regional differences in the adoption of hybrid maize as a new technology in the USA. One of his primary assumptions was that adoption had three distinct characteristics: an origin, a slope and a ceiling. His goal was to learn something about the ways in which technological change is generated and propagated in US agriculture. His data were best fit by a logistic model in

Table 12.1. Adoption of GM crops in major crop producing states. (Source: Fernandez-Cornejo *et al.*, 1999.)

Field crop	Year of first introduction	% of Estimated planted acreage		
		1996	1997	1998
Soybean				
Herbicide-resistant	1996	7.4	17.0	44.2
Maize				
Bt	1996	1.4	7.6	19.1
Herbicide-resistant	1996	3.0	4.3	18.4
Maize totals		4.4	11.9	37.5

which observations were not points of equilibrium that may or may not change over time, but points on an adjustment path, moving consistently towards a new equilibrium position (Griliches, 1957). In his work, the new equilibrium position to be achieved was the ceiling. This ceiling varied from state to state, as he saw different potential maximum adoption rates based on economic value of different soil types in different states. Griliches presented a logistic growth curve expressed as:

$$P = K/(1 + e^{-(a + bt)}) \quad (12.1)$$

where, P = percentage of maize planted with hybrid seed; K = ceiling or equilibrium value; t = time variable; b = rate of growth coefficient; and a = constant of integration.

While at first this appears to be a good model to follow, there are some key differences. Griliches viewed hybrid seed as not being a single invention available to all at once. He made a clear distinction between availability and adoption. Availability started in the Corn Belt and spread outwards (Dixon, 1980). With genetically modified organisms (GMOs), supply was available throughout the USA. Another significant difference between the work of Griliches and GMO researchers of today is that Griliches did his work in the 1950s and had almost 20 years of data with which to conduct a time series analysis. Today, researchers only have 4–5 years of data on which to make estimations. While we do now have more complex methods of estimation, 20 years' worth of data would make analysis a simpler task.

More recently, the USDA has undertaken an extensive survey of GMO adoption through the Agricultural Risk Management Study (ARMS). Their survey is a three-part task involving: screening, obtaining production practices and cost data, and obtaining financial information. Their 1997 survey for soybeans covered 19 states representing 93% of all soybeans grown in the USA. They used a total of 1444 observations from 17 states after further screening of the data. In the paper by Fernandez-Cornejo *et al.* (1999) a per acre profit function was constructed from the collected data. The profit function included probabilities estimated through a probit analysis conducted to calculate the probability of a farmer using genetically enhanced seed varieties. They ran this model for herbicide-resistant soybeans, herbicide-resistant cotton and Bt cotton. While they found no significant change in profit resulting from adoption of GMOs, they did find that elasticity of yield with respect to probability of adoption of herbicide-resistant soybeans increased 0.03. McBride and Brooks (1999) found similar results from the same data set.

Data Collection

For the purpose of this research, we worked with the Ohio Corn Growers Association (OCGA) in developing a sample from which to conduct a survey of Ohio farmers. The National Corn Growers Association in conjunction with numerous agricultural magazines developed the OCGA database. The database contains a vast array of demographics that cover virtually all aspects of production agriculture. The grower information that we used to develop our sample was a random sample of 1922 Ohio maize and soybean producers.

From the randomly selected 1922 producers, 600 of these were randomly chosen to receive a mail survey that included general attitude questions regarding GMOs, previous history and future intentions regarding GMOs, and acreage, cost and price information for 1998, 1999 and 2000. We designed two versions of the survey and divided the sample into two groups of 300. Both editions of the survey included identical questions regarding attitudes about GMOs and general questions about benefits of and concerns regarding GMOs. In the long edition, producers were asked to fill out a chart with data regarding acreage, seed cost, herbicide cost, insecticide cost, commodity price and yield for both GMO and conventional varieties for crop years 1998, 1999 and 2000. For the short edition, we only asked growers about the acreage for GMO and conventional maize and soybeans in 1999 and 2000. As an incentive for recipients to complete and return the survey, they were asked to fill out a card to enter them in a raffle for Ohio State University (OSU) football tickets (a large incentive in Ohio). Also, the envelopes were marked with labels noting that GMO and Ohio State Football information was inside the envelope.

Of the 600 surveys that were sent in May 2000, 130 were returned giving a response rate of 21.5%. This response rate was close to our goal as the survey was sent in the middle of planting season in Ohio (Pennings *et al.*, 1999). Due to time constraints for both the recipients and the researchers, limited follow-up was completed in order to raise the response rate.

Table 12.2. Farm size and operator age from different sources.

Variable	OSU	OCGA	ODA
Average farm size (acres)	1218	>500	186
Average age of operator (years)	47.9	52	53

Source: ODA, OCGA.

Table 12.3. Survey summary statistics (%).

	Maize		Soybeans	
Questions	Yes	No	Yes	No
Did you plant any GM varieties in spring of 1999?	52.4	47.6	84.3	15.7
Did you encounter any elevators that would not accept GM varieties last autumn?	4.3	95.7	0	100
Did you receive premiums for non-GM varieties last autumn?	7.0	93.0	21.3	78.7
Do you plan to segregate GM from non-GM varieties this autumn?	18.6	81.2	35.5	64.5
Do you expect premiums for non-GM varieties this autumn?	7.0	93.0	22.3	77.7

Of the surveys that were returned, the following descriptive statistics describe the demographics of the sample. The average age of the respondents was 47.9 years, with 76 (60.3%) under 50 and 46 (39.7%) over age 50. As for education, 67 (52.8%) had a high school degree or less while 57 (47.2%) had at least some college education. Forty-seven respondents (42.3%) reported income levels over US$50,000 year^{-1}. One demographic area in question involves the amount of farmland dedicated to crops. As stated earlier, the average farm in Ohio was reported to be 186 acres in 1999; from the respondents to this survey, the average farm size is 1218 acres with 43 (39%) over 1000 acres. Overall, these data represent farmers who are younger, more educated and who farm more land than the average. Table 12.2 compares average farm size and average farmer age among our survey, OCGA and the Ohio Department of Agriculture (ODA).

Table 12.3 shows a summary of statistics gathered concerning farmers' production in 1999 and their responses to a series of questions regarding premiums and segregation of GMOs from non-GMOs. As is evident from Table 12.3, a strong majority of respondents have adopted GMO on some level. Also notable are the percentages of respondents who expect premiums and plan to segregate soybeans in the hope of finding a niche for their product.

With reference to farmers who have planted some or no GMOs, Table 12.4 gives a summary of data from the survey we conducted. The results show that the adoption rates in terms of number of farmers were higher for soybeans than for maize. It is also evident that while the number of farmers

Table 12.4. Sample statistics of GMO adoption.

Crop	Item	1998	1999	2000
Soybeans	Number of farmers adopting GM varieties	27 (77.1%)	90 (86.5%)	92 (87.6%)
	Number of farmers not adopting GM varieties	8	14	13
	Total sample	35	104	105
Maize	Number of farmers adopting only Bt maize	9 (32.1%)	41 (42.7%)	28 (29.8%)
	Number of farmers adopting only herbicide-resistant maize	0	1 (1%)	6 (6.4%)
	Number of farmers adopting both Bt and herbicide-resistant maize	0	11 (11.5%)	10 (11.2%)
	Number of farmers not adopting GM varieties	19	43 (44.8%)	50 (53.2%)
	Total sample	28	96	94

adopting GM soybeans and herbicide-resistant maize have steadily increased, the number of farmers using Bt maize decreased in 2000.

Model

The model chosen for this study loosely follows previous work by Fernandez-Cornejo *et al.* (1999). While they used a probit model to estimate probabilities of adoption and then developed a profit function as a result of the adoption probability, we will use a Tobit model to estimate probability, number and share of acres planted with GMOs. This type of model is a censored regression model as first studied by Tobin in the 1950s. The Tobit model is a two-stage model in which the result of the first stage regression is identical to that of probit.

Through the Tobit model, we were able to estimate acreage and share of acreage for herbicide-resistant soybeans, Bt maize and herbicide-resistant maize for 1998, 1999 and 2000 planting seasons. Independent variables can be grouped into six categories: cost and profitability, price risk, demographics, safety, environmental and other. Within each group, there are questions regarding producer attitudes, potential and realized benefits and concerns, previous planting history and future intentions.

Based on the frequency of response and distribution of answers, questions that required the

Table 12.5. Variable description and expected signs for 2000 soybean acreage estimates.

Variable	Description	Expected sign
Cost and profitability		
COST2	Farmers are realizing cost savings by using GM varieties	Positive
PROF	Higher profitability as an advantage	Positive
TILL	Less tillage as a benefit	Positive
YIELD	Yield increase as a benefit	Positive
HERBCOST	Decreased herbicide cost as a benefit	Positive
WEEDCON	Improved weed control as a benefit	Positive
Price risk		
MKT	Ability to market GMOs as a concern	Indeterminate
CNTRCT	Percentage of grain that is contracted	Indeterminate
PREM99	Premiums received in 1999	Negative
Demographics		
SOY99	GMO soybean planted in 1999	Positive
CORN99	GMO maize planted in 1999	Positive
INCM2	Income greater than US$100,000 year^{-1}	Indeterminate
EDU2	Education greater than high school	Positive
NE	Farm in north-east Ohio	Indeterminate
NW	Farm in north-west Ohio	Indeterminate
OWNS	Percentage of farm that is owned	Indeterminate
LVSK	Livestock on farm	Positive
Safety		
SCIE2	Scientists have not studied the long-term risks of eating GM foods	Negative
BABY1	I would not be hesitant to feed babies with GM food	Positive
Environmental		
RESIS	Resistant weeds as a concern	Negative
HERBUSE	Lower herbicide use as a benefit	Positive
INSCTUSE	Decreased insecticide use as a benefit	Positive
Others		
KNOW2	I consider myself knowledgeable about genetic modification	Indeterminate
WORM2	I would adopt maize resistant to root worms if it becomes available	Positive

Table 12.6. Estimated R^2 of regression.

Crop	Acre00	Acre99	Share00	Share99
Soybeans	0.63	0.524	0.619	0.398
Bt maize	0.45	0.556	0.556	0.476
Herbicide-resistant maize	0.783	0.962	0.719	0.875

respondent to answer on a six-point agree–disagree scale were translated into a series of dummy variables. Coding was done at three points (1, 2, 3) for most questions. This typically included a dummy variable if the respondent answered strongly disagree or disagree, a dummy variable if the respondent answered strongly agree or agree, and a dummy variable for the more neutral responses disagree somewhat and agree somewhat. The only agree–disagree question not coded this way was the question that asked, 'I would be hesitant to feed babies with GM food.' This question, asked as a measure of perceived safety of GMOs was coded with two dummy variables, one for disagree strongly, disagree and somewhat disagree and one for strongly agree, agree and somewhat agree.

A majority of other variables were measured through the use of dummy variables in order to account for a particular characteristic or practice. For example, under the environmental category, if the respondent marked that lower herbicide use was a benefit of herbicide-resistant soybeans, a dummy variable was used to indicate this. An abbreviated list of variables used, short descriptions of them and expected signs are included as Table 12.5. This list includes some variables that are statistically significant in the 2000 soybean acreage equation. A positive expected sign reflects that possessing this characteristic would increase GMO acreage or share of production. A complete list of variables, including expected signs and sample descriptive statistics, is given in Appendix Table A1. This includes only the dummy variables used in the model. Dropped dummy variables are not included. Using the question 'Farmers are realizing cost savings by using GM varieties' as an example, the dummy variables are as follows: 1 if the respondent signified they agreed or strongly agreed to the statement, 0 otherwise; 1 if the respondent signified they disagreed or strongly disagreed to the statement, 0 otherwise; and 1 if the respondent disagreed somewhat/agreed somewhat to the statement, 0 otherwise. The variable for disagree somewhat/agree somewhat is not included in the model.

One independent variable of interest is the region of Ohio in which the farm operates. We divided the Ohio grain producing areas into three sections, the north-west, north-east, and south-west. Average farm size was also computed for each region in order to look for causes for differing adoption rates between regions. While the average farm size overall was 1218 acres for the 107 respondents who reported farm size, 63 farmers from the north-west had an average of 1118 acres, 19 farmers from the north-east had an average of 1277 acres, and 25 farmers from the south-west had an average farm size of 1432 acres.

As a measure of 'goodness of fit', R^2 is used to describe how well our data fitted the model. R^2 tells us what percentage of variation of the dependent variable our model explains. Table 12.6 gives the R^2 for 1999 and 2000 models. The regressions for 1998 are not credible due to a small sample and therefore are not included in the reporting of results. As evident from this table, our models do an acceptable job of explaining variations of acreage and share, considering that cross-sectional data were used. The R^2 ranged from 0.398 all the way to 0.962. This can be interpreted as meaning that on the lower range, we can explain 39% of the variance in the model and in the high range we can explain 96% of the variation in the model.

When working with cross-sectional data, one typically expects R^2 to be lower than our estimate for herbicide-resistant maize models. If one just looks at the soybean and Bt maize estimates, the R^2 lie in a more acceptable range for this type of research. It is believed that the R^2 for herbicide-resistant maize is higher than expected due to the low number of adopters of this technology. In the 2000 herbicide-resistant maize estimates, there were 79 observations and only 16 were uncensored in the Tobit model. The uncensored observations represent adopters of the technology. In the 1999 estimates for both acreage and share of production, there were 82 observations of which 11 were uncensored. This leads to the conclusion that only a few observations are responsible for explanation of the variance in the

Table 12.7. Sample statistics of GM acreage and share of production.[a]

Crop	Variable	1998 Sample size	1998 Mean	1998 Max	1999 Sample size	1999 Mean	1999 Max	2000 Sample size	2000 Mean	2000 Max
GM soybeans	Acreage	25	209	850	85	311	1200	83	372.5	3000
	Share	25	0.41	1.00	85	0.57	1.00	83	0.66	1.00
Bt maize	Acreage	25	33.68	500	82	76.72	800	79	64.08	1600
	Share	25	0.10	1.00	82	0.17	1.00	79	0.15	1.00
Herbicide-resistant maize	Acreage[b]				82	16.64	700	79	29.83	800
	Share				82	0.05	1.00	79	0.07	1.00

[a] Sample size refers to the number of observations used for regression analysis.
[b] Blanks indicate data not available.

model. If there are similarities in the uncensored observations, it will lead to a high proportion of error explanation.

Soybeans

Estimates for soybean acreage and acreage share were run for 1998, 1999 and 2000. Over this time period, the mean acreage of GM soybeans increased from 209 acres in 1998 to 372.53 acres in 2000. Increases have also been seen in acreage share as the mean value has risen from 0.41 to 0.66 this year. Table 12.7 gives data on average acres and share for all soybeans and maize. For this table, the maximum values were used as an indication of range. This is possible because for all estimates, the lower bound was zero.

The sample sizes for this table were obtained from the regression results. For example, the GM

Table 12.8. Significant variables in soybean estimation.

Variables	Acre00	Acre99	Share00	Share99
Cost and profitability				
COST2	+	+	+	+
PROF	+	+		
YIELD			+	
TILL			+	
Price risk				
CNTRCT	–			
Demographics				
AGE				
INCM1	+		+	
INCM2	–	–		
NE				+
CROPS	+	+		
Safety				
BABY1	+			
Environmental				
HERBUSE	+		+	
Other				
KNOW1	–	–		–
WORM1				–
WORM2		+		

soybeans 1998 sample size reflects that there were 25 usable observations for estimating acreage and 24 usable observations for estimating share. 1998 sample sizes are much lower because only those respondents completing the long version of the survey were asked to give data from 1998. Since the small sample in 1998 would not yield credible results, we present and discuss only the regression results for 1999 and 2000.

When analysing the results of estimation over the 3-year period, it is necessary to observe which variables are statistically significant. Of the variables used, COST2, the respondent believing that farmers are realizing cost savings by adoption of GMOs, was significant in the estimation of acreage in 1999 and 2000 as well as for share in 1999 and 2000. All of the coefficients were positive implying that a belief in cost savings increases adoption. Table 12.8 summarizes variables that are statistically significant at the 10% level or lower in estimating the adoption of GMO soybeans over this period of time.

Interpretation of these results can be simplified to the following, if a farmer does not consider him or herself to be knowledgeable about GMOs, or has income below US$20,000 year^{-1}, he or she will be less likely to adopt GMO soybeans. If a farmer believes that cost savings can be achieved, or that GMO soybeans lead to higher profitability, they will be more likely to adopt. It is noted that cost savings have stronger impacts than profitability, as the profitability is not significant at all in the acreage share equations. While there are conflicting results when interpreting the effect of the number of acres of total crops on GMO adoption, it is proposed that the relationship is positive despite the 1998 estimated coefficients. Complete regression results from the soybean estimation for 2000 are included in the Appendix Tables A2 and A3.

Table 12.9. Significant variables in Bt maize estimation.

Variables	Acre00	Acre99	Share00	Share99
Cost and profitability				
COST1		+	+	+
COST2			+	
YIELD	+	+	+	+
INSCTCON	+		+	
Price risk				
MKT	+			
PREM			–	
CNTRCT	+		+	
Demographics				
HERBPROD	+		+	
AGE		+		
INCM1	–	–	–	–
NE			+	
NW			+	
CROPS	+	+		
Safety				
RISK2		+		+
SCIE1			+	
SCIE2		–		–
Environmental				
HERB2		+		+
RESIS		+		+
INSCTUSE			+	
Other				
WORM1				–
WORM2	+	+	+	

Bt Maize

Similar to GMO soybeans, estimates of Bt maize adoption were obtained for 1999 and 2000, again using both acreage number and share as dependent variables (see Appendix Tables A2 and A3 for complete results). Unlike GMO soybeans, which have seen steady growth in average acreage and share, Bt maize average acreage and share, in parenthesis, increased from 33.68 (0.10) to 76.72 (0.17) from 1998 to 1999, but then decreased in 2000 to 64.08 (0.15). While some of this decrease can be related to decreased expected infestation levels of the ECB, we wanted to investigate what factors also influenced the farmers' decision to adopt. When compared with GMO soybeans, it is possible to find more variables that are significant across years (Table 12.9).

Some key differences can be noted between factors that affect adoption of Bt maize when compared with GMO soybeans (Table 12.8). Most important of these is the significance of income. While for soybeans, income under US$20,000 (INCM1) was negatively related to GMO adoption, for Bt maize, income above US$100,000 (INCM2) was negatively related. This result implies that farmers with a very high income would not adopt as much Bt maize as those with lower income levels. Also of significance is the strong relationship between adoption of Bt maize and the respondents' answer to the question, 'I would adopt maize resistant to root worms [another maize pest] if it becomes available.' There is a strong positive relationship between the respondent agreeing to this question and adopting Bt maize.

While not a significant variable, it is worth mentioning that the more grain that the farmer forward contracts, the more Bt maize he is estimated to grow. When the level of forward contracting is used as a measure of risk, with high levels of contracting representing a farmer being risk adverse, this sign is consistent with theory. Bt maize is often looked at as an insurance policy for the farmer. In the case of high levels of infestation, they are protected from the pest; if low infestation occurs, the technology fee is seen as the insurance premium. Therefore, the

Table 12.10. Significant variables in herbicide-resistant maize estimation.

Variable	Acre 00	Acre99	Share00	Share99
Profitability				
WEEDCON			+	
Price risk				
MKT		+		
CNTRCT	–	–	–	–
Demographics				
BTPROD	+	+	+	
INCM1				–
NE	+		+	
NW			+	+
CROPS	+			
Safety				
RISK1				–
SCIE1		+		+
SCIE2		+		+
BABY1		+		
Environmental				
HERB2	–	–		
RESIS		–		
Other				
KNOW1	–	+		+
KNOW2		+		+
WORM2		+	+	+

more risk averse a farmer is, the more likely they should be to adopt Bt maize in cases where infestation levels of the ECB are expected to be a problem for the crop.

The results also show that cost savings were significant in Bt maize equations. Regarding this, it is important to mention that COST1 and COST2 represent change in comparison with the dropped variable, not necessarily a direct change in cost. The profitability variable was not significant at all. This corresponds to similar results achieved by USDA researchers (Fernandez-Cornejo *et al.*, 1999). These results imply that cost or profitability play less important roles in the adoption of Bt maize than for GM soybeans.

Herbicide-resistant Maize

It is interesting to analyse similarities and differences between adoption of herbicide-resistant maize and Bt maize. As evident from Table 12.10, adoption of herbicide-resistant maize has increased, but not at the rate of either Bt maize or GMO soybeans, reaching only an average acreage of 29.83 and average share of 0.07 in our sample (see Appendix Tables A2 and A3 for complete sets of regression results).

As is evident from Table 12.10, adoption of herbicide-resistant maize is positively related to the adoption of Bt maize. The variable BTPROD measures the relationship that Bt adoption has on herbicide-resistant adoption. The signs of HERB2, strongly agree or agree that GMOs lead to reduced herbicide and insecticide use, appear to contradict the theory that reduced herbicide use would lead to greater adoption.

When compared with Bt maize, herbicide-resistant maize is similar in that adoption is positively related to willingness to adopt root worm tolerant maize if it comes on to the market. This variable was used to measure willingness to adopt technologies that are still in development stages. Of the 124 respondents who answer this question, 77 either agreed or strongly agreed that they would adopt root worm tolerant maize.

Being located in the north-east part of Ohio is another variable that is significant for both Bt and herbicide-resistant maize in 2000. For this variable, NE, being located in this demographic area results in higher adoption of both maize varieties. On the other hand, the level of forward contracting is inversely related to the amount of herbicide-resistant maize that is grown. This is in contrast to the positive relationship that was seen with Bt maize. Again using forward contracting as a measure of willingness to accept risk, this distinction is theoretically correct. While adoption of Bt maize was seen as a risk adverse decision, adoption of herbicide-resistant maize can be viewed as risk seeking. Herbicide-resistant maize allows the farmer a greater level of convenience of pest control. They are able to use a broad-spectrum herbicide to rid the field of pests. In this light, they are risk seeking in the hope that the value of convenience is greater than the associated technology fee.

Conclusions

This chapter presents preliminary work being done on GMO adoption in Ohio. Our objective is to analyse factors that are significant in the estimation of GMO adoption. From the results presented, it is apparent that several factors are significant over time in estimating adoption. The work being conducted for this project is very new. Attempts have been made to find other works that have attempted to estimate GMO adoption as a function of attitudes, perceptions and behaviours, but none has been found. Again, due to the fact that our research is ongoing, the results should be regarded as preliminary.

Our preliminary estimations have shown that there are consistent factors affecting adoption across years for the same crops as well as across crops during the same year. It has also been found that while the adoption of GM soybeans and herbicide-resistant maize has continued to increase, the adoption of Bt maize decreased in 2000. Monsanto originally predicted that the convenient GM seed would eventually replace conventional soybeans. Questions for the future on this topic will be primarily concerned with attempting to predict adoption patterns in the future. In order to do this, another survey is being proposed in January 2001 after the 2000 harvest season. This edition would be sent to another 1000 farmers in Ohio in order to gain an even better understanding of what drives adoption of GMOs.

References

Carpenter, J. and Gianessi, L. (1999) Herbicide tolerant soybeans: why growers are adopting Roundup Ready varieties. *Agbioforum* 2(2), archived at: www.agbioforum.org

Dixon, R. (1980) Hybrid corn revisited. *Econometrica* 48(6), 1451–1461.

Fernandez-Cornejo, J., Klotz-Ingram, C. and Jans, S. (1999) Farm-level effects of adopting genetically engineered crops in the U.S.A. Paper presented at the *International Consortium on Agricultural Biotechnology Research.* Rome, Italy, 17–19 June.

Griliches, Z. (1957) Hybrid corn: an exploration in the economics of technological change. *Econometrica* 25(4), 501–522.

Hyde, J., Martin, M.A., Preveckel, P.V. and Richard, C. (1999) The economics of Bt corn: valuing protection from the European corn borer. *Review of Agricultural Economics* 21(2), 442–454.

McBride, W.D. and Brooks, N. (1999) Survey evidence on producer use and costs of genetically modified seed. Paper presented at the *International Consortium on Agricultural Biotechnology Research.* Rome, Italy, 17–19 June.

Pennings, J.M.E., Irwin, S.H. and Good, D. (1999) *Surveying Farmers: a Research Note.* AgMAS Project Research Report 1999-04, October.

Appendix

Table A1. Variable description and signs for 2000 acreage estimates.

Variable	Description	Soybeans[a]		Maize[b]		Expected sign
		No. of respondents	%	No. of respondents	%	
Cost and profitability						
COST1	Farmers are not realizing cost savings by using GM varieties	15	15	14	15	Negative
COST2	Farmers are realizing cost savings by using GM varieties	49	48	40	44	Positive
PREM99	Premiums received in 1999	24	23	6	7	Negative
PROF	Higher profitability as an advantage	37	36	n/a	n/a	Positive
HERBCOST	Lower herbicide cost as a benefit	22	21[c]	13	14[d]	Positive
WEEDCON	Improved weed control as a benefit	n/a	n/a	36	40[d]	Positive
YIELD	Higher yield as a benefit	5	5[c]	27	30[e]	Positive
TILL	Less tillage as a benefit	8	8[c]	n/a	n/a	Positive
INSCTCON	Improved insect control as a benefit	n/a	n/a	20	22[e]	Positive
Price risk						
MKT	Ability to market GMO as a concern	69	67	55	60	Negative
PREM	Premiums for non-GMOs as a concern	30	29	25	27	Negative
CNTRCT	Percentage of grain that is contracted	n/a	20	n/a	20	Indeterminate
Demographics						
SOY99	GMO soybean planted in 1999	90	87	n/a	n/a	Positive
CORN99	GMO maize planted in 1999	n/a	n/a	51	56	Positive
BTPROD	Bt maize planted	n/a	n/a	38	42	Positive
HERBPROD	Herbicide-resistant maize planted	n/a	n/a	16	18	Positive
AGE[f]	Age in years	(48.50)	n/a	(48.50)	n/a	Negative
INCM1	Income less than US$20,000 $year^{-1}$	11	11	19	21	Indeterminate
INCM2	Income greater than US$100,000 $year^{-1}$	12	12	13	14	Indeterminate
EDU2	Education greater than high school	50	49	46	51	Positive
NE	Farm in north-east Ohio	14	14	13	14	Indeterminate
NW	*Farm in north-west Ohio*	50	49	45	49	Indeterminate
OWNS[f]	Percentage of farm that is owned	(0.29)		(0.30)		Indeterminate
CROPS[f]	Acres of land in crops	(1176.00)		(1267.00)		Indeterminate
LVSK	Livestock on farm	51	50	46	51	Positive

Safety						
RISK1	Eating GM food is not highly risky	84	82	72	79	Positive
RISK2	Eating GM food is highly risky	4	4	4	4	Negative
SCIE1	Scientists have studied the long-term risks of eating GM foods	27	26	23	25	Positive
SCIE2	Scientists have not studied the long-term risks of eating GM foods	24	23	19	21	Negative
BABY1	I would not be hesitant to feed babies with GM food	78	76	66	73	Positive
Environmental						
HERB1	Farmers are using less herbicide and insecticide by using GM varieties	3	3	2	2	Negative
HERB2	Farmers are not using less herbicide and insecticide by using GM varieties	71	69	62	68	Positive
RESIS	Resistant weeds as a concern	45	44	36	40	Negative
HERBUSE	Lower herbicide use as a benefit	17	17	10	11	Positive
INSCTUSE	Decreased insecticide use as a benefit	n/a	n/a	18	20	Positive
Other						
KNOW1	I do not consider myself knowledgeable about genetic modification	8	8	7	8	Indeterminate
KNOW2	I consider myself knowledgeable about genetic modification	45	44	42	46	Indeterminate
WORM1	I would not adopt maize resistant to root worms if it becomes available	6	6	5	5	Negative
WORM2	I would adopt maize resistant to root worms if it becomes available	62	60	55	60	Positive

[a] Based on the total sample of 103 soybean-producing respondents.
[b] Based on the total sample of 91 maize-producing respondents.
[c,d,e] The responses to these questions are mutually exclusive.
[f] The figures in parentheses are sample means.
n/a, not applicable.

Table A2. Variable description, expected signs, and estimated coefficients for 2000 acreage equations.

Variable	Description	Expected sign	Estimated coefficients[a]		
			Soybeans	Bt maize	Herbicide-resistant maize
Cost and profitability					
COST1	Farmers are not realizing cost savings by using GM varieties	Negative	**–216.98**	239.23	–144.71
COST2	Farmers are realizing cost savings by using GM varieties	Positive	**240.06**	**145.78**	174.55
PROF	Higher profitability as an advantage	Positive	**106.5**		
HERBCOST	Lower herbicide cost as a benefit	Positive	48.41		129.72
YIELD	Higher yield as a benefit	Positive	34.12	**418.48**	
TILL	Less tillage as a benefit	Positive	130.93		
INSCTCON	Improved insect control as a benefit	Positive		**368.34**	
WEEDCON	Improved weed control as a benefit	Positive			195.47
Price risk					
MKT	Ability to market GMO as a concern	Negative	–70.15	**180.58**	88.78
PRE	Premiums for non-GMOs as a concern	Negative	12.34	–106.2	88.98
CNTRCT	Percentage of grain that is contracted	Indeterminate	**–3.57**	4.87	**–17.4**
Demographics					
SOY99	GMO soybean planted in 1999	Positive	**150.86**		
CORN99	GMO maize planted in 1999	Positive	42.63	–8.998	7.74
AGE	Age in years	Negative	–0.6365	**4.59**	–2.23
INCM1	Income less than US$20,000 $year^{-1}$	Indeterminate	**152.34**	**–368.68**	–88
INCM2	Income greater than US$100,000 $year^{-1}$	Indeterminate	**–232.74**	**–212.93**	–271.2
EDU	Education greater than high school	Positive	**90.86**		
NE	Farm in north-east Ohio	Indeterminate	124.71	142.07	**443.78**
NW	Farm in north-west Ohio	Indeterminate	51.43	152.56	132.66
OWNS	Percentage of farm that is owned	Indeterminate	–86.4	174.48	–70.38
CROPS	Acres of land in crops	Indeterminate	**0.1723**	**0.051**	**0.09**
LVSK	Livestock on farm	Positive	15.95	–13.49	–64.95
Safety					
RISK1	Eating GM food is not highly risky	Positive	116.61	9.75	97.56
RISK2	Eating GM food is highly risky	Negative	12.79	**439.19**	–1233.47
SCIE1	Scientists have studied the long-term risks of eating GM foods	Positive	21.32	18.85	–6.08
SCIE2	Scientists have not studied the long-term risks of eating GM foods	Negative	39.22	–7.8	–150.56
BABY1	I would not be hesitant to feed babies with GM food	Positive	**145.29**	90.54	–2.32

Environmental					
HERB1	Farmers are not using less herbicide and insecticide by using GM varieties	Negative	171.45	−1331.4	−571.31
HERB2	Farmers are using less herbicide and insecticide by using GM varieties	Positive	−78.1	146.3	**−285.06**
RESIS	Resistant weeds as a concern	Negative	7.79	−109.14	2.81
HERBUSE	Lower herbicide use as a benefit	Positive	**153.6**		−1152.8
INSCTUSE	Decreased insecticide use as a benefit	Positive		246.996	
Other					
KNOW1	I do not consider myself knowledgeable about genetic modification	Indeterminate	−238.7	125.61	**−406.93**
KNOW2	I consider myself knowledgeable about genetic modification	Indeterminate	−57.8	−37.65	−18.93
WORM1	I would not adopt maize resistant to root worms if it becomes available	Negative	−37.2	−1482.11	−834.74
WORM2	I would adopt maize resistant to root worms if it becomes available	Positive	36.71	**283.37**	122.45

[a] The coefficients in bold indicate that the variable is statisitically significant at the 10% level or lower.

Table A3. Variable description, expected signs, and estimated coefficients in 2000 share equations.

Variable	Description	Expected sign	Estimated coefficients[a]		
			Soybeans	Bt maize	Herbicide-resistant maize
Cost and profitability					
COST1	Farmers are not realizing cost savings by using GM varieties	Negative	–0.175	**0.7732**	0.2239
COST2	Farmers are realizing cost savings by using GM varieties	Positive	**0.4133**	**0.3504**	–0.0169
PROF	Higher profitability as an advantage	Positive	0.0554		
HERBCOST	Lower herbicide cost as a benefit	Positive	0.1142		0.143
YIELD	Higher yield as a benefit	Positive	**0.3247**	**0.7347**	
TILL	Less tillage as a benefit	Positive	**0.3147**		
INSCTCON	Improved insect control as a benefit	Positive		**0.7119**	
WEEDCON	Improved weed control as a benefit	Positive			**0.4151**
Price risk					
MKT	Ability to market GMO as a concern	Negative	0.0084	0.1482	0.0108
PRE	Premiums for non-GMOs as a concern	Negative	–0.0729	**–0.5106**	0.355
CNTRCT	Percentage of grain that is contracted	Indeterminate	0.001	**0.0096**	**–0.0272**
Demographics					
SOY99	GMO soybean planted in 1999	Positive	0.027		
CORN99	GMO maize planted in 1999	Positive	–0.0554	–0.0771	0.0689
AGE	Age in years	Negative	0.0024	0.001	–0.0071
INCM1	Income less than US$20,000 year^{-1}	Indeterminate	**0.2493**	**–0.4386**	–0.0264
INCM2	Income greater than US$100,000 year^{-1}	Indeterminate	–0.1599	–0.1905	–0.3271
EDU	Education greater than high school	Positive	0.0139		
NE	Farm in north-east Ohio	Indeterminate	0.0681	**0.3915**	**0.7687**
NW	Farm in north-west Ohio	Indeterminate	–0.0718	**0.4225**	0.2988
OWNS	Percentage of farm that is owned	Indeterminate	–0.1413	**0.4009**	–0.3572
CROPS	Acres of land in crops	Indeterminate	–0.0000234	–2.13E-05	0.0000595
LVSK	Livestock on farm	Positive	0.0735	0.007	0.0492
Safety					
RISK1	Eating GM food is not highly risky	Positive	0.1454	0.0292	0.271
RISK2	Eating GM food is highly risky	Negative	0.1666	0.5848	–2.7106
SCIE1	Scientists have studied the long-term risks of eating GM foods	Positive	0.0223	**0.3601**	0.02888
SCIE2	Scientists have not studied the long-term risks of eating GM foods	Negative	–0.0142	0.2064	–0.2644
BABY1	I would not be hesitant to feed babies with GM food	Positive	0.1185	–0.1524	–0.2214

Environmental					
HERB1	Farmers are not using less herbicide and insecticide by using GM varieties	Negative	–0.1061	–2.9229	–1.5481
HERB2	Farmers are using less herbicide and insecticide by using GM varieties	Positive	–0.1243	0.3411	–0.1883
RESIS	Resistant weeds as a concern	Negative	0.1021	–0.2122	0.0118
HERBUSE	Lower herbicide use as a benefit	Positive	**0.2156**		–2.3138
INSCTUSE	Decreased insecticide use as a benefit	Positive		**0.5004**	
Other					
KNOW1	I do not consider myself knowledgeable about genetic modification	Indeterminate	–0.1122	0.0609	–0.4248
KNOW2	I consider myself knowledgeable about genetic modification	Indeterminate	0.0417	–0.1007	0.056
WORM1	I would not adopt maize resistant to root worms if it becomes available	Negative	**–0.564**	–3.2716	–1.9602
WORM2	I would adopt maize resistant to root worms if it becomes available	Positive	0.0903	**0.3973**	0.2379

[a]Coefficients in bold indicate that the variable is statistically significant at the 10% level or lower.

13 *Ex ante* Economic Assessment of Adopting Genetically Engineered Crops in Finland

Jyrki Niemi and Meri Virolainen
Agrifood Research Finland, Economic Research (MTTL), Helsinki, Finland

Introduction

Adoption of genetically modified (GM) crops has risen dramatically since their commercial introduction in the mid-1990s. Cultivation is rapidly expanding, especially in the USA, Mexico, Argentina and China. The amount of land under GM crops totalled approximately 60 Mha in 2000. Europe accounts for less than 1% of this as the European authorities have only given conditional approval to the new technology. The attitude of the European regulatory authorities reflects widespread public concern about the use of GM crops in food production. The strength of feeling varies between different countries, but even among many people who accept the value of GM technology for producing new medical treatments there is uneasiness about the 'unnatural' character of GM food.

However, with the increasing uptake of GM technology and the significant number of new GM developments being prepared for the market place, GM crops are set to become a major influence on competitiveness through the entire food chain. Competition will be intensified, particularly in terms of production cost and quality of the food products. While the controversy and dispute over granting approval to GM food in Europe means that there is very little current GM production in Europe, specific concern has been expressed about the future competitiveness of European agriculture. Some proponents claim that rapid acceptance and adoption of GM technology is necessary if European producers are to retain their share of the world market.

The introduction of GM technology would trigger changes and produce challenges for farmers as well as for the rest of the food chain. For a Finnish farmer, this naturally gives rise to some questions: What are the possibilities/barriers for commercial introduction of GM crops on Finnish farms? What are the economic impacts of GM technology? How will GM crops influence the profitability and international competitiveness of Finnish agriculture? Who benefits from the new technology, and who faces the disadvantages? What experiences have there been in Finland? These questions form the basis for this chapter, the purpose of which is to identify the specific impact of GM technology on Finnish agriculture.

The information given in the chapter is based on literature review, mostly from the USA, and due to a lack of a sufficient domestic literature, on interviews with key persons directly or indirectly involved in Finnish production, trade or consumption of GM food.

The chapter is divided into three main parts. First, certain elements that determine the economic effects resulting from the adoption and diffusion of agricultural biotechnology are examined. This will involve tracing the key productivity gains delivered by GM technology, and analysing how such gains will be shared between the farmers and the rest of the food chain. The next section examines the special character of Finnish agriculture, and thus observes the position of Finnish farmers and the competitiveness of Finnish agriculture should gene technology be strongly adopted in the EU as well. The main thrust of the discussion will be on the key economic and strategic issues and implications of GM technology on Finnish farms. The final section summarizes the findings.

Economic Effects of Gene Technology at the Farm Level

The first generation GM crops

Innovations of gene technology in the agricultural sector can be divided into two separate waves: the first and the second generation innovations. The first generation includes plants developed to resist pests, diseases and herbicides; or plants with the ability to tolerate adverse environmental conditions, such as drought and frost.[1] Benefits of the first generation technology for a single farmer can be summed up as follows (Kalaitzandonakes, 1999): increased yields, decreased pest management costs, better risk management, improved insurance against pests, time savings, more efficient use of land and reductions in equipment outlays.

According to Salo *et al.* (1998) yields can be increased by 5–20 % using new GM crops instead of the old conventional species. Because in traditional plant breeding an improvement in crop yield amounting to a few percentage units is considered large, the effect is appreciable in terms of increasing productivity in agriculture. Higher yields are simply based on the improved resistance of GM crops against pesticides and insects, due to the fact that the GM crops do not suffer from spraying as much as traditional crop species, or that sprayings are carried out less frequently.

Thus the first generation technology is cost-

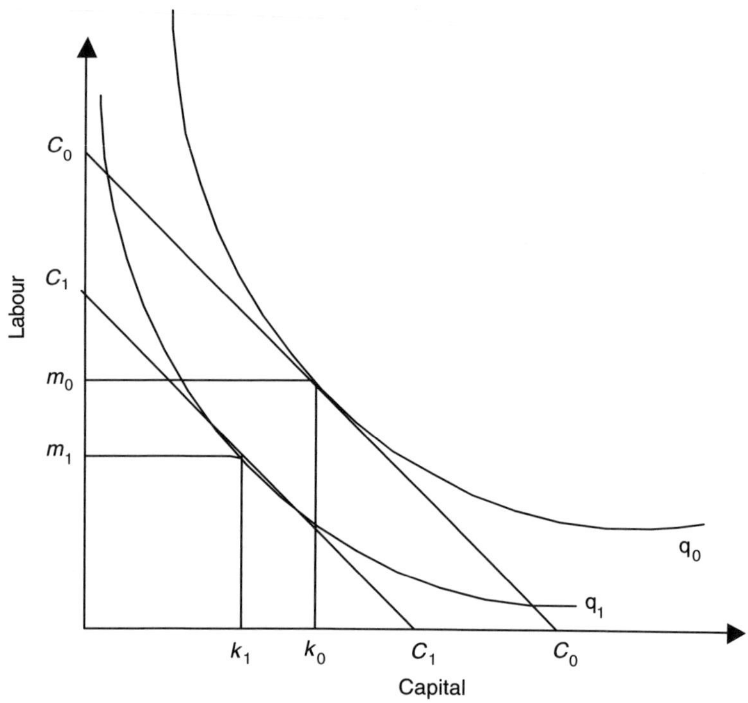

Fig. 13.1. The case of cost-saving technology as an example of technological change.

[1] Early commercial products include *Bacillus thuringiensis* (Bt) maize. These genetically altered hybrids contain a naturally occuring soil bacterium, Bt, that kills European corn borers.

reducing technology, which allows a producer to offer a greater or equal quantity of commodities at lower price. Cost-saving technology reduces the producer's unit production costs by increasing yields or reducing input costs. The first wave of biotech innovations, such as herbicide-tolerant cotton or soybean, can be easily classified as output-increasing technology and cost-saving technology. Figure 13.1 illustrates the case of cost-saving technology. Output q is assumed constant between time period 0 and 1 ($q_0 = q_1$), when technological change reducing firm cost from C_0 to C_1 has taken place. The prices of capital and labour determine the slope of the isocost line. Factor-neutral technology implies that the need for labour m and capital k diminishes by the same proportion from m_0 to m_1 and from k_0 to k_1.

Many studies have analysed the agronomic and economic effects of adopting GM crops, especially in the USA (for example Culpepper and York, 1998; Marra *et al.*, 1998; Klotz-Ingram *et al.*, 1999; Traxler and Falck-Zepeda, 1999). Although cultivation experiences of GM species have so far been mostly positive, there have been some disappointments. For example, herbicide-tolerant cotton was associated with significant increases in yields, but disappointment was experienced in herbicide use, which did not decrease according to expectations. Cultivation tests with herbicide-tolerant soybeans, on the other hand, provided just the opposite result. The use of herbicides decreased significantly, but no major improvements in yields took place (USDA, 1999). Klotz-Ingram *et al.* (1999) showed that the benefits of GM crops depend largely on regional and yearly differences, including input costs (pesticide and insect costs, elite seeds used or not, etc.) and other production attributes.

The distribution of benefits from the new technology is likely to be uneven. Based on the US experience, the adoption of GM seeds has raised farmers' income in the short run, but revenues have vanished in the long run. Kalaitzandonakes (1999) points out that early adopters of new technology realize increased profits, but as more farmers adopt, the increase in aggregate supply causes agricultural prices to fall. The magnitude and speed of the supply change will depend on the rate of adoption and diffusion of the technology. Changes in consumer and producer benefits then depend on the extent to which changes in price affect demand. If demand is fairly price-inelastic, then supply increases would result in some increase in the quantity demanded, but a considerable drop in price. This is characteristic of the demand for many crops. First wave GM technology is, therefore, not able to solve the problem concerning inelastic demand of farm commodities. The prices of farm commodities are likely to stay low and consumers will gain at the expense of farmers.

The process described above is known as the 'treadmill theory' (Cochrane, 1965), describing how new profitable and productive technological innovations are rapidly adopted by some farmers, leading later to wider adoption of new innovations and chronic surplus production and decreasing prices. Those who were slow to adopt or did not adopt would lose. Hence, the new technology might involve unhappy consequences for farmers. In the end, the impacts of the first stage GM technology seem to have been negative for farmers.

The second generation GM crops

In the second wave, still at a development stage, gene technology will be improved to better meet customers' needs. The second generation technology results in improvement in the attributes of the food product. Some examples of potential second generation products include crops with added value output, such as improved flavour, texture, shelf-life and nutritional content, or improved process properties for later processing stages. The quality-enhanced food products have the potential to increase producer profits through increased demand for the improved food (Caswell *et al.*, 1994). These products can be sold at a higher price if consumers value the quality change. The higher prices may be an incentive to agricultural producers to adopt these technologies even if production costs remain unchanged or increase. Up to the present only a few value-added commodities have entered the markets, such as 'golden rice' developed for children suffering from blindness and lack of vitamin A in the developing world.

Determining the economic effects of second generation crops can be more complex than in the first generation case. Speciality crops that contain consumer-desired traits must be separated from the commodity supply chain in order to preserve the added-value component. This could be accomplished by either 'crop segregation' or 'identity preservation' (IP). Effective segregation or IP – which begins at the farm level from isolated storage to different transport systems – requires that the entire value-chain be coordinated in order to preserve the value trait in a food product until it reaches the consumer, leading to more segmented commodity markets.

Segmentation of a commodity market into speciality sub-markets, and a residual commodity market, should, in turn, result in higher total market value. In addition to creating value, such segmented markets are less vulnerable to the technological treadmill phenomenon as demand is typically more elastic (Kalaitzandonakes, 1999). The confluence of strategies of the various players along the agro-food chain will determine how value from GM technology will be distributed among them and the way the chain will be restructured. To be viable players in the value-added chain, farmers have to develop closer relationships along the entire chain.

As a result, the value-enhanced crops could eventually change the entire scope of production in agriculture, moving it towards contract production, similar to that found in the arrangements for speciality grains and oilseeds. An increased use of contractual arrangements leads to greater vertical integration and coordination in the whole food chain (Caswell *et al.*, 1994). Interdependence will replace independence. Both supply and demand factors underlie the trend towards vertical integration. On the demand side, the need for efficient consumer response is positioning agribusiness to serve consumers' wants and needs more accurately. For processors, integration ensures predictable supply and consistent quality

For farmers, contracts are likely to offer premiums over average market prices for agricultural commodities, greater access to new technology and inputs, and new sources of capital (Caswell *et al.*, 1994). The enhanced value of the GM product will not, however, automatically jump into the farmers' wallet, without extra efforts. GM technology also demands new skills from farmers who have to change their traditional production patterns. Contracts may erode a tradition of farmer independence in production decisions and management. Farmers are expected to plant, grow, harvest, store and deliver according to the specific needs of end-users in food manufacturing, the livestock sector and even the pharmaceutical industry (Riley and Hoffman, 1999).

Despite the extra efforts caused by higher requirements for a value-enhanced GM commodity, production of a value-added crop may turn out to be a profitable business for farmers. For growing tailored traits, farmers can earn premiums on each kilogram produced. Value-enhanced GM commodities include the same properties as luxury commodities, that is more elasticity of demand. This may be a key for farmers to gain from new biological innovations and produce premium commodities instead of bulk commodities; eluding the vicious circle of producing more and more at a lower price. The relative negotiating position of individual farmers in each supply chain will determine their share of value-added from second generation agro-biotechnology.

Risks of gene technology for farmers: who gains?

There is no doubt that gene technology is able to generate concrete productivity gains at the farm level, in the form of cost reductions, higher yields and improved risk management. However, the new technology also introduces new risks. Uncertainties exist because there is only short-term experience of their large-scale cultivation. Firstly, some authors are concerned that over the long term, harmful environmental effects – such as the evolution of resistant insect populations – in the case of certain plants could reduce the benefits derived from the characteristics bestowed by the transferred gene.

Another well-known risk, often referred to by opponents of biotechnology, is the potential for too great a role for big corporations, jeopardizing farmers' independent role in the food chain (Hayenga, 1998). In the 'darkest scenario' farmers will lose their independence, innovation firms extract all of the economic benefits generated by their products, and vertical integration will lead to unlimited power for multinational companies. The ability of the input sector to receive the largest part of economic benefits is, however, a contradictory issue. Are they able to abuse their monopoly position in the market? According to some recent studies, there is no straightforward answer to this.

Bijman and Enzing (1995) showed that adopting biotechnology in the Dutch potato food chain raised significantly potato processing firms' importance in the chain. Processing companies became larger and fewer, because biotechnological research is costly and risky, leading to the search for economies of scale. Traxler and Falck-Zepeda (1999), on the other hand, demonstrated in their study carried out in the USA that firms in the input sector (innovators) receive only a fragment of the total profit, because they have to offer a clear incentive for farmers to adopt the cultivation of GM seeds. In fact, the innovators' share of total profit seemed relatively modest and a significant share of the benefits was transferred to the other parts of the food chain.

Traxler and Falck-Zepeda (1999) were, however, quick to point out that it is very difficult to identify the monopoly profits in empirical work because of the difficulty in estimating marginal production costs, making it a quite difficult task to untangle the exact proportions in which benefits from an innovation are shared.

One crucial issue concerning the position of farmers in the chain is the number of contract partners. The adoption of the first wave technology involves a risk for farmers if there is only a limited number of firms in the agricultural input market, as in Finland. The new technology is not expected to generate changes that would raise the number of players in the input sector. In fact, at present the movement is exactly the opposite, with the increasing consolidation and concentration of the agricultural input market.

The introduction of second wave technology complicates the picture as it generates market segmentation, new niche markets and so on, leading to greater vertical coordination. None the less, it should be kept in mind that a highly vertically integrated system will not automatically raise the farmers' position in the food chain. If a single farmer has only a limited number of contract partners and conditions are weak, contract farming may turn against farmers. In addition, there is the risk that new actors may appear in the food chain who try to obtain a part of the economic value added. New members in the chain could reduce the farmers' share of the total gains.

Interestingly, the majority of the GM experts in Finland have argued forcefully that it is very unlikely that an 'exploitation' system would take place (Virolainen and Niemi, 2000). Vertical coordination is based on an idea of creating and sharing the extra value of a value-enhanced commodity. It is, however, recognized that farmers may lose their independence if contract farming is implemented without farmers' collective negotiation organization to ensure their share of value-added commodities.

Finally, Hillyer (1999) has pointed out that the second stage technology involves other risks for farmers that are difficult to control. The second wave products face several challenges on their way from laboratory to consumers, including a risk of 'empty promises' if new inventions cannot be concretized. Costs of segregating are an unknown element, such as large-scale building of new warehouses. Kalaitzandonakes (1999) asserts that the adoption and acceptance of second stage commodities by consumers will be even slower than the first generation commodities, creating a remarkable risk as well.

An additional obstacle for the adoption of GM farming in Europe is the Common Agricultural Policy (CAP) of the EU. Oversupply has been a major concern for European policy makers from the late 1970s. There is no doubt that more productive GM plants would lose their attractiveness in the eyes of the EU if it should increase the excess supply and therefore diminish farmers' revenues and increase the need for export subsidies.

Challenges Faced by Finnish Agriculture over GM Technology

On the basis of in-depth interviews with leading actors in the Finnish agro-food and agro-biotechnology firms, Virolainen and Niemi (2000) concluded that the EU will eventually approve the new technology. Hence, in the future GM food will appear on the Finnish markets, and this will fundamentally affect the agro-food industry. The advent of GM crops is expected to trigger structural changes in the entire agricultural industry, dramatically changing the ways farmers produce and market their products. For policy makers interested in assessing the economic effects of adopting the new technology, it is important to identify the specific market conditions of the agro-food industry in question. Therefore, the effects of GM technology must be considered in terms of competitiveness prior to adoption, and the ability of the industry to adjust to the new technology.

The competitiveness of Finnish agriculture is weakened chiefly through the unfavourable climatic conditions and predominantly small-scale farm structure. Although Finnish livestock farms rank among the most productive in Europe and increasing farm size could cut down production costs, the poor cost-effectiveness of plant production constitutes a permanent disadvantage. Cost comparisons indicate that unit costs in grain production are notably higher in Finland than in the other EU countries. The productivity of plant production is poor primarily due to the geographic location. This is most clearly visible in the low yield level with respect to the amount of production inputs used. Finnish farmers have to use crop species providing only a medium yield level, because higher yielding elite seeds cannot be adjusted to the demanding climate of Finland.

Overall, the development of productivity in Finnish agriculture has been much slower than in the leading agricultural countries in Northern Europe. In the first few years in the EU the growth in productivity was probably slowed down by the costs due to increasing the unit size (i.e. adjustment costs), but growth accelerated as a result of the rapid structural development in agriculture. In the past couple of years there have been indications that growth in productivity might be somewhat more rapid compared with the long-term average. However, it is obvious that the difference in productivity between Finland and the rest of the EU has not been essentially reduced during Finland's first years as an EU member (Myyrä and Pietola, 1999).

In many cases, GM crops would probably increase the productivity of Finnish agriculture. According to some Finnish GM experts (for example E. Pehu, personal communication, 19 January 2000; K. Raininko, personal communication, 10 January 2000) agronomic gains from gene technology might be even greater for countries like Finland, suffering from difficult circumstances, than countries with more favourable farming conditions. In these 'optimistic scenarios' Finland will benefit considerably from the new biotech innovations, like improved resistance to temporary cold periods and improved ability to adjust to lower temperatures, for example in the early spring.

Naturally, the possibility of applying higher yielding crop species sounds very attractive, if inventions can be concretized. One important gain is the possibility of maintaining the crop yield on a sufficient level through improved pest and insect management. Field trials in Finland have been undertaken in connection with, for example, transgenic sugarbeets, indicating an approximately 2–5% larger crop. Of course, the cost of sugarbeet seeds would be higher. At the same time, pesticide costs would decrease by 60–75%, however. According to Raininko (personal communication, 2000), farmers could earn an additional FIM 600 ha^{-1} (10% of gross returns) using transgenic sugarbeets.

On the other hand, concerns have been raised as to whether large farms in good agricultural regions can make better use of the GM technology and reduce their unit production costs more than the small Finnish farms in remote areas. In fact, it is difficult to conceive that the difference in productivity between Finland and the leading agricultural countries in northern Europe, which exists due to natural disadvantage, could be essentially reduced with the help of GM technology. For example, Salo *et al.* (1998) have suggested that the benefits derived from GM crops in Finland could be somewhat lower than elsewhere, since disease, weed and pest problems are less serious than they are in countries further south.

Although GM crops may improve productivity in Finnish agriculture and production costs could be cut by the use of GM crops, the yield constraints of crop farming are expected to remain an essential disadvantage. This disadvantage could leave small Finnish farms less competitive and eventually force them out of the market. An important issue in this context is whether farm size/location is related to the ability to use the new technology more efficiently (Caswell *et al.*, 1994). The present literature suggests that there seems to be little empirical support for the view that GM technology is particularly well suited for large farms. However, if the information costs associated with adoption are high, then 'information bias' may exist, where large farms find it easier to adopt new technology, since they can spread the fixed cost of adoption over a greater level of production (Kinnucan *et al.*, 1990).

Some concern has been expressed about the costs of R&D efforts to develop GM crop varieties that would provide real productivity gains for Finnish farmers. Many of the crop varieties grown in Finland have been traditionally bred there, owing to the fact that the large global corporations do not set out to develop varieties intended for Finland's boreal climate and small markets. These varieties have to be developed either as a result of domestic R&D work, or by purchasing genes and GM plants from foreign companies for use in domestic plant breeding (Salo *et al.*, 1998).

The capability of adjusting to the biotech competition facing Finnish agriculture varies a great deal in different production lines. According to Salo *et al.* (1998) there are certain Finnish conventional crops whose competitive ability is low if GM species enter the markets. Expanding imports of GM rape and GM barley will be a real threat for the cultivation of conventional barley and oilseed crops in Finland. In particular, imported GM rape, including an optimal GM modification for the processing of oil, could jeopardize domestic oilseed production. For years Finnish barley has been successfully used as raw material for the Finnish brewing industry. Barley purchases may, however, be targeted abroad instead of domestic GM-free raw material due to cost savings obtained (Virolainen and Niemi, 2000).

Conclusions

The cultivation of GM crops is going to increase in the near future and the EU will eventually take part in this development. However, the rate and direction of this development and the use of new technology depend on a multitude of factors, including EU policies, consumer demand, and advances in science. There is little doubt, drawing on the literature, that the adoption of GM technology will alter the farmer's position in the food chain, creating new threats as well as new options. The first wave of biotechnology is expected to provide tangible productivity gains at the farm level, such as higher yields, lower pest management costs and greater cropping practice flexibility. The second wave technology promises to enhance the value of crops from the farmer to the consumer.

However, at this point it remains relatively uncertain what specific impact GM technology will have on Finnish agriculture, and on rural Finland in general, for that matter. On the one hand, there are expectations that farmers will benefit from increased profits for their GM crops. Among other things, the new technology promises to deal better with the problematic climatic conditions of Finland with an option to choose higher yielding crop species. On the other hand, concerns have been raised as to whether large farms in the leading agricultural countries in northern Europe can make better use of the GM technology and reduce their unit production costs more than the small Finnish farms in remote areas. In a large, sparsely populated country with few alternative sources of income, this could be a threat to the viability of certain regions. The relationship between new technology and farm size/location may, therefore, determine the structure of rural communities and who wins and loses from technological change.

Because both the efficacy and the price of GM products are still unknown, it is difficult to make any quantitative assessment of the economic impact. However, it is evident that the introduction of GM crops would reinforce vertical integration in the Finnish food chain. GM varieties will only be grown under the approval of the customer (trading firm, processing company, retailer or consumer). Approval will only be given if the GM variety holds an evident qualitative improvement compared with non-modified varieties. The need for approval strengthens the integration between different stages of the product chain. Farmers must become more specialized and need to work more closely together if they are to maintain their share of the market. In future they must also become much better informed and react more flexibly to market changes.

The evolution of EU regulations concerning food safety and environmental protection will naturally have a significant influence on the type of GM technology innovations that are applied by farmers. Ultimately, the success of the new technology will be decided in the market place by a wider set of consumer interests. Concerns about the nature of genetic engineering and risks to human health have led retailers to consider carefully their response to the prevalence of products which might contain genetically modified organisms. The uncertainty of consumer reactions is the largest impediment to assessing the future potential of GM technology in Finnish agriculture.

References

Bijman, J. and Enzing, C. (1995) *Biotechnology and Vertical Co-ordination in the Agrofood Chain. A Case Study of the Dutch Potato Chain,* Science and Public Policy 22. Beech Tree Publishing, Guildford, UK, pp. 391–398.

Caswell, M., Fuglie, K. and Klotz, C. (1994) *Agricultural Biotechnology: an Economic Perspective,* Agricultural Economic Report 687. Economic Research Service, United States Department of Agriculture, Washington, DC.

Cochrane, W. (1965) *The City Man's Guide to the Farm Problem.* University of Minnesota Press, Minneapolis, Minnesota.

Culpepper, A.S. and York, A.C. (1998) Weed management in glyphosate-tolerant cotton. *Journal of Cotton Science* 4, 174–185.

Hayenga, M. (1998) Structural change in the biotech seed and chemical industrial complex. *AgBioForum* 1, 43–55. www.agbioforum.org

Hillyer, G. (1999) Biotechnology offers U.S. farmers promises and problems. *AgBioForum* 2, 99–102. www.agbioforum.org/vol2no2/hillyer.html

Kalaitzandonakes, N. (1999) A farm level perspective on agrobiotechnology: how much value and for whom? *AgBioForum* 2, 61–64. www.agbioforum.org

Kinnucan, H., Upton, H., Molnar, J. and Venkateswaran (1990) Scale neutrality of bovine somatotropin: ex-ante evidence from Southeast. *Southeastern Journal of Agricultural Economics* 1–12.

Klotz-Ingram, C., Jnas, S., Fernandez-Cornejo, J. and McBride, W. (1999) Farm-level production effects related to the adoption of genetically modified cotton for pest management. *AgBioForum* 2, 73–84. www.agbioforum.org

Marra, M., Carlson, G. and Hubbell, B. (1998) Economic impacts of the first crop biotechnologies. www.ag-econ.ncsu.edu/faculty/marra/firstcrop/img001.gif

Myyrä, S. and Pietola, K. (1999) Tuottavuuskehitys Suomen maataloudessa vuosina 1987–97. [Abstract: total factor productivity in Finnish agriculture in 1987–1997]. Research report 234, Agricultural Economics Research Institute, Helsinki.

Riley, P. and Hoffman, L. (1999) Value-enhanced crops: Biotechnology's next stage. *Agricultural Outlook* March, s. 18–23.

Salo, A., Kauppinen, V. and Rask, M. (1998) Loppuraportti kasvitekniikasta: Kasvigeenitekniikka ravinnontuotannossa. [Abstract: plant gene technology in food production]. Publication 1998/4, Committee for the future, Parliament of Finland, Helsinki.

Traxler, G. and Falck-Zepeda, J. (1999) The distribution of benefits from the introduction of transgenic cotton varieties. *AgBioForum* 2, 94–98. www.agbioforum.org/vol2no2/traxler.html

USDA (1999) *Impacts of adopting genetically engineered crops in the U.S. – preliminary results.* www.econ.ag.gov/whatsnew/issues/gmo

Virolainen, M. and Niemi, J. (2000) Geenitekniikan ja muuntogeenisen ruuan taloudelliset vaikutukset Suomen elintarvikeketjussa. [Abstract: economic effects of genetic engineering and genetically modified food in the Finnish food sector], Working papers 7/2000, Agricultural Economics Research Institute, Helsinki.

14 Biotechnology, Farm Management and Local Agricultural Development[1]

Maria Teresa Gorgitano and Valeria Sodano
Department of Agricultural Economics, University of Naples – Federico II, Portici, Italy

Introduction

The chapter analyses the potential effects of biotechnology innovations in agriculture, referring to the case of the processed tomato sector in southern Italy. To date, there are no commercially available biotechnology products, but the spreading life-science revolution will soon offer new opportunities in this sector. After a brief presentation of the current competitive position of the examined industry, the analysis of biotechnological impact is carried out in two stages.

Primarily, an analysis of the potential demand for biotechnologies is conducted, stemming from the results obtained in a wide study on the demand for innovation by farmers in southern Italy. Moreover, perception and acceptance of biotechnologies by different operators along the tomato 'filiera', are investigated through focused interviews.

In the second stage, by drawing on the current research effort and comparing the potential innovation with expressed needs on the demand side, the economic impact of biotechnologies is described, taking into account the complex institutional problems affecting the sector. At this stage the analysis is limited to a qualitative basis and does not face the environmental and ethical problems related to genetically modified (GM) food.

In the concluding section, the results are connected to the current discussion on genetically modified organism (GMO) regulatory policies, demonstrating that in some weak institutional frameworks, like in southern Italy and in developing countries, the social control of risks and benefits associated to the new technologies is very hard to achieve, regardless of the chosen policy instrument.

The Italian Processed Tomato Sector

Italy is the second largest world producer of processed tomato and supplies about 50% of the European Union (EU) total demand. The processing industry is concentrated in two regions, one in central Italy (Emilia) and the other in southern Italy (Campania). As opposed to Emilia, Campania is characterized by the three following elements: (i) agricultural production is located out of the region and is concentrated in Puglia, which is nearby; (ii) firms are small, (while Emilia accounts for 15% and 32% of total domestic firms and production,

[1] The chapter is the outcome of the work of the two authors with financial support from the MURST (Ministero dell'Università, della Ricerca Scientifica e Tecnologica) project 'Innovative Pest Control Systems and Quality Improvement in Mediterranean Agricultural Area'. M.T. Gorgitano wrote the first half of the chapter (to 'Survey at the Farms') and V. Sodano wrote the second half.

Campania supplies 50% of production with 60% of firms); and (iii) Campania is strongly specialized in traditional products, and is the world leader in the peeled tomato, covering 80% of the total supply.

The tomato industry expanded quickly during the 1970s and 1980s, driven by the increasing demand and by some important innovations, such as: (i) the development of hybrids with enhanced yield and quality; (ii) the diffusion of picker and seeding machines; and (iii) the modernization of the processing plants.

During the 1990s, demand increased very slowly and industry tried to keep the sector lively by introducing new products, such as prepared flavoured sauces and spiced tomatoes, added to the traditional ones (i.e. peeled tomato, concentrated sauce and pulp sauce).

The most important event by far to influence the sector's evolution during the last 20 years is the introduction of an income support scheme by the EU (reg. 1152/78 and 2200/96), which grants subsidies to processors up to fixed quotas. To receive the subsidies, processors must pay the farmer a price not lower than the target price annually set by the Commission. Moreover, processors are constrained to make purchasing contracts with the farmers' associations rather than with single farmers. The EU intervention allowed the sector to survive in a framework of very low competitiveness with respect to the world market, where prices are about 40% lower than the domestic minimum guaranteed price.

To date, the sector lies in the maturity stage of its life cycle. Profits are still positive, but the operators are seriously concerned about negative signals from price trends. The situation can quickly worsen if the EU decides to cut the support and this eventuality negatively affects expectations and investments among the operators.

Empirical Analysis of the Innovation Demand

The analysis of innovation demand in the southern Italy processed tomato sector was conducted through a field survey organized in two stages. In the first stage, we preliminarily explored innovation problems by interviewing exponents of different bodies acting in the 'filiera', such as processing firms, producer associations and regional agricultural development services. In the second stage, we studied the relationships between farmers' structural and managerial characteristics and their need for innovation. To this aim we developed a specified questionnaire and submitted it to a large sample of growers from Puglia and Campania, which sell to the local processors.

The perception of innovation needs by the agriculture-related sectors

Processing firms

Processors are mainly interested in product innovation to achieve two goals: (i) to make use of plant machinery throughout the year; and (ii) to diversify the final product to match new consumers' attitudes.

Firms selling through supermarket private labels are concerned with quality control and guarantee. They also highlight the increasing demand of retailers for tomatoes obtained by following integrated techniques of production.

Firms selling branded products generally focus on product differentiation in niche markets, such as regional or organic ones. Some people are attracted by functionally engineered food, such as tomatoes with a higher concentration of antioxidants (lycopene and flavonoids), which are supposed to prevent cancer.

All of the companies are interested in innovations that raise the productivity of the cycle of transformation, through either an increase in the Brix degree, or a very high resistance to peeling.

Biotechnologies are generally disregarded, firstly because of the current mistrust shown by European consumers, secondly because the applications that have been proposed up to now, have not responded to the specific needs of firms operating in Campania.

Producer associations

Producer associations are made up of producer cooperatives that are very heterogeneous in sizes and behavior. The associations are predominantly from the Campania region, while cooperatives are equally distributed between Campania and Puglia. The distance of the associations' management from farmers accounts for the low interest they exhibit in agricultural innovation problems. Many managers of the associations have a clear perception of the most urgent technical problems, such as virus prevention.

Nevertheless, they tackle farmers' needs in a very lazy way. Very few cooperatives provide technical assistance for their members and more often the job is given to technicians from the processing companies. In the same way, even though there are many cooperatives that have joined in the request by the companies to have production obtained by integrated techniques, their role is still limited to resolving bureaucratic and administrative problems that come from the EU and do not extend to technical assistance.

With regard to biotechnologies, the associations' most frequent statement is that the current information on risk and opportunities is too low to allow a responsible judgement on the effectiveness of their introduction in the sector.

Public bodies for agricultural extension

The directors of public extension bodies in Puglia and Campania are faced with diverse agricultural problems. Independent farmers have varying amounts of capital and land, as well as diverse future prospects. In Puglia, the cultivation of tomatoes does not pose any particular problems and the main job is to help farms with a high level of technical efficiency to grow. The situation in Campania is quite different. Farms are generally quite small in size, and the cultivation of tomatoes was abruptly reduced by the spread of viruses and then aggravated by the competition from Puglia. Public extension deals with the problem by teaching farmers to operate in market niches with more added value, for example by instituting collective brands (DOP, IGP) for the valorization of regional products such as San Marzano.

The position towards biotechnology is uncertain: moderately favourable if aimed at resolving problems associated with viruses, and negative if directed at increasing yield.

The survey at the farms

We used a questionnaire based on the following sets of questions: (i) general information about farm structure and farmer; (ii) production processes; (iii) relationships within the cooperatives; (iv) coordination with upstream and downstream related sectors; (v) innovation patterns; (vi) future perspectives; and (vii) GMO perception and acceptance.

Of the farms in the sample, 40% are located in Puglia (all outside Foggia) and 60% in Campania, in the Nocerino-Sarnese area, the traditional territory for tomato cultivation. Farms in Puglia are large; 20% of them cultivated an area larger than 100 ha. In contrast, in Campania small sizes prevail and more than 80% of the farms do not extend to 2 ha. In Campania farmers crop 50% of their cultivated area to the tomato while in Puglia tomatoes constitute no more than 10% of the land.

Many direct questions were used in order to identify the most serious problems faced while cultivating tomatoes. The answers given show that the choice of and the quality of the propagation material (plants and seeds), and other technical means do not represent a problem given that the processing firms suggest the variety and the cooperatives provide technical inputs.

The most frequently declared problems are: pest disease (80%), lack of manpower (50%), yield reduction due to soil impoverishment (27%), harvest timing according to delivery organization (20%) and irrigation (18.5%).

Moreover, 80% of farmers highlighted institutional and economic problems stemming from the regulation of the relationship with producer associations and processors within the framework of the EU intervention. Problems arise from the strong contractual power the intervention gives to processors and from the lack of loyalty and competence in the associations' conduct. Negative effects are low prices and quality and excessive market risk.

In order to identify more precisely the farmers' potential demand for innovation, we analyse the different weights they give to each one of the problems described above, their acceptance of GMOs and the motivations supporting these choices.

We performed a cluster analysis in order to divide the sample into homogeneous groups. This classification was made in two steps. In the first one, through a principal component analysis, we obtained few variables (factors) after having identified relationships among a larger set of interrelated variables (elementary variables). This analysis permits us to identify non-directly observable variables (factors) as linear combinations of the observed variables (elementary variables). The goal is to reduce the number of variables while keeping the ability to explain most of the observed variability and identifying (through the factors) the underlying dimensions that can be useful to explain the complex

Table 14.1. Factor loading of elementary variables,[a] explained variance and communality.

Variables	Communality	Factor 1	Factor 2	Factor 3
	% of the variance	26.94	26.30	25.30
	Cumulated variance	26.94	53.24	78.54
X1	0.78	0.89		
X2	0.77	0.83		
X3	0.84		−0.87	
X4	0.79		0.82	
X5	0.72			−0.91
X6	0.81			0.80

[a]Coefficients lower than 0.30 in absolute value are dropped.
X1: Mechanized production process degree; *X2*: Processes enhancing quality; *X3*: Share of tomato acreage; *X4*: Cropped farm area; *X5*: Age of the farmer; *X6*: Initial year of tomato cultivation.

phenomena analysed.[2] In the second step, they were utilized as variables for a cluster analysis to obtain homogeneous groups of farmers.[3]

The elementary variables we used are:

1. Mechanized degree of production process;
2. Production methods enhancing quality;
3. Share of tomato acreage;
4. Cropped area of farms;
5. Age of the farmer;
6. Initial year of tomato cultivation.

The results of the principal component analysis are shown in Table 14.1. Three factors have an eigenvalue greater than 1, and are responsible for 78.5% of the total variation. The communality range of the elementary variable set is between 0.72 and 0.84.[4]

The first factor explains more than 26.9% of the total variation. It identifies the mechanization degree of the tomato production process and the level of 'quality' in tomato production methods. A high score for this factor identifies a farm that has a high degree of mechanization and uses integrated pest management techniques.

The second factor is responsible for more than 26.3% of the total variation. It is defined by two variables which enter the linear combination with opposite signs: share of tomato acreage and cropped farm acreage, which identifies the different farm size. A farm with a high score for this factor is a large farm with only a small part of its land devoted to tomato production.

The last factor considered (25.3% of the total variation) includes two elementary variables with a

[2] The operational steps of the factor analysis have been: (i) to compute the correlation matrix for all elementary variables; (ii) to extract the factors using the principal component analysis; and (iii) to perform an orthogonal rotation of the factors.
After the correlation analysis, in order to verify whether the elementary variables are related to each other or not, and to what extent, we extracted factors. They are linear combinations of the elementary variables, so that any of the observed variables is closely related to one or more factors when its weight in the linear combination (factor loading) in absolute value is large. Each factor is extracted by maximizing the explained variability, under the constraint of being uncorrelated with the other factors. In this way, the hope is for the first few factors to be able to account for the largest amount of total variance. In fact, to explain all of the variance in the sample, a number of factors equal to the number of variables must be extracted. Each factor explains part of total variance, and altogether account for total variance. In order to reduce the factor number, we sorted the factors according to the percentage of explained variance, and selected only those factors which explain more variance than a single variable on average, by selecting the factors with an associated eigenvalue greater than 1. Finally the factors were rotated (using the Varimax method) to transform them into other factors that are easier to interpret.

[3] The method applied for joining clusters is the Ward's method, and the distance among cases was computed with the squared Euclidean distance index.

[4] The communality of a variable is the proportion of variance explained by the factors equal to the sum of the square of the factor loading of the variable (the proportion of variance accounted for by each factor).

coefficient larger than 0.3 in absolute value: the farmer's age and the initial year of tomato cultivation; these are highly correlated to the factors, but with opposite sign. A farm with a low score for this factor is managed by an old farmer who has cultivated tomato for a long time.

The cluster analysis identified four groups of farms. The first group includes 31% of the farms and is made up of farms with traditional production characteristics: (i) small farms; (ii) use of family labour only; (iii) labour intensive production process; and (iv) intensive farming system without clearly defined crop rotation. Tomato cultivation started in the 1970s and is still based on local traditional production processes. It is necessary to stress that 94% of these farmers did not finish primary school, so their formal education level is lower than other groups. This is not surprising because these farmers are the oldest among interviewed farmers. Often they produce tomato crops without contracts. Generally, their market integration is very low, even though the relationship between farmer and producer association is very strong and reinforced by the procurement services and credit supplied to the farmer.

The second group identifies the 'small and medium farms' and consists of more than 27% of the farmers. This group and the former are similar with respect to the farm structural aspects. Nevertheless main differences are related to some of the farmers' characteristics (such as age and formal education level) and to the part-time organization of labour, with farmers also working outside the agricultural sector. They are not efficiently market integrated and the producer association works to some extent as an exchange facilitating agency.

The third group includes the 'large farms' (19% of the sample) which alternate tomato production with extensive crops (such as wheat). They use crop rotations, which allow the repetition of tomato on the same land only one year out of four. The formal education level is the highest among interviewed farmers, in fact more than 60% of the farmers have a high school or a college degree. The links between the farmers and the producer associations are strong, but only for services supplied to the farmer (economic and technical advice).

Finally, the fourth group puts together more than 23% of the farmers. It can be defined as the 'high quality producers' group. Its peculiar aspects are: young farmers with a medium level of formal education, medium farm size, and family farm with a large amount of hired labour. The tomato production process is oriented towards an efficient use of farm labour and is fully mechanized, from planting to harvesting. All farms supply high-quality tomatoes, using formally integrated production methods. The mechanical level of the productive process is the

Table 14.2. Farmers' opinions on general problems and GMO acceptance (%).

	1st group	2nd group	3rd group	4th group
General problems				
Pest control	88	80	70	75
Labour	35	66	60	41
Soil impoverishment	29	46	46	25
Irrigation	11	20	40	8
Contracts	47	66	60	91
Associations	23	20	90	58
EU policy	11	13	30	42
Marketing	—	13	30	33
Market opportunities	17	33	40	75
GMO acceptance				
No	83	54	60	67
Unknown	93	63	17	25
Consumers' dislike	7	37	33	50
Mistrust	—	—	16	25
Ethical reasons	—	—	34	—
Yes	17	46	40	33
Useful	35	42	75	50
Experiment	65	58	25	50

highest. Relationships with producer association are very strong because the latter provide farmers with all kinds of extension, including credit, managerial, technical and marketing services.

In Table 14.2, we present the farmers' opinions on technical and economic problems and their inclination to make use of GMOs. The farms in the first group (traditional producers) perceive the pest control problem as the most important and identify the strong contractual position of the industry as the main cause of their economic problems. Most of them (83%) are not interested in GMOs, either because they do not have enough information (93%), or because they do not believe in market opportunities (7%). Among the farmers who would make use (17%), 65% are interested in preliminary experiments, and 35% trust the effectiveness of the new technologies.

The farmers in the second group (small and medium farms) declare three important technical problems (labour organization, decreasing yield and viruses) and many among them (66%) complain about the strong power of processors. About 54% dislike GMOs (lack of information, 63%; no market opportunities, 37%), while among those in favour, 58% would like to experiment and 42% believe in the effectiveness.

The farmers in the third group (large farms) do not suffer from technical problems, while they are particularly aware of the institutional distortions induced by the EU intervention. Most (90%) hold the producer associations responsible for low prices and marketing inefficiencies. Biotechnologies are generally disagreed with (60%), not only because of the lack of information (17%) and of market opportunities (33%), but also because of ethical principles (34%). Farmers who like GMOs are mostly convinced (75%) that the new technologies could have large positive economic effects.

The farmers in the fourth group are much more worried about institutional problems. They criticize the behavior of producer associations (58%) and above all the contractual power of processors (91%). They also express the need for a wide reform of the EU policy. The interest in GMOs is quite limited and 67% of those interviewed would not employ it, mainly (50%) because of marketing problems. The favourable farmers would experiment with GMOs (50%) and believe in the potential success.

Biotechnology Research in the Processed Tomato Sector

At present there are only a few commercial transgenic tomatoes, those with the fruit ripening altered, which have had a discrete diffusion in the US fresh market since the beginning of the 1990s. We can suppose that in a few years new products will be available. Transitions in the life science industry (Enriquez and Goldberg, 2000) accelerate the introduction of GMOs into as much of the market as possible and this is proved by the large number of experimental fields present in the EU. In Italy tomato is the second product (by number of experiments) in the list of EU deliberate field trials (European Commission–JRC, 1999) and it is in first place for the variety of transgenic characteristics tested. The 30 trials included in the Commission's complete list of allowed field trials (see Appendix) are divided among the following seven trait innovations (numbers in brackets refer to the number of experiments per group):

- glufosinate tolerance (2);
- Bt-derived insect resistance(1);
- improvement of processing quality (6);
- increased yield (3);
- virus resistance (13);
- drought resistance (1);
- others (4).

The analysis of the groups which carry on the trials gives rise to the following remarks.

1. The research is mainly supported by the public sector. This could positively affect the distribution of potential benefits and environmental risk control, but could compromise the commercial success of the eventual new products.
2. Multinational life-science companies are scarcely involved and this may be a result of two issues: (i) tomato is not yet considered a commercially interesting market; and (ii) Italy does not provide enough research facilities.
3. Seed companies are much more interested in the sector. This is because the strong direct relationship with local customers they have had over the past 20 years created major market opportunities while also giving them the possibility of exploiting the organizational economies created by the existing commercial network.
4. The processing industry does not seem to be so fascinated by biotechnologies, in part because of the

low consumer acceptance in the EU market (that is the main market for the sector) and in part because current innovative strategies focus on different aims such as regional and organic production, quality control and distribution organization. Moreover, biotechnological research requires strong financial and knowledge resources while the tomato industry is a traditional sector with a low strategic and economic effort in R&D. Manufacturers could develop innovative projects only by making agreements with biotech companies interested in speciality and local markets, like start-up related to the public research network, but these are just the kind of firms that are not found in the weak socio-economic environment of southern Italy.

The Potential Effects of Private Biotechnology Innovations in the Southern Italy Tomato Sector

The economic effects of biotechnology innovations in agriculture are generally evaluated looking at the user's improvement in technological efficiency (McBride and Brooks, 2000) and at the welfare effects. In the case of a cost reducing innovation, assuming no negative external factors, the welfare effect is measured by the change in Marshallian surplus plus the monopoly profit captured by the discoverer of the innovation through the IPR protection (Moschini and Lapan, 1999; Falck-Zepeda *et al.*, 2000). While these approaches correctly capture the main effects for standardized commodities, they hardly address problems stemming from biotechnologies in speciality and processed agricultural products. When more than one sector operates in the marketing channel for the transgenic product, it is difficult to estimate the final effect on the consumer, and to evaluate the distribution of the producer's surplus among the different vertically related sectors. Monopsonistic practices in the primary market can lead to the exploitation of the grower's benefits of innovation from the processing industry, likewise monopolistic power in the final market can lower the consumer's surplus. The analysis framework is even more complicated when vertical coordination is not achieved through the market but through contractual agreements, and when competitive strategies are based both on cost advantage and on differentiation.

We try to evaluate the potential effects of biotechnologies in the processed tomato sector by considering the strategic and organizational implications of the innovations. The analysis is strictly qualitative because of the lack of both available data and of a formal model that is able to capture the complexity of the suggested scenarios. Considering five main vertical stages in the tomato sub-sector (biotech industry, plant nurseries, growers, processors and retail distribution) we analyse the effects of the innovations with respect to the following three issues: (i) the vertical distribution of cost and benefits; (ii) the competitiveness of southern production; and (iii) the evolution of structures and strategies within the sector. Innovations are grouped into two categories: growing improvements and processing improvements.

Growing improvement innovations

The current research focuses on two characteristics, yield increase and virus resistance. A higher productivity, although technologically more efficient, could lead to economic losses for the growers. In the actual situation with the EU support, crop abundance, instead of pulling the sector towards lower consumer prices and agricultural modernization (by the exit of marginal farmers), could worsen the contractual weakness of growers. As we have already seen, the producer associations do not try to plan production to respect community goals and quotas but try to depress the market to permit the maximum opportunistic exploitation from the processors. When an innovation is introduced, the growers should pay the technology fee and should accept a price lower than the guaranteed one.[5] Also, consumers probably would not see any advantages because the final price is imposed by the retailers according to their own cost structures and strategies.[6] Furthermore, extra profits reached by the greater market power on the procurement side could distract the industry from carrying out optimal interventions in product differentiation and quality

5 As we will discuss in the next section, legally processors should pay the guaranteed prices but in the system there exist many informal and quite illegal ways by which the industry and the association set a different price.

6 As in most processed food, the agricultural component's share of the total cost is very low and there is a low downward price transmission elasticity along the marketing channel.

control. Therefore the results could be: gains for the innovation suppliers and for processors; losses for the growers and maybe for the consumers who would pay for the total organizational efficiency loss. Moreover, the competitive strength of the southern 'distretto' could be worsened, because of an aggravation of the institutional problems affecting the sector and the loss of the actual southern competitive advantages.

In the case of virus-resistant plants, in the current situation with little disease loss, the effects could be compared with those of the previous case. Higher protection from virus would raise productivity. The only additional effect would be the lower cultural risk, but under the hypothesis of risk-adverse growers this would not lead to relevant economic effects. Obviously in the case of a wide disease diffusion as happened in Campania during the 1980s, the innovation would be positive for all the components of the 'filiera', simply avoiding an economic disaster. Nevertheless, this dramatic scenario is very unlikely even without biotechnologies, because we need only a few correct agronomic practices, such as the use of certified seed and good crop rotation, to prevent it.

Processing improvement innovations

We must distinguish the case in which the innovation is introduced by the biotech industry from the case in which it is introduced by some processing firm.

In the first case, if the innovation produces cost and quality advantages we can suppose it will soon spread over the whole processing industry, causing an increase in competitiveness and a consumers' surplus improvement (either through a lower price, or through higher quality assurance and variety). The processor's ability to exploit some of the benefits will depend on its relative market power. Surely the main benefits will occur for the discoverers and the downward suppliers of the innovation, such as the plant nursery. The main cost will be borne by growers, who will pay the technology fee and will lose strategic autonomy within the 'filiera'.

In the second case (innovation stemming from processing firm research activity), the effect could be a less competitive sector due to the competitive advantages achieved by the innovating firms. Probably before imitation effects are produced, consumers could lose with regard to the price but gain in product differentiation either in the horizontal or in the vertical dimension. The gains from the technology fee will be shared between inventors (processors) and suppliers (the seed industry and plant nurseries) according to the specific contracts that the parties will be able to design and enforce. Growers will have to accept higher cost; nevertheless the increased strategic role of procurement activities will probably lead processors to a cooperative relationship with the farmers. In certain cases it could happen that industry would be forced to transfer part of the innovation gains to the growers to give them the incentives to cultivate the transgenic product.

Biotechnologies, Local Development and Institutional Problems in Southern Italy

One interesting result from the empirical research was that the farmer's basic need is to improve vertical coordination, with the aim of lowering price risk and balancing the buyer's contractual power.

This result demonstrates that even though it has been working for more than 20 years, the EU intervention did not achieve the following two goals:

1. Assuring farmers about production placement and price (paying the subsidy to processors should assure placement; moreover constraining the receipt of the subsidy to the payment of the target price should assure price);
2. Promoting collaborative relationships between farmers (the obligation to make contracts through the producers association should help to achieve this goal) to improve production planning and the farmers contractual position.

What really happens in the southern Italian regions we examined is that EU rules only are formally enforced, while exchanges are regulated by informal and implicit contracts, as the following farmers' statements demonstrate:

1. Price is set at the end of the harvest, according to the supply volumes.
2. Processors use different ways to pay a price lower than the legal one. They can declare lower weights and quality for delivered production, or impose high rates on loans informally granted to farmers (with the producer associations acting like brokers).

3. Payment timing and procedures are defined according to informal agreements that aim at maximizing brokers' (i.e. producer association) margins.

These informal exchange mechanisms, sometimes quite illegal, are enforced by the behaviour of the associations which act as agents of processors instead of the formal principal (the farmers). This distortion in incentives can be explained noting that present directors of the associations were previously dealers and have been keeping the same economic behaviour over time. Moreover it is a fact that the EU intervention gave them a strong monopsonistic power towards farmers.

Here we find most of the institutional problems that negatively affect southern Italy's development.[7] The presence of informal relationships between processors and associations, enforced by traditional social conventions, to some extent constitutes a form of social capital (according to the Coleman's definition). The exploitation of this social capital allows both parties to reach personal advantages by the implementation of opportunistic behaviours that damage the farmers and the public operator. Surplus stemming from these behaviours deviates the system from efficiency. Competition is no more based on the achievement of competitive advantages through managerial efforts, but on the exploitation of social capital in an opportunistic way. These kinds of institutional problems can affect biotechnology innovations in the processed tomato sector.

The weakness of the legal system hampers investments by life-science companies, especially those made jointly with local bodies. Conversely, it can give rise to incentives for unfair behaviour, like a low level of risk control in field trials or the opportunistic exploitation of local biological resources.

The contractual handicap moves the agricultural sector from efficient innovative patterns, increasing the risk of adopting biotech innovations when more suitable competing technologies could be implemented.

Finally the described vertical coordination forms, which lead to unfair conduct, could negatively affect any effort of the public sector to adequately inform and assure consumers about the use (or the non-use) and the risk of GMOs.

Summary and Conclusions

The empirical investigation into the possible development of biotechnologies in the processing tomato sector in southern Italy gave rise to three main findings.

1. To date, the demand for GM products is very low in the sector either because of the lack of information or because of the need for alternative innovations. Growers look for organizational innovations that are able to increase their contractual power over the buyers. Processing firms pursue product innovations in niche markets like those for organic and regional products, and judge GM products to not be appealing enough for the consumers.
2. The interest in biotechnologies is also limited on the supply side. The life science companies are actually more interested in standardized crops than in speciality products like processed tomato. Public bodies that could achieve research projects increasing welfare, seem to be too weak both on managerial and financial grounds. The processing industry, while having some interest in GM products with added value output traits, does not invest enough in R&D activities.
3. Given the actual competitive structure, the institutions acting in the sector, and the local socio-economic peculiarities, biotechnological innovations would positively affect mainly the innovators and the operators owning the most market power along the 'filiera'. Growers could even suffer from innovation leading to yield increase because of the depressing market effects along with the payment of the technology fee. The most negative effect could occur in Puglia, where the agricultural competitive advantage is actually based on the natural higher productivity the region exhibits.

The results suggest some discussion about two important issues. The first topic, for which the available literature is very scant, concerns the evalu-

[7] In the 1980s many studies explained the underdevelopment in southern Italy from an institutional perspective. Some of them widely made use of Coleman's concept of social capital (Putnam, 1993; Fukuyama, 1995), classifying the southern regions among those with a low social capital and a strong family's cultural rule. Some critics (Mutti, 1994; Bagnasco, 1999) underlined that it is not the 'quantity' of the social capital, but its 'quality', that seriously hampers the development. This interpretation is consistent with some theoretical studies (Mutti, 1998; Piselli, 1999; Trigilia, 1999) which attempt to make the concept of social capital better fit development problems.

ation of the relationships between the food system organizational structure and the innovative patterns in one or more vertically related stages. Some examples are given by the recent radical innovations in the food retail sector that led to revolutionary management concepts like ECR (efficient consumer response) or CM (category management) and dramatically changed the competitive framework in the backward vertically related sector (Lanciotti, 1999; Senauer and Kinsey, 1999).

In the case of the processed tomato sector, we argued that different vertical coordination forms can give different incentives to biotechnology innovation and strongly influence the distribution of derived costs and benefits among the components along the 'filiera'.

When vertical coordination is achieved by informal and implicit contracts rather than market or formal contracts, the effects on innovating activity are even more difficult to understand. To explore this issue we need to improve the traditional efficiency-based theory (like the theory of contracts or the transaction cost economy) using analytical tools better able to tackle problems, such as the power and evolution based incentives or the social embeddedness of economic activity. Some analytical frameworks successfully used in local development studies such as the New Economic Sociology stemming from Granovetter or the New Sociological Institutionalism (Rizza, 1999), have also recently been recommended for the food system vertical coordination analysis (Boon, 1999; Galizzi and Venturini, 1999).

The second issue is related to the EU regulatory options about the use of biotechnology in agriculture. The mandatory labelling option and the precautionary principle accepted in risk assessment and management, is irreconcilably opposite to the USA, which instead recognizes zero-risk for GMOs (until science-based proof of dangerousness is acquired) and allows voluntary labelling of products that do not use GMOs (Caswell, 2000).

Even if people who are really worried about the negative environmental impact of GMOs would be happy to believe that EU policy is driven by the Jonas' ethical 'responsibility principle' (Jonas, 1990), it is quite clear that EU policy is principally influenced by economic and political consensus considerations. Consensus is related to the recent interest of public opinion in food safety problems. Economic considerations refer to the EU aim of protecting domestic production, specialized in speciality and regional products, from the new transgenic products, for which the USA has strong competitive advantages.

The results on the potential effects of biotechnologies in the southern Italy tomato sector allow for at least two critiques of the EU position. Firstly, the presence of numerous field trials, in spite of the low interest in biotechnology emerging from interviews, demonstrates that, in time, biotechnologies will widely affect world agriculture. Then the problem is no longer to ban or to label GMO products but to responsibly and successfully participate in the life science revolution. Only if all the components of the food system are involved in the discovery and in the introduction of innovations, will it be possible to have an effective risk monitoring of the new technologies. The best aims of the public sector should then be research support and the promotion of collaborative relationships among different bodies involved in biotechnologies.

Secondly, the research demonstrates that in some weak institutional frameworks (like we highlighted in southern Italy) differences between voluntary and mandatory rules are rather more theoretical than practical. When there are no market incentives to guarantee the effectiveness of the voluntary approach, and when the mandatory rules cannot be enforced by an efficient legal system, both the approaches will fail to reach their objectives (Segerson, 1999). In southern Italy and in many underdeveloped countries with similar institutional problems (Shiva, 1993), public intervention should aim at improving human and social capital, by promoting knowledge and trust.

References

ANICAV (1999) *Rapporto sull'industria italiana delle conserve di pomodoro.* ANICAV, Napoli.

Bagnasco, A. (1999) Teoria del capitale sociale e political economy comparata. *Stato e Mercato* 57, 351–372.

Boon, A. (1999) Capabilities, transaction costs, and vertical co-ordination in the food system. In: Galizzi, G. and Venturini, L. (eds) *Vertical Relationships and Co-ordination in the Food System.* Physica-Verlag, New York.

Caswell, J.A. (2000) An evaluation of risk analysis as applied to agricultural biotechnology (with a case study of GMO labeling). *Agribusiness* 16(1), 115–123.

Coleman, J.S. (1990) *Foundation of Social Theory.* Harvard University Press, Cambridge, Massachusetts.

Enriquez, J. and Goldberg, R.A. (2000) Transforming life, transforming business: the life-science revolution. *Harvard Business Review* 4, 96–104.

EU Commission-JRC (1999) Field trials, Italy. www.foodjrc.it

Falck-Zepeda, J.B., Traxler, G. and Nelson, R.G. (2000) Rent creation and distribution from biotechnology innovations: the case of Bt cotton and herbicide-tolerant soybean in1997. *Agribusiness* 1(16), 21–32.

Fonte, M., Salvioni, C. and D'Ercole, E. (2000) L'analisi economica delle biotecnologie applicate all'agricoltura. *La Questione Agraria* 1, 54–74.

Fukuyama, F. (1995) *Trust.* The Free Press, New York.

Galizzi, G. and Venturini, L. (1999) Towards a theory of successful vertical cooperation in the food system. In: Galizzi, G. and Venturini, L. (eds) *Vertical Relationships and Co-ordination in the Food System.* Physica-Verlag, New York.

INEA (2000) *Le Proposte di Riforma del Sostegno al Pomodoro da Industria e Agli Agrumi Destinati alla Trasformazione.* INEA, Roma.

Jonas, H. (1990) *Il principio responsabilità.* Einaudi, Torino.

Lanciotti, C. (1999) Supply chain integration, category management and the Italian food retail sector. In: Galizzi, G. and Venturini, L. (eds) *Vertical Relationships and Co-ordination in the Food System.* Physica-Verlag, New York.

McBride, W.D. and Brooks, N. (2000) Survey evidence on producer use and costs of genetically modified seed. *Agribusiness* 16(1), 6–20.

Moschini, G. and Lapan, H. (1999) Intellectual property rights and the welfare effects of agricultural R&D. In: Galizzi, G. and Venturini, L. (eds) *Vertical Relationships and Co-ordination in the Food System.* Physica-Verlag, New York.

Mutti, A. (1994) I sentieri dello sviluppo. *Rassegna italiana di sociologia* 73, 109–120.

Mutti, A. (1998) *Capitale sociale e sviluppo.* il Mulino, Bologna.

Pelt, J.-M. (2000) *L'orto di Frankestein.* Feltrinelli, Milano.

Piselli, F. (1999) Capitale sociale: un concetto situazionale e dinamico. *Stato e Mercato* 57, 395–417.

Putnam, R. (1993) *Making Democracy Work.* Princeton University Press, Princeton, New Jersey.

Rizza, R. (1999) Le istituzioni tra economia e sociologia. *Sociologia del Lavoro* 73, 119–143.

Segerson, K. (1999). Mandatory versus voluntary approaches to food safety. *Agribusiness* 1(16), 53–70.

Senauer, B. and Kinsey, J. (1999) The efficient consumer response initiative: implications for vertical relationships throughout the U.S. food system. In: Galizzi, G. and Venturini, L. (eds) *Vertical Relationships and Co-ordination in the Food System.* Physica-Verlag, New York.

Shiva, V. (1993) *Monoculture of the Mind. Perspectives on Biodiversity and Biotechnology.* Zed Books, London.

Trigilia, C. (1999) Capitale sociale e sviluppo locale. *Stato e Mercato* 57, 419–440.

Appendix

Italy: Field trials (EU Commission-JRC) (source: www.food.jrc.it/gmo).

- Tryptophan-2-monoxygenase synthesis (Istituto Sperimentale per l'Orticoltura)
- Virus resistance (tomato spotted wilt virus) (Istituto Sperimentale per l'Orticoltura)
- Tryptophan-2-monoxygenase synthesis (Istituto Sperimentale per l'Orticoltura; Section of Montanaso Lombardo)
- Virus resistance (cucumber mosaic virus) (Istituto Sperimentale per la Patologia Vegetale; Monsanto Europe)
- Virus resistance (cucumber mosaic virus) (Istituto Sperimentale per la Patologia Vegetale; Peto Italiana s.r.l.)
- Virus resistance (cucumber mosaic virus); Virus resistance (potato virus y) (Istituto Sperimentale per la Patologia Vegetale; Key Gene)
- Virus resistance (cucumber mosaic virus) (Istituto Sperimentale per la Patologia Vegetale di Roma)
- Ac/Ds two components transposon system; Gene tagging; Glufosinate tolerance (Metapontum Agrobios s.c.a.r.l.)
- Ac/Ds two components transposon system; Glufosinate tolerance (Metapontum Agrobios s.c.a.r.l.)
- Virus resistance (cucumber mosaic virus) (Metapontum Agrobios s.c.a.r.l.)
- Downregulation of pectin esterase; Improvement of processing quality (Peto Italiana s.r.l.)
- Improvement of processing quality; Polygalacturonase synthesis (Peto Italiana s.r.l.)
- Improvement of processing quality; Polygalacturonase synthesis (Peto Italiana s.r.l.; Stazione Sperimentale per l'Industria delle Conserve Alimentari; Zeneca Plant Science)
- Virus resistance (cucumber mosaic virus) (Peto Italiana s.r.l.)
- Virus resistance (cucumber mosaic virus) (S&G Sementi s.p.a.)
- Drought tolerance; Levan sucrase synthesis (Sementi Nunhems s.r.l.)
- Increased cell wall thickness; Pyrophosphate synthesis (Sementi Nunhems s.r.l.)
- Increased yield; Kinase synthesis (Sementi Nunhems s.r.l.)
- Increased yield; Sucrose transporter protein synthesis (Sementi Nunhems s.r.l.)
- Virus resistance (cucumber mosaic virus) (Sementi Nunhems s.r.l.)
- Virus resistance (cucumber mosaic virus) (SME Ricerche s.c.p.a)
- Bt-derived insect resistance (SME Ricerche s.c.p.a)
- Downregulation of pectin esterase; Improvement of processing quality (Peto Italiana s.r.l.)
- Improvement of processing quality; Polygalacturonase synthesis (Stazione Sperimentale per l'Industria delle Conserve Alimentari)
- Improvement of processing quality; Polygalacturonase synthesis (Stazione Sperimentale per l'Industria delle Conserve Alimentari; Zeneca Plant Science)
- Virus resistance (cucumber mosaic virus) (Tecnogen s.c.p.a.)
- Osmotic synthesis; Pathogenesis related proteins synthesis (Università degli Studi della Tuscia; Dipartimento di Produzione Vegetale)
- Virus resistance (tomato yellow leaf curl virus) (Vilmorin Italia)
- Virus resistance (tomato yellow leaf curl virus) (Vilmorin SA)

15 Public Acceptance of and Benefits from Agricultural Biotechnology: a Key Role for Verifiable Information

Wallace E. Huffman[1]* and Abebayehu Tegene[2]*
[1]*Department of Economics, Iowa State University, Ames, Iowa, USA;* [2]*Resource Economic Division, ERS, USDA, Washington, DC, USA*

The on-going genetically modified (GM) food and genetically modified organism (GMO) controversy threaten to destroy the near-term market for ag-biotech food and inputs and alter greatly the net social benefits that are potentially attainable from agricultural biotechnology. For example, consider:

1. In February 2000 Greenpeace filed a lawsuit challenging the US Environmental Protection Agency's decision to allow the release of GM insect resistant (Bt) crops. In their news release they stated 'The EPA should stop [GE] polluters before the environment is threatened' (Greenpeace, 2000a).
2. During the week of 23 March 2000 Greenpeace joined over 50 other organizations in a petition to the US Food and Drug Administration (FDA) calling on the agency to remove genetically engineered (GE) foods from the market because it failed to require safety testing or labelling (Greenpeace, 2000b).
3. In 1998, the European Union set out to update Council Directive 90/220 covering the deliberate release of GMOs. In the spring of 2000, the EU decided not to approve any more releases until the directive is revised.
4. During 1999, more than 50% of the crop biotech field experiments in the UK were disrupted by anti-GMO activists.

Nettleton (1999) states that the anti-biotech activists have achieved a masterful feat in communication, subverting the purpose of biotechnology, whipping up public alarm and feeding political agendas to protect agricultural markets.

Advances in science enable new technologies and advances in technology increase the demand for science. Advancing science and technology are uncertain and costly activities (Holmstrom, 1989; Huffman and Evenson, 1993; Huffman and Just, 2000). Although some new technologies have benefited society greatly, much uncertainty surrounds most new technologies. For example, little accurate information or knowledge exists about the attributes, including effects, of new ag-biotechnologies,

* The authors are Professor of Economics, Iowa State University, and Economist, Resource Economics Division, ERS, USDA, respectively. Matt Rousu provided research assistance, and the Economic Research Service (cooperative agreement no. 43-3AEL-8-80125) and Iowa Agriculture and Home Economics Experiment Station provided financial support. Views presented in this chapter are the authors' and do not represent ERS or the USDA. Journal paper number J-19000 of the Iowa Agriculture and Home Economics Experiment Station, Ames, Iowa, Project 3077, supported by Hatch Act and State of Iowa funds.

and some of the knowledge/information that does exist is public information and some private.[1] Additional research can be undertaken to increase the knowledge about the beneficial and harmful effects of new technologies, some of which will reduce the uncertainty over future irreversible catastrophes. Advances in communication and information networks make possible rapid worldwide dissemination of public scientific discoveries and other information.

Private information is the source of asymmetric information, and it leads to an informational advantage to the party possessing the information. In two party interactions with one party possessing private information, the informed party can be expected to exercise their informational advantage whenever they can expect to gain from it, and the other party loses. When experienced parties develop intuition about situations potentially leading to opportunistic behaviour of others, asymmetric information can destroy trade/exchange between parties where the potential gains from trade/exchange are large.

The objective of this chapter is to examine the welfare effects of information from communication (by interested parties) on the decisions of producers of agricultural products and potential users of biotechnical inputs and on consumers of agro-biotechnical products. Interested parties are considered to be private biotechnical companies whose profits depend positively on sales of GM inputs to agricultural producers and 'environmental groups' whose utility depends positively on the quantity and quality of environmental stocks. The ag-biotech companies provide or distribute information primarily in advertising, news releases, informational brochures and web sites, personal contacts and demonstrations. The environmental groups provide or distribute information primarily in news releases, informational web sites and demonstrations. Final consumers of agricultural products and agricultural producers that might use GM inputs are assumed to be approximately 'neutral parties' in this communication process, but they face important decisions relating to their own welfare.[2] We acknowledge that other possibly more trusted sources of information exist for producers and consumers, and the chapter focuses on the importance of these sources to good decision making.[3] We will show that verifiable information plays a central role in socially good decision making and that an independent agency should store and make freely available verified information. This agency might also engage in research needed to refute or confirm claims made by interested parties and others.

The Model

Many decision makers must or choose to rely on information provided by individuals or groups who are affected by their decisions. Furthermore, these decision makers may not know the alternatives available and have no control over the information provided to them by interested parties. These interested parties may manipulate by distorting or concealing information. For example, farmers rely on information, including advertising, provided/distributed by biotech companies about the expected performance of new biotech inputs, which is quite selective. These firms are constrained somewhat by an interest in repeat sales, but their communication may not reflect accurately all known impacts. Consumers rely on information and advertising distributed by food companies and environmental groups which seem likely to be tinged with self-interest. For example, communications by Greenpeace and Friends of the Earth opposing GM foods may exaggerate the potential harm to the environment and distract from other important issues.

Greenpeace, Friends of the Earth and other environmental groups are interest groups. Individuals self-select into these groups because of a common interest or goal focused on the environ-

[1] See Frey (2000) for a discussion of some of the prospects and problems associated with the use of biotechnology in plant breeding.

[2] Hoban (1997, 1999) shows that food safety concerns from biotechnology are rated relatively low by consumers in the USA, Western Europe and Japan, relative to microbial contamination, pesticide residuals and antibiotics or hormones.

[3] Although Hoban (1996) reported that ag-biotech companies and activist groups are not ranked high by US consumers as a trusted information source, this does not mean that they never use any of the information in decision making. In The Netherlands, a survey of the general public showed that environmental and consumer organizations were seen as the most reliable sources of information on biotechnology (Heijs and Midden, 1996). In the UK, Martin and Tait (1992) found that in a local community reaction to a GMO release, the public chose to support the perspective provided by Greenpeace.

ment, and achieving the group's goal is a public good to its members. Hence, free-riding by one member on the efforts of other members is a major organizational problem (Olson, 1965; Sandler, 1992; Cornes and Sandler, 1996: 324–326). Each of these groups has resources – largely members' time and financial contributions – and their impact is affected by organizational efficiency. By choosing relatively narrow objectives, these groups reduce coordination and decision making costs over organizations that have diverse goals. Advances in communication and information technologies have greatly reduced organizational costs of interest groups and have undoubtedly increased their productivity. They are now able to construct low cost web sites for displaying their objectives, news releases, short articles and other information. For example, the web site of Friends of the Earth has been used to display the locations or addresses of large-scale plantings of GMO crops in England (Friends of the Earth, 2000). These groups can also use e-mail to rapidly distribute communications among members, for example dealing with demonstrations, and others, such as letters opposing GMO use and policies.[4]

In our model, the two interested parties provide information in the form of communications attempting to affect agricultural producers' and consumers' decisions. The communications are signals which reflect the self-interest and private information of each party. Communication is cheap because it requires little action on the part of the sender, and new information technologies, such as e-mail and web sites, have reduced its cost and greatly increased the swiftness of delivery. There remains some modest fixed cost of preparing a communication, but the marginal cost of distributing it has become approximately zero (Shapiro and Varian, 1999: 19–51). Misinformation can be as easily distributed as useful information. Computer viruses, like 'I Love You', are one example of misinformation which can be sent swiftly around the world and clog the information network. In particular, the new information technology has greatly expanded the possibility of individuals communicating with others whom they do not know personally and from whom they may have greatly different norms and values. Hence, new problems with assessing the quality of information obtained from web sites and e-mail have arisen (World Bank, 1999: 72–81). Because signalling with communications is so cheap, one possible outcome is that they degrade the quality of information to the point that communications from interested parties are ignored.

Although the marginal cost of distributing information is approximately zero, it remains costly for decision makers to interpret this information, especially contradictory information. Consumers and agricultural producers, however, differ in the long-term consequences of using bad information. Consumers maximize utility subject to a resource constraint. When they fail to use objective information in decision making, their utility or well-being decreases, but this does not generally cause them to exit the economy (except when the consequence is death). Producers on the other hand can be described as long-run profit maximizers. If they do not use good information, their profits are reduced. If they become negative over the long term, most likely they will be forced to exit the industry. Thus, there is selectivity operating among farmers that is generally different from that of consumers. For consumers and producers, the ability to process information and make good decisions is a valuable skill in the sense that it can be welfare or profit increasing, and this ability seems most likely to be related to their years of schooling and accumulated experience as decision makers (Huffman, 2002). Given some outlay on information, a decision maker might choose to rely on only one interested party in making their decisions, or they might choose to rely on several interested parties possessing different points of view. There are potential costs and benefits of each of these actions.

Although Milgrom and Roberts (1986) have shown that it is possible for a decision maker to make fully informed decisions when the decision maker relies on one interested party for information, the necessary conditions seem quite restrictive and are unlikely to be fulfilled. For example, the interested party's preferences must be known to the decision maker, the information must be freely verifiable, the decision maker must know the factors about which the interested party has information to be able to detect situations in which information is being withheld, and the decision maker must be

[4] Furthermore, advances in communication and information technologies may have facilitated coalition formation among interest groups wishing to oppose ag-biotechnology for whatever reason (see Greenpeace, 2000b and the comment by Nettleton, 1999).

able to draw the appropriate inference when information is withheld (i.e. he or she must be a sophisticated, sceptical decision maker). These are demanding, but perhaps not impossible attributes for successful agricultural producers to possess, but they seem to exceed the attributes of most consumers of food. The implication is that for agricultural producers (or consumers) to rely only on information provided/disseminated by biotech supply firms or environmental groups is unlikely to lead to fully informed decisions. Hence, good reasons exist for society to be sceptical of claims made by both the suppliers of ag-biotech inputs and the environmental interest groups.

Adding (having or allowing) competition among interested parties in providing information greatly reduces the restrictions or assumptions necessary for good decision making. The interested parties must be able to convey their information to the decision maker, and the decision maker must listen to all interested parties who want to convey information, that is, there must be an opportunity for the different interests to come out. Ag-biotech firms, environmental groups and other interested parties seem likely to differ in their ability and effectiveness in conveying information that they have. This ability might be associated, among other things, with training, communication skills, personalities, organizational objective/philosophy and information technologies.[5] Rapid advances in communication and information technologies are widely available today, and with the dramatic fall in the cost of sending messages and storing information through new networks (Shapiro and Varian, 1999; World Bank, 1999), accessibility to technologies seems minimally constrained by capital or credit. Biotech companies are primarily private companies interested in long-term profits associated with the sales of their products and the value of the company, and the information that they distribute can be expected to be consistent with this long-term objective and to be constrained by it. Environmental groups are pursuing non-monetary goals which seem likely to be less constraining on their actions and possibly on the objectivity of the information they distribute.

However, the decision makers can now be unsophisticated, having little or no idea of available options, of issues bearing on the decision or preferences of the interested parties. He or she must, however, be able to process the information that he or she receives, and the information must be verifiable. Under these conditions, fully informed decisions are possible. The implication is that agricultural producers and consumers can make better decisions when they use information from diverse and possibly interested parties, provided the information is verifiable.

Much information being distributed these days about ag-biotechnology, however, is not currently verifiable. First, biotechnology is advancing rapidly so many effects and impacts of new products are unknown. Second, a coalition of anti-biotechnology interests has been formed to slow the acceptance of ag-biotechnology. These groups have raised new questions about both the short-term and the long-term effects on health and the environment of using ag-biotechnology and consuming GM foods. Third, some of the activities of the anti-biotech groups seem to be focused on disrupting the experiments that might lead to important and useful advances in the stock of knowledge about ag-biotechnology.

When information is not verifiable, the reliability of information provided by an interested party (or parties) depends on the congruity between the objectives of the decision maker and those of the interested party (parties). When objectives diverge, decision making is difficult, and these complex problems have been the topic of optimal incentive schemes in principal-agent or agency theory literature (e.g. see Holmstrom, 1979; Holmstrom and Milgrom, 1987; Gibbons, 1998). These models are, however, well suited only to decision problems with few, e.g. two, players.

When information is not verifiable, communications by interested parties might lead to unduly restrictive public policies being adopted (e.g. banning GM food production or imports) or it might degrade the information content to the extent that sophisticated decision makers ignore it.[6] This will, however, be generally welfare reducing relative to

[5] In dealings with strangers of unknown credibility and no binding contract, reputation based on providing accurate and reliable information is valuable. See Sobel (1985) for one perspective on how a concern about credibility can be expected to affect the quality of information provided to a decision maker over time.

[6] The 4-year moratorium on processing patent applications on transgenic plants and animals by the European Patent Office, which ended in December 1999 with a decision by the enlarged board of appeals, seems to have been a matter of interpretation of European Union directives and not of verifiable information (Scheermeier and Dickson, 2000).

fully informed decision making. For example, social cost-reducing inputs for crop production might not be used by farmers, or socially beneficial GM foods might not be consumed.[7] More generally, long-term delays in adopting GM technologies because of the time required to verify or refute claims by the environmental groups about ag-biotechnical products will reduce the expected social and private payoff to R&D in this area. This has implications for where the private sector places its future R&D investments (Frey, 2000), meeting future food needs and for economic growth.

When a large number of decision makers can use or need the same verified information, research to produce this information produces a public good, which may be of great social value. The knowledge once produced is non-rival, that is, use by one decision maker does not affect the quantity or quality available to others, and it is not (or may not be) economically feasible to exclude users. Because the opportunity cost of an added user of the information is zero and each user's valuation of the information is private information, private decisions will lead to under provision unless some organizational device is used to internalize externalities associated with free-riding. The price system is of no (or little) aid in extracting information on the social value of verified information or a system to manage it.

Institutionalizing a Verified Information System

An ideal verified information system for ag-biotechnology must provide a mechanism for disclosing private information (i.e. information or knowledge that exists but is not available to everyone), establish a process for refuting or confirming claims of interested parties, and advance the stock of knowledge about short- and long-term effects of biotechnology.[8] The size of the problem is large because ag-biotechnology is a global intergenerational public good. Biotech knowledge is non-rival and non-excludable on a global basis, although particular techniques and products have been converted into impure public goods through the institution of patenting and international patent agreements. Furthermore, the impacts (benefits and hazards) are multi-generational – in terms of both potential benefits and hazards.

Although most of the currently available ag-biotechnologies have been developed for large-scale agriculture and high income consumers, the potential exists for ag-biotechnology to help low income countries to meet future food needs (Serageldin, 1999; National Research Council, 2000) and many of the low income countries want access to or to have the opportunity to use ag-biotechnology to help meet their future food needs (OECD, 2000). Hence, reliable information on ag-biotechnology is a public good with potentially large global value.

Because new ag-biotechnology has the potential to produce benefits and realize hazards over the long term, it has an intergenerational dimension. This means that sequencing of generations becomes important in setting policies and determining benefits; that is, both equity and efficiency dimensions are important. The current generation has a 'first-mover' advantage because it can choose an agenda or pathway that best serves its own purposes, even though these purposes may be at odds with later generations (Sandler, 1997). This is an especially important issue for actions that are irreversible; for example, reducing the earth's biodiversity, and for some transfers of genes across unrelated species. Far-sighted decisions can be promoted by including individuals/representatives distributed over wide age and standard of living ranges in decision making on ag-biotech policy and by evaluating costs and benefits of biotechnology in real terms without discounting.[9]

Some mechanisms for revealing private information lead to better decisions than others. When a product's 'quality' is at issue, 'untruthful' advertising

7 As an indication of benefits from agricultural biotechnology, Falck-Zepada *et al.* (2000a) estimate the first-year worldwide economic surplus from the introduction of Bt cotton in the USA was US$240 million. The economic surplus to herbicide tolerant soybeans is larger (see Moschini *et al.*, 1999; Falck-Zepeda *et al.*, 2000b). In the surplus computations, biotech inputs are treated as having neutral effects on human health and the environment. See Alston *et al.* (1995) for more information on methods of social cost-benefit analysis dealing with research.

8 Avery *et al.* (2000) are concerned with a much narrower issue. They propose a computerized market for the collection and distribution of subjective evaluations of a product of uncertain quality purchased by consumers.

9 Discounting seems to be a questionable practice for social cost–benefit analysis that spans several generations because with discounting at any positive rate, the current weight to a distant generation's disaster is very small (and frequently negligible). See Sandler (1997: 62–69).

tends to lead to a breakdown in the market for the good and no trade occurs because of adverse selection; for example Akerlof's 'market for lemons' (Akerlof, 1970; Molho, 1997: 19–26). With 'truthful' advertising, monitored effectively by an independent body, good market performance is obtained. Privately provided information on the route to reputation building and repeat sales is less effective for simple 'quality' issues (Molho, 1997: 52–53), and it can be expected to be quite deficient in the biotech area where the scientific issues are frequently complex, the quality dimensions are frequently changing and the stock of knowledge is steadily advancing.

Private information is best revealed publicly through an independent agency which has the authority and responsibility to independently verify information from interested parties. An independent agency can go about making objective assessments of information and claims made by interested parties and others. These assessments and evaluations are costly to make because they use scarce resources, but once verified the information is a pure public good. New information and communication technologies have greatly reduced the cost of storing and rapidly transferring this information and greatly increased the potential accessibility to a global scale.

Advances in the stock of knowledge are important to a successful information verification system. It expands the topics, issues or dimensions of the knowledge base on which verification can be made. The primary contributors to this activity seem likely to be scientists employed by commercially independent institutions and funded primarily by the public sector, that is, scientists in 'open universities' and possibly government agencies. An open university is one where the direction and funding of research is not driven primarily by commercial interests or any other narrow interest group and where high scientific-control standards are in place (David, 2000; Huffman and Just, 2000). Universities where the research agenda and/or funding has been captured by a single interest group, or one or more large private companies (e.g. a life science company), does not meet this criterion. Also, if the direction of the research programme of a government agency is driven heavily by commercial considerations or has low scientific-control standards, it will not meet the criterion of open and objective science either. For scientific discoveries to be highly supportive of the information verification system, they must originate from an institutional process that signals openness and objectiveness to disinterested parties.

Good research requires considerable time to undertake and to verify itself. When the frontiers of biological and related sciences are advancing rapidly, a significant period may exist where considerable scientific uncertainty exists about outcomes, effects, impacts or the quality of the information (see, e.g. Frey, 2000: 61–79). Interested parties may attempt to exploit this information lag which can be to the detriment of producers and consumers of agricultural products and to society generally. An efficiently functioning knowledge generation and verification system can, however, shorten this lag. By doing so, it creates an environment where interested parties have a strong incentive to reveal voluntarily more (rather than less) of the private information that they possess and where verified information is freely available and easily accessible. The cooperation of interested parties is achievable primarily because sophisticated decision makers would infer even worse outcomes (Milgrom and Roberts, 1986).

No supernational body is likely to be created to provide verifiable information which is a global public good. Nations have been unwilling to empower such bodies with the authority to collect taxes for such purposes (Sandler, 1997). For an individual country to provide the good, the social benefits must exceed the costs. The greater public good nature of verified information created by advances in information technologies means savings for some countries whose decision makers free-ride and weaker incentives for any one country to undertake the verification activity. Only one good biotech verifiable information system is needed; or given one system, the marginal product of another system is zero. In the public economics literature, this has sometimes been labelled as 'best shot' technology of public good supply aggregation (Sandler, 1998). Thinking about a global coordination game across countries for provision of verifiable information, only one country needs to act. The country which has the largest expected net social benefit from action, or largest stake, can be expected to provide a biotech verifiable information system. This seems likely to be a country with a strong research system, a large ag-biotech industry, a technically advanced large-scale agriculture, large population, high price of time (for acquiring and evaluating information) and high income.

Trust in Public Institutions

Because a verifiable biotechnology information system must be financed by public tax collections, it most likely will be operated by one or more government agencies.[10] For this institutional framework to be successful, it must have the trust of the public. Currently agencies of national governments dealing with similar issues in Western developed countries vary in the amount of trust or confidence that the public places in information that they provide.

In the USA, public trust is high, but in Western European countries public trust is low. Gaskell *et al.* (1999) reports on a survey of the US public showing a high level of trust in information dealing with the safety of biotechnology provided by the USDA and FDA. Also, Hoban (1997) reports on a 1994 survey of the US public where respondents were asked to rank 15 different sources of information on agricultural biotechnology for trustworthiness. The National Institutes of Health and Food and Drug Administration ranked very high: second and third.[11] (The USDA and EPA were not included in the reported rankings.)

For Europe, Gaskell *et al.* (1999) report on a survey of the public in 17 European countries showing low trust in national public bodies 'to tell the truth about GM crops grown in fields'. European governments have accumulated a bad record with the general public on food safety issues because of their past experiences dealing with governments that mishandled information on UK BSE meat and dioxin contamination of dairy and poultry products in Belgium and The Netherlands. Decision making by producers and consumers is made more difficult when government agencies cannot be trusted as a verifiable information source.

Current public information systems in place in the USA and Europe dealing with information on agricultural biotechnology are primarily focused on regulation for environmental and health safety and secondarily focused on proving information to consumers through labelling. In the USA, the regulation of biotechnology products is through the Coordinated Framework established in 1986 (National Research Council, 2000a) which ties together the US Department of Agriculture (USDA), Food and Drug Administration (FDA) and Environmental Protection Agency (EPA). The framework is based on the principle that techniques of biotechnology are not inherently risky and that biotechnology should not be regulated as a process, but rather that the products of biotechnology should be regulated in the same way as products of other technologies. Responsibility and jurisdiction over transgenetic products were assigned as follows: (i) plants came under the jurisdiction of the Federal Plant Pest Act (FPPA) administered by the USDA; (ii) food and feed under the Federal Food, Drug, and Cosmetic Act (FFDCA) administered by the FDA; and (iii) microorganisms and substances used for pest control under the jurisdiction of the Federal Insecticide, Fungicide, and Rodenticide Act (FIFRA) and part of FFDCA administered by the EPA. As new biotech products have been developed, environmental and consumer groups have expressed concerns that EPA rules do not adequately cover all the relevant risk issues (e.g. oral toxicity, potential for allergenicity) and the USDA should examine more thoroughly for risks of new plants outcrossing

[10] The exchange of information and clearing house mechanism under the 1993 Convention on Biological Diversity have a somewhat different focus. The 1992 Convention on Biological Diversity has as objectives the conservation of biological diversity, the sustainable use of its components and the fair and equitable sharing of the benefits arising out of the utilization of genetic resources. The Convention went into effect in December 1993 with 168 signing countries (the USA being absent). Article 17 requires that members facilitate the exchange of information relevant to the conservation and sustainable use of biodiversity, including results from technical, scientific and social research and other information. Article 18 requires the establishment of a clearing house mechanism to promote and facilitate technical and scientific cooperation (Convention on Biological Diversity, 1992; www.biodiv.org/chm/conv/cbd_text_e.html). Both of these articles are to facilitate access to and exchange of information on biodiversity around the world, especially information needed to implement provisions of the Convention. Although the exchange of information and clearing house activities facilitate voluntary sharing of information on biodiversity, they are not actively involved in quality control or verification. The Cartagena Protocol on Biosafety was adopted in January 2000 to address environmental (but not food safety) impacts of bio-engineered products that cross international borders. The Protocol establishes an Internet-based Biosafety Clearing-House to help member countries exchange scientific, technical, environmental and legal information about living modified organisms.

[11] However, in 1999, US environmental activists and some consumer groups intensely criticized and demonstrated against the FDA's GM food policies (Macilwain, 1999).

with wild relatives to produce unusually hardy weeds or adversely affect biodiversity (National Research Council, 2000a).

Labelling for GMO content has not been adopted in the USA, but indirectly information will be provided in a new USDA standard for 'organic food' labelling (Golan *et al.*, 2000). For crops, the standard means that the use of genetic engineering, irradiation and sewage sludge in the production or processing stages is prohibited. Although all organic farmers and handlers would be expected to abide by the standard, it remains to be seen whether the standard can be effectively enforced.

For countries in the European Union (EU), the EU has established polices as directives to member nations on environmental legislation. The EU Directive on the Contained Use of Genetically Modified Organisms (Directive 90/219/EEC) and on Deliberate Release (Directive 90/220/EEC) attempt to establish a system for controlling the use of GM organisms (European Commission, 2000). The directives were modelled after the EU's chemical notification directives. Since these EU directives are not implemented uniformly across member countries and no central monitoring authority exists, the system is somewhat loosely controlled.

Directive 90/219/EEC provides common rules throughout the EU for the use of GM microorganisms in research laboratories and industrial facilities and provides appropriate measures to protect human health and the environment from any risks arising from activities using GM microorganisms. The Directive outlines appropriate procedures for risk management. Microorganisms and activities using them are to be classified by their potential for risk and to containment and control measures. Each country must designate 'competent authorities' to receive information from commercial companies and research institutes. These authorities must organize inspection and other control measures. They must also examine the conformity of notifications received with the requirements of the directives. Effective risk management is expected, and it means that a careful risk assessment of contained use must be made, the appropriate level of containment must be exercised and suitable preventive measures must be taken (European Commission, 2000).

Directive 90/220/EEC covers deliberate release of GM organisms into the environment for research and development purposes and the placement on the market of products containing GMOs. The directive takes a preventive approach, emphasizing prior assessment and approval. The main elements are: (i) an environmental risk assessment must be carried out before any experimental or commercial releases of GMOs into the environment; (ii) no release can be carried out without the consent of the competent authority; (iii) an approval procedure by a nation's competent authorities should limit experimental releases to at most 90 days; and (iv) EU Community approval is required for commercial releases of GMOs (European Commission, 2000).

The latter procedure has been implemented as follows (Maurer and Harl, 2000). On receipt of the notification, the competent authority in a member country has 90 days to either forward the notification dossier to the European Commission with a favourable opinion or inform the notifier that the proposed commercial release does not fulfil the requirements of the directive. After receiving a notification dossier, the Commission immediately forwards it to the competent authorities of all other member states. If no objection is raised by the competent authorities of these states within 60 days after the commission forwards the notification, the competent authority that first received the notification issues a written consent to the applicant and informs the other member countries and the Commission of the consent. When member countries do not reach agreement on a notification, the Commission draws on relevant scientific committees for information and opinions. After weighing this information, the Commission makes a decision which is binding on all EU members.[12] Individual countries are also urged to have effective penalties for improper release of GMOs, but none has been established.

Public confidence in EU GMO policies has been undermined by recent information that unapproved GMOs have been sold in some EU countries and planted by farmers; for example, oilseed rape in several countries and cotton in Greece and Spain (Greenpeace, 2000c,d). The establishment of governmental policies that are unenforceable or weakly enforced does not build public confidence or trust in government. Directive 90/220/EEC is currently under revision.

GMO labelling was adopted by the European Commission in 1997. It requires each member

[12] For a company attempting to obtain commercial approval for selling GMOs, this is a relatively complicated and time consuming procedure in comparison to having one central EU agency make a decision.

country to enact a law requiring labelling of all new products containing GM organisms approved under Directive 90/220/EEC (i.e. where there are safety reasons). Under this policy, a GMO is defined as 'an organism in which the genetic material has been altered in a way that does not occur naturally by mating and/or natural recombination'. Also, mixtures with non-GMOs would indicate the possibility that GMOs may be present.

The EU's labelling policy has been controversial. First, proven safety risk evidence is generally lacking. Second, the scientific meaning of 'none' has been controversial in an era where rapid advances in scientific instrumentation has made it possible to detect ever smaller units, and accidental physical mixing of GMO and non-GMO commodities can easily occur. Third, the requirements for restaurants and caterers is so low as to be virtually useless to interested consumers.

Conclusions

This chapter has addressed the economics of information as it affects the acceptance of and benefits from agricultural biotechnology. We have shown that the producers and consumers can make good decisions on acceptance and use of GM products if there is freely accessible, verifiable information and competition in the provision of information by interested parties. However, when information is not verifiable and decision makers must rely on interested parties, achieving good decisions is much more difficult. Ag-biotech companies, environmental groups and others seem likely to have interests that diverge from those of consumers and producers and to use private information strategically, especially when the supply of new GM products is advancing rapidly. Furthermore, rapid advances in communication and information technologies have greatly reduced the cost and increased the speed with which information can be distributed. The private information that these groups have can cause the market for GM products in one way or another to collapse. This may be at considerable loss in social welfare.

We have argued that the services provided by an institutionalized information verification system operated by an independent body would be a mechanism for producing good public information services that would have large social value. This institution would reveal private information, establish a process for refuting or confirming claims of interested parties and advance the stock of knowledge on short- and long-term effects of biotechnology. Advances in the stock of knowledge are important to a successful information verification system because it can expand the topics, issues or dimensions of the knowledge base on which verification can be made. The primary contributors to this activity seem likely to be scientists employed by commercially independent institutions, funded primarily by the public sector and working to meet scholarly and scientific standards.

The relevant categories of information would cover topics consistent with a broad range of income and intergenerational interests. To obtain and maintain public trust, it must be 'consumer driven' and broader than pure scientific issues, although they would be one important component. It would, however, include scholarly presented and summarized information on ethical, social, economic, environmental, food safety, scientific and trade issues dealing with GMOs.

Verified information is costly to produce, but once provided, it has international public good attributes. We have argued that the advances in information and communication technologies have increased the free-rider problem by weakening the incentives for any one country to undertake such activities. We suggest that the provision will most likely be by some large country that stands currently to receive a significant share of the net social benefits from a verifiable information system for ag-biotechnology.

Currently, public institutions dealing with regulating GMOs and labelling in the USA have relatively high public trust but none fulfils our conceptual view of a verifiable information system. Public bodies in Europe are experiencing low public trust as an information source, and this is undoubtedly making good decision making more difficult. An effective information system dealing with biotechnology must be managed well in order to provide large social benefits.

References

Akerlof, G. (1970) The market for lemons: quality uncertainty and the market mechanism. *Quarterly Journal of Economics* 84, 488–500.

Alston, J.M., Norton, G.W. and Pardey, P.G. (1995) *Science Under Scarcity: Principles and Practice for Agricultural Research Evaluation and Priority Setting.* Cornell University Press, Ithaca, New York.

Avery, C., Resnick, P. and Zeckhauser, R. (2000) The market for evaluations. *American Economic Review* 89, 564–584.

Convention on Biological Diversity (1993) User's guide to the clearing house mechanism. www.biodiv.org/chm/userguide/CHM_User_Guide.html

Cornes, R.C. and Sandler, T. (1996) *The Theory of Externalities, Public Goods and Club Goods,* 2nd edn. Cambridge University Press, New York.

David, P.A. (2000) The political economy of public science. In: Smith, H.H. (ed.) *The Regulation of Science and Technology.* Macmillan, London.

European Commission (2000) *Guidelines to the Approximation of European Union Environmental Legislation.* Part 2G. *Chemicals and Genetically Modified Organisms.* European Commission, Brussels.

Falck-Zepeda, J.B., Traxler, G. and Nelson, R.G. (2000a) Surplus distribution from the introduction of a biotechnology innovation. *American Journal of Agricultural Economics* 82, 360–369.

Falck-Zepeda, J.B., Traxler, G. and Nelson, R.G. (2000b) Rent creation and distribution from biotechnology innovations: the cause of Bt cotton and herbicide-tolerant soybeans in 1997. *Agribusiness.*

Frey, K.J. (2000) *National Plant Breeding Study-IV. Future Priorities for Plant Breeding.* Special Report 102, Iowa State University, Iowa Agri. Home Econ. Experiment Station.

Friends of the Earth (2000) Genetically engineered food: farm scale trial locations. Friends of the Earth, GM sites. www.foe.co.uk/camps/foodbio/sites/index.html

Gaskell, G., Bauer, M.W., Durant, J. and Allum, N.C. (1999) Worlds apart? The reception of genetically modified foods in Europe and the U.S. *Science* 285, 384–387.

Gibbons, R. (1998) Incentives in organizations. *Journal of Economic Perspective* 12, 115–132.

Golan, E., Kuckler, F. and Mitchell, L. (2000) *Economics of Food Labeling.* US Department of Agriculture, Agricultural Economics Report No. 793.

Greenpeace (2000a) EPA responds to Greenpeace lawsuit defends rubber stamping of bud science?' Greenpeace USA Media Center, 20 April.

Greenpeace (2000b) Kellogg's tiger targeted in action condemning use of Franken Food. Greenpeace USA Media Center, 23 May.

Greenpeace (2000c) Illegal GE crops found in several European countries: Greenpeace calls for urgent action to prevent further imports. Greenpeace Genetic Engineering Highlights, 19 May. www.greenpeace.org/-geneng/highlights/gmo/00_5_19.html

Greenpeace (2000d) Greenpeace and farmer decontaminate oil seed rape field in Germany. Greenpeace Genetic Engineering Press Release, 29 May. www.greenpeace.org/pressreleases/geneng/2000may29.html

Heijs, W.J.M. and Midden, C.J.H. (1996) *Biotechnology. Attitudes and Influencing Factors: Fourth Survey.* Eindhoven University of Technology, Department of Philosophy and Social Science, Eindhoven, Netherlands.

Hoban, T.J. (1997) Consumer acceptance of biotechnology: an international perspective. *Nature Biotechnology* 15, 232–234.

Hoban, T.J. (1999) Consumer acceptance of biotechnology in the U.S. and Japan. *Food Technology* 53, 50–53.

Holmstrom, B. (1979) Moral hazard and observability. *Bell Journal of Economics* 10, 74–91.

Holmstrom, B. (1989) Agency costs and innovation. *Journal of Economic Behaviour and Organization* 12, 305–327.

Holmstrom, B. and Milgrom, P. (1987) Aggregation and linearity in the provision of intertemporal incentives. *Econometrica* 55, 303–328.

Huffman, W.E. (2002) Human capital: education and agriculture. In: Gardner, B.L. and Rausser, G.E. (eds) *Handbook for Agricultural Economics,* Vol. 1. Elsevier Science, Amsterdam (in press). (Also, Iowa State University, Dept. Economics Staff Paper No. 222, August 1999.)

Huffman, W.E. and Evenson, R.E. (1993) *Science for Agriculture: a Long Term Perspective.* Iowa State University Press, Ames, Iowa.

Huffman, W.E. and Just, R.E. (2000) Setting efficient incentives for agricultural research: lessons from principal-agent theory. *American Journal of Agricultural Economics* 82, 828–841.

Macilwain, C. (1999) U.S. Food-safety body hears protests over genetically modified food. *Nature* 402, 571.

Martin, S. and Tait, J. (1992) *Release of Genetically Modified Organisms: Public Attitudes and Understanding.* The Open University, Centre for Technology Strategy, London.

Maurer, M.T. and Harl, N. (2000) *Obtaining Approval to Market GMOs in the European Union.* Working paper, Iowa State University, Department of Economics.

Milgrom, P. and Roberts, J. (1986) Relying on the information of interested parties. *RAND Journal of Economics* 17, 18–32.

Molho, I. (1997) *The Economics of Information.* Blackwell, Malden, Massachusetts.

Moschini, G., Lapan, H. and Sobolevsky, A. (1999) *Roundup Ready Soybeans and Welfare Effects in the Soybean Complex.* Staff working paper No. 324, Iowa State University, Department of Economics.

National Research Council (2000a) *Transgenic Plants and World Agriculture.* National Academy Press, Washington, DC.

National Research Council (2000b) Committee on Genetically Modified Pest-Protected Plants. *Genetically Modified Pest-Protected Plants: Science and Regulation.* National Academy Press, Washington, DC.

Nettleton, J.A. (1999) The Frankenfood myth. *Food Technology* 53, 20.

OECD (2000) Rapporteurs' Summary: GM food safety: facts, uncertainties, and assessment. From *OECD Edinburgh Conference on the Scientific and Health Aspects of GM Foods,* 28 February–1 March.

Olson, M. (1965) *The Logic of Collective Action.* Harvard University Press, Cambridge, Massachusetts.

Sandler, T. (1992) *Collective Action.* The University of Michigan Press, Ann Arbor, Michigan.

Sandler, T. (1997) *Global Challenges: an Approach to Environmental, Political, and Economic Problems.* Cambridge University Press, New York.

Sandler, T. (1998) Global and regional public goods: A prognosis for collective action. *Fiscal Studies* 19, 221–247.

Scheermeier, Q. and Dickson, D. (2000) Europe lifts patent embargo on transgenic plants and animals *Nature* 403, 3.

Serageldin, I. (1999) Biotechnology and food security in the 21st century. *Science* 285, 387–389.

Shapiro, C. and Varian, H.R. (1999) *Information Rules: a Strategic Guide to the Network Economy.* Harvard Business School Press, Boston, Massachusetts.

Sobel, J. (1985) A theory of credibility. *Review of Economic Studies* 52, 557–573.

World Bank (1999) *Knowledge for Development.* World Development Report 1998/99. Oxford University Press, New York.

16 Science and Regulation: Assessing the Impacts of Incomplete Institutions and Information in the Global Agricultural Biotechnology Industry

Stuart Smyth and Peter W.B. Phillips
Department of Agricultural Economics, University of Saskatchewan, Saskatoon, Saskatchewan, Canada

Introduction

This chapter outlines some of the economic, regulatory and social costs of incomplete information. Incomplete scientific experiments have contributed to government regulations and standards that impose possibly excessive costs on biotechnology firms, producers and consumers. Classical economic theory posits that markets operate with perfect information; however, it is readily apparent that the marketplace for genetically modified (GM) agrifood products has an abundance of imperfect information. Concerns that have risen regarding GM products have had a global impact on government policy. This chapter offers an examination of the actual policy changes resulting from two incomplete and contentious research results.

Starting in the summer of 1998, biotechnology came under attack regarding the safety of the GM agrifood products that the industry was producing. This attack started with the public announcement by Dr Arpad Pusztai that GM agrifood products could be harmful to human health. Critics of biotechnology immediately began to advance this research as evidence that the long-term implications of biotechnology are not known and that restrictions should be immediately placed on further biotechnology developments. The speculative nature of Dr Pusztai's announcement resulted in a large, public backlash against the biotechnology industry that is still being felt today. The pressure on the biotechnology industry increased in the spring of 1999 with the announcement by Dr John Losey that GM *Bacillus thuringiensis* (Bt) maize was harmful to monarch butterfly larvae. The media presented this as evidence that multinational corporations are irresponsibly producing products that can unintentionally kill valued insects, such as the monarch butterfly. As in the first case, this research was flawed and subsequent research contradicted all of Dr Losey's claims. The damage, however, was already done. Many consumers now believe that biotechnology kills butterflies and the cost to the biotech industry to inform and educate society as to the facts will be enormous.

In each case, consumer concerns forced governments to respond. In the UK, the relevant regulatory agencies, which did not enjoy great public confidence, responded with mandatory labelling that imposes significant monetary and opportunity costs on consumers and the economy. Additionally, it is not clear that the policy prescription can adequately address consumer concerns. In the USA, in contrast, the Environmental Protection Agency

(EPA), which is a well established, relatively credible and respected institution, responded to the concerns about the monarch butterflies, but in a more measured way that appears to have minimized the costs relative to the benefits.

This chapter investigates the role of imperfect information and institutions in biotechnology markets. The following section describes the background and consumer attitudes towards GM products. The third section outlines the theoretical approach used to conduct the analysis. The fourth section presents the scientific experiments, the scientific community's reaction, the government regulations and standards that resulted from two scientific experiments and some of the costs that have developed. The final sections provide a short analysis of the developments using the model followed by some conclusions.

Background

Molecular genetics has revolutionized the agrifood business, converting the industry from a largely supply push, evolutionary system towards a demand-pull, engineered system. Biotechnology innovations entered global agrifood markets in 1995 with the introduction of recombinant bovine somatotropin (rbST) and herbicide-tolerant maize, cotton, canola and soybean varieties. Since then the rate of new product introductions has risen sharply, with more than 13 crops already transformed by more than 40 traits. James (1999) estimates that in 1999 GM crops were being produced in 12 countries and that more than 40 new varieties involving input and output traits are in the R&D pipeline and will probably be ready for commercialization before 2005. Given the potential for stacking both input and output traits, the potential permutations and combinations leading to new products are enormous.

A large number of consumers are or may be consuming these products. Maize, soybeans and canola are extensively traded throughout the world and their processed elements are essential ingredients incorporated into an estimated 60% of the processed foods on grocery store shelves. Consumer attitudes and preferences about GM products vary widely both within and between key markets, reflecting the differing views about human health, environmental safety and product quality. People with different experiences and interests have widely different perspectives on these products. *The Economist*/Angus Reid World Poll (1999) revealed that consumers in Germany and the UK were the most aware of GM foods (95% and 94%) while US consumers were least aware (66%). Related to that, 82% of German consumers and 67% of UK consumers but only 57% of US consumers would be less likely to buy GM labelled products, while 5% of German, 25% of UK and 37% of US consumers were indifferent to GM elements in foods. The rest of the countries surveyed (Australia, Canada, France and Japan) fell in the middle.

These differing consumer attitudes are both a reflection of and reflected in national regulatory systems. Parallel to the introduction of these new products, government regulatory systems in key producing and consuming regions began to diverge. Canada and the US adopted regulatory standards and processes that relatively efficiently reviewed the objective, science-based risks of the products and approved products for environmental release, feed use and human consumption (Isaac and Phillips, 1999). In contrast, regulatory failures in the European agrifood system heightened concerns about biotechnology based agrifood products and destroyed confidence in existing regulatory authorities, forcing the European Union (EU) to develop a new, purpose-built regulatory structure for biotechnology products. Initially this system appeared to have the potential to deliver decisions consistent with North America, but no new products have been approved since October 1998 and none is likely to be approved until recently proposed revisions to the EU system are approved, now expected in 2002.

Against this backdrop, two new studies were released in 1998 and 1999 that suggested agricultural biotechnology might have significantly higher risks than previously anticipated. The UK and US systems responded significantly differently, partly based on the degree of public confidence in the regulators. This was not the first time the regulatory systems have been challenged. There are numerous instances in the USA and Europe (and other countries as well) where safety systems have failed. In the USA there have been a number of environmental failures, including the Three Mile Island nuclear accident of 1979, the environmental damage that resulted from long-term use of DDT and the pollution problem surrounding the 'Love Canal'. In Europe there have been numerous food safety failures, including the scare surrounding the use of hormones in livestock production in Italy in the 1970s, the presence of listeria in unpasteurized cheese, the evidence of antifreeze in certain types of wine in the 1980s, and the BSE disaster of the 1990s. The long-term impacts of these failures of risk analysis systems have differed between Europe and the USA.

The result in Europe is that the general public does not trust the institutions responsible for food safety. This lowered trust has developed over time due to previous food safety concerns and culminated with the bovine spongiform encephalopathy (BSE) problem. At the time, in the early stages of the BSE problem the British government and scientists told consumers that the beef was safe to eat and there were no concerns regarding linkages to human illnesses. As more information became available it became clear that there were linkages to human illnesses and consumers were advised not to eat beef, which only served to discredit the ability of the food safety institutes in the eyes of consumers. Europeans have developed a risk aversion when faced with the choice of consuming products that are safe where there is minimal factual information. The low level of trust that Europeans have in their food safety institutions forced the EU to legislate a tolerance level of 1% in an attempt to prevent GM agrifood products from entering Europe. In the UK the level of trust is even lower than that of other European nations and public pressure surrounding GM foods forced the British government to enact regulations that forced the food industry to label all ingredients as GM if they contained greater than 1% GM content.

The general populace in the USA has higher levels of trust in the institutions responsible for public safety. This level of trust has been fostered by the actions that followed from the above-mentioned failures. Following the Three Mile Island accident, regulations and standards were thoroughly reviewed and revised to ensure an even greater level of safety. Much of this was conducted in the public domain, which allowed those concerned to witness that the institutions were responding in a prudent manner to the concerns of the nation. The result of this is that the American public believes that their food safety institutions have the ability to ensure that products in the marketplace are safe and that there is minimal risk to consumers when they purchase these products. Even so, when concern about the impact of Bt maize on non-target organisms developed in 1999, the US Government responded by strengthening the standards regarding planting of Bt maize. While the new rules are arguably somewhat beyond what might be needed to manage the risks, the incremental costs were controlled. The American public responded favourably to these new standards due to the high level of trust that they have in their institutions.

This chapter discusses the institutional challenges, their responses and the different financial and opportunity costs of the two regulatory responses.

The Theory of Risk and Markets

Economic theory has tended to ignore many of the issues related to product quality. The Marshallian demand curve is usually specified as a function of income and the product price, occasionally supplemented with cross price elasticities for substitutes and complements. Tastes and preferences, which are fundamental factors in determining demand, are often assumed to be homogeneous and/or static.

Product quality is often ignored or assumed to be reflected in price and income elasticities. Quality of new products is especially problematic. Irrespective of the fundamental features of any new products, all new products almost by definition incorporate experiential and credence elements (Tirole, 1988). Trust is a vital element in creating and managing markets for those products.

Firms are often not able to create, by themselves, the conditions of trust that generate the socially optimal quantities of goods and services produced and consumed. Publicly managed risk analysis systems are vital to creating that trust. Van den Daele *et al.* (1997) identifies three types of risk that affect the safety of product and consumer perceptions of those risks:

- Probabilistic risks involve those theoretically grounded and empirically demonstrated risks related to the product or its technology. The methods and much of the evidence is available in peer-reviewed journals or public records.
- Hypothetical risks, in contrast, involve those possibilities that are grounded in accepted theory but lack empirical experience or evidence that can establish probabilities. Most of these can be identified in academic literature.
- Speculative risks, in contrast to the other two areas, have neither established theory nor experience to back them up. Those speculative risks that have much basis can often be found in working papers or other developing literature.

Beyond that, almost any correlations can be made to show the potential for risk, irrespective of whether there is any theoretical basis for the possibility.

New products can raise concerns about all three types of risk. Effective risk analysis systems encompass elements that work to assess, manage and communicate the risks to consumers and citizens:

- Risk assessment, which involves an *ex ante* evaluation of risks and benefits of GM foods, is usually based on an objective examination of the science on a case by case basis.
- Risk management, which involves the human system of rules and procedures, is designed to contain products during the research and commercialization phase and undertake post-release monitoring and surveillance.
- Risk communication, which involves dissemination of information and messages in an effort to narrow the gap between actual risks and the perceptions of risk.

As one might imagine, most risk analysis systems should be able to effectively handle risk assessment, which is relatively mechanistic and should also be able to handle risk management. Risk communication, however, is extremely complex for issues such as GM foods because of the wide array of speculative issues that science cannot adequately respond to. In the absence of complete certainty about, or at least significant experience in, a new technology or product, risk analysis systems must inherently be based on trust. In turn, trust is instilled as a result of reactions to risk assessments that are carried out by institutions. Trust grows as risk management procedures address public concerns. Finally, trust is firmly embedded in the public via the success of the risk communication process that develops over time. Strong institutions are crucial for successful and trustworthy food safety systems.

Frequently when food safety institutions attempt to deal with risk they adopt the approach of risk elimination, which results in costly, unnecessary regulations and/or standards. There will always be risk associated with everything individuals do. The key is to lower the level of risk until it is within an individual's comfort zone. This is the case when dealing with food safety. The utmost care is taken to ensure that products on store shelves are as safe as possible; however, there is still a minimum level of risk that must be faced when purchasing food.

Traditional risk assessment theory suggests that risk is a combined measurement of the length of exposure multiplied by the level of adverse effects of the agent to other organisms, or hazard. This can be expressed in the following formula:

$$\text{risk} = \text{hazard} \times \text{exposure}$$

If the time of exposure is brief (fractions of a second) and the level of hazard is a high dosage, the level of risk would be low or minimal. Science has used this formula to evaluate whether initial research findings should proceed or be halted. If the assessment was conducted and the level of risk was determined to be high, then government agencies would not approve the technology or product for release. Experts will naturally have a different view about the level of risk associated with a new product or technology from that of the public. As a result, experts are often confused by consumer reactions to new products and technologies.

Sandman (1994) has argued that regulators should instead use the following formula for understanding consumer perceptions of risk:

$$\text{risk} = \text{hazard} \times \text{outrage}$$

Sandman believes the old formula underestimated the actual level of risk because it ignored outrage. Public concern is focused on whether the risk is acceptable rather than on the scientific perceived level of public risk. This has important implications for risk communication, as food safety institutions must address consumer outrage in their response to the risk assessment.

Craven and Johnson (1999), in writing about food scares, discuss the new approach that risk communicators must take when faced with consumer concerns.

> The science of risk communication is still relatively new, though valid and effective precepts have been clearly defined. Mitigating a hazard itself does not mitigate the outrage about the hazard. To defuse a scare, outrage must be addressed, that is, the public's particular concerns must be addressed and dealt with in a way responsive to their emotional needs regarding the issue.
>
> (Craven and Johnson, 1999)

The difficulty for risk communicators is to determine which scares, whether they are food related or not, will provoke an outrage response from consumers. It is clear that those responsible for risk communication in the biotech industry and food safety institutions were not anticipating the consumer outrage that developed in the UK and Europe over the introduction of GM agrifood products.

Snow (1997, cited in Craven and Johnson, 1999) suggests four critical elements necessary in any approach for risk communicators when dealing with consumer outrage. First, the communications must be open and fully disclose the threat and issues. Second, the communicator must acknowledge responsibility for the issue. Third, the communicator must be courteous (even if the public response is angry and impolite). Fourth, the communicator must demonstrate compassion, by respectfully recognizing and addressing people's fears and apprehensions.

This chapter examines the US and UK regulatory responses to a changed view of risk resulting from incomplete, speculative science. The reason for the failure of British and European food safety systems to respond to consumer fears over GM foods becomes clearer when compared with these four approaches: the British and European food safety institutions acted in the complete opposite manner, as is described above. Had these food safety institutions been better risk communicators, it might not have been necessary to implement the costly labelling regulations described below. In contrast, the US risk communicators responded more to the outrage than the hazard, effectively lowering the cost of maintaining trust in the products and the regulator.

Hypothetical and Speculative Science and Risk Analysis

An extensive set of standards and regulations governs the food sector nationally and internationally, but the rules for biotechnology-based products are still evolving. The incomplete institutional structure for managing risks of new products has been unable to effectively handle new, often incomplete scientific information. Recently the biotechnology community had its foundation shaken by the release of scientific data claiming that biotechnology may be producing products that are not as safe as the biotech industry was claiming. The research on rats fed GM potatoes and monarch larvae resulted in new regulations and standards for biotechnology.

Regulations and standards that affect biotechnology have been implemented by several national governments. These regulations and standards have been implemented based on the results of often-incomplete scientific research. The focus of this portion of the chapter will be to examine the impacts that have resulted from Dr Pusztai's rat/GM potato research and Dr Losey's monarch butterfly larvae/Bt maize pollen research. Neither of these scientific research papers was peer reviewed prior to being publicly released and both created media sensations. This carried over to consumer concerns, which resulted in governments developing and implementing regulations and standards that were designed to show consumers that domestic governments were doing something to address the potential problems being caused by biotechnology. The resulting regulations and standards were implemented based on incomplete information.

Dr Pusztai's GM potato research

The Scottish Office of Agriculture, Environment and Fisheries Department (SOAEFD) commissioned the Rowett Research Institute in Scotland to study the effects of two specific strains of transgenic potatoes. The principal investigator was Dr Pusztai, who was on a 3-year research grant at the Institute. The potatoes had been engineered to produce lectins that are toxic to insects that feed on potatoes. The research was conducted by feeding GM potatoes and non-GM potatoes to different groups of rats. The study comprised four experiments: (i) a control group fed raw potatoes for 10 days; (ii) a group fed raw GM potatoes for 10 days; (iii) a group fed cooked GM potatoes for 10 days; and (iv) a group fed raw and cooked GM potatoes for 110 days. Each experiment group had five or six rats and at the end of each experiment the rats were killed and their organs were weighed and examined for differences.

After the organs had been weighed and examined, Dr Pusztai concluded that he 'is of the opinion that the existing data fully support our suggestion that the consumption by rats of transgenic potatoes... has significant effects on organ development, body metabolism and immune function' (SOAEFD, 1998). He went on to say that in all four experiments, the rats fed raw and cooked GM potatoes experienced 'major and in most instances highly significant changes in the weights of some or most of their vital organs'. The vital organs that he refers to are the liver, spleen and thymus. Dr Pusztai concludes by saying that GM potatoes are not substantial equivalents to conventional potatoes.

In August 1998, Dr Pusztai appeared on a British television programme and stated that the consumption of GM potatoes caused rats to grow at a slower rate and impaired their immune system. Three days later the Rowett Institute suspended him, prohibited him from accessing any of his data and requested that he not speak publicly about his experiments. The Rowett Institute set up an audit committee to provide a detailed analysis of Dr Pusztai's experiments and this committee reported in October 1998. The report stated that the genetic modification 'did not have any deleterious effects on the growth of rats in three short-term and one long-term feeding experiments even when added at 100-times the concentration expressed in the tubers of the transgenic products' (Nuffield Council on Bioethics, 1999). In February 1999, Dr Pusztai attended a media conference at the Houses of Parliament in London that was organized by Friends of the Earth. At this news conference, Dr Pusztai claimed it was not the lectin gene that had been transformed that caused the damage but rather the viral promoter that was used. By making this claim, Dr Pusztai effectively condemned GM foods due to the fact that most genetic transformations use viral promoters. In May 1999, the UK's Royal Society released their analysis of Dr Pusztai's experiment, which found no evidence linking the consumption of GMO-potatoes in rats to harmful effects. For the next few months the debate between both sides on this issue was carried out in the British newspapers, with the main complaint being that the paper had never been peer reviewed.

In October 1999, *The Lancet,* a well respected scientific journal, announced that it would publish Dr Pusztai's research. *The Lancet* had the paper peer reviewed by six scientific experts, who all found 'the study to be defective in design, execution and interpretation (*Independent,* 1999)'. A majority of the six reviewers, which is twice the usual number of reviewers, recommended that Dr Pusztai's research not be published. *The Lancet* overruled all of this advice and decided to publish the results of the experiments in the 16 October 1999 edition.

The scientific reaction

Those who have had the opportunity to review Dr Pusztai's experiments have determined that the experiments were so poorly conducted that it was virtually impossible to make any factually based conclusions. 'In these control experiments the most striking result was how poor a diet raw potatoes was for the animals ... the animals fed raw potatoes did not gain any weight. They seemed actually to be close to starvation' (Braun, 1999, personal communication). It was discovered by one of the reviewers that Dr Pusztai started the long-term experiment with raw potatoes and when they began to show signs of starvation, he switched to boiled potatoes. This switch meant that the results of the long-term trial are inconclusive in terms of revealing useful scientific data. Additionally it was revealed that the potatoes Dr Pusztai was working with were never intended to be used as a table potato, rather they were developed for research purposes. The research was part of standard toxicology tests that are routinely conducted when developing new varieties.

In the course of its audit of Dr Pusztai's experiments, the Rowett Institute found that when he spoke on TV in August 1998 and said that GM potatoes lowered the responsiveness of rat's immune systems, the reality was the relevant experiments had not yet been concluded. In fact, when the vital organs were examined, it was found that in all experiments but one, the weights were actually well within the standard deviation range for those organs. One expert that analysed the research states in his overview that:

> there are no significant differences in growth rate and organ weights which could reasonably be attributed to feeding transgenic potatoes to the rats. No two experiments in the series were done in the same way, so a detailed comparison of the reliability of the data is impossible or at least very difficult.
> (Braun, 1999, personal communication)

The Royal Society in the UK reviewed the experiments and concluded that they were so poorly structured, executed and analysed that no conclusions could be drawn. This review expanded on the poor nutritional value of potatoes by stating that the 'GM potatoes used contained almost 20 per cent less protein than unmodified potatoes, and rats in the long-term feeding study were given additional protein to avoid starvation' (*Nature,* 1999). In addition to this they point out that the results of one potato study cannot be imposed on to other GM agrifood products. Each individual GM agrifood product needs to be examined and analysed on its own merits.

The final reaction, and indeed possibly the largest, was the outrage expressed when *The Lancet* decided to reject the advice of its peer review committee and publish Dr Pusztai's research. Highly respected research scientists in the UK openly condemned this and questioned the value of peer review if journal editors decide to publish material based on consumer interest. Typically the peer review process is an anonymous one that allows experts to make critical, unbiased assessments of experiments. Professor John Pickett, one of the peer reviewers for *The Lancet*, was so infuriated by what the journal's editors did that he broke with anonymity and publicly stated that '[i]t is a very sad day when a very distinguished journal of this kind sees fit to go against senior reviewers' (*Independent*, 1999).

Resulting government regulations

The media coverage this research received encouraged consumers to reject biotechnology. Due to the earlier government and scientific fiasco regarding mad cow disease and British beef, many consumers have lost confidence in what government and scientists tell them about food safety. When Greenpeace and Friends of the Earth used Dr Pusztai's research to proclaim that all GM agrifood products are unsafe to eat, they generated a huge consumer backlash against GM foods and biotechnology in general. In an attempt to calm a nervous public, the British Government announced that it would require mandatory labelling of all products regarding GM content. The British Government did this at the same time as Prime Minister Tony Blair, senior cabinet members and government officials were publicly advocating the safety of GM foods.

The British regulation requires all food products that contain more than 1% GM content to be labelled, including all restaurant meals, and there are financial penalties for non-compliance (Phillips and Foster, 2000). The result of this legislation has forced many British supermarkets to drop products that do not comply, thus reducing product availability to consumers. The products that can prove there is no GM content are labelled as such but consumers pay between 10 and 15% more for these products (BBC Online Network, 30/10/99: www.news2.thls.bbc.co.uk). The early indication of this legislation is that consumers have fewer food choices at a higher cost.

The incomplete information that flowed from Dr Pusztai's experiments in part contributed to the creation of mandatory labelling regulations. The unfortunate aspect of this legislation is that it not only negatively affects British consumers but also restricts the ability of other nations to export products to the UK. Both Canadian and American firms have lost British market access because the ability to segregate to the level of less than a 1% GM tolerance level is not presently available. In this instance, Dr Pusztai's decision to circumvent the traditional peer review process and use the media to release his findings has truly had global implications.

The financial and economic cost of the regulations

There are two approaches to estimating the impact of the UK labelling system. In the first instance, one can look at the financial costs of running the system. Alternatively, one can look at the opportunity cost of the new rules.

When the British Government enacted its legislation that forced labelling on the food industry, the policy was developed quickly to address consumer fears and as a result, very little thought was given to the financial cost of implementation. The Ministry of Agriculture, Fisheries and Food (MAFF) was responsible for the labelling legislation and it stated that the

> majority of businesses affected by the new Regulations will already be labelling food under EC Regulation 1139/98. The additional cost, for reprinting menus etc. is unlikely to be any higher than similar costs incurred in respect of the implementation of EC Regulation 1139/98, which was estimated at £1M–£2M. (UK MAFF, 2000)

The British Hospitality Association (BHA) has calculated the cost of reprinting menus to be as high as £80 million. One small business expects that annual costs for keeping information up to date could reach £50,000 and larger firms are expecting annual update costs of £2 million. The BHA reported that MAFF officials were surprised to discover how high the costs would actually be and further surprised to learn that the costs would be passed on to consumers.

Testing for compliance became the responsibility of individual counties or boroughs. On 1 April 2000 the Food Standards Agency (FSA) became the accountable institution for food safety. Previously, this was the responsibility of MAFF. Due to the low level of trust in MAFF, a new institution had to be established to deal with the issue of labelling GM agrifood products. Each county or borough has the responsibility to fund its own Standards Office. The result of this is that the level of testing and enforcement of the new labelling regulations is directly linked to the amount of funding that a local borough is able to allocate to its Standards Office. Boroughs that do not have adequate financial resources to fully staff a Standards Office may face the situation where the Standards Office conducts no tests and is only capable of minimal regulation enforcement.

The principle reason for this is the cost of conducting tests on processed food products. At present in the UK, the cost is £117 per test. It takes on average 25 days to obtain results from the test and the test is only capable of determining whether there is, or is not, GM material. This is problematic as the 1% tolerance level will allow some GM content and when a business is notified that there is GM material in its processed food, the immediate question is about the level of GM content. One borough, Reading, conducted 52 sample tests between 1 January 1999 and June 2000 and found that about 10% of samples did not meet the requirements of the UK regulations. While each borough has the power to take legal action for non-compliance, what is happening is that a due diligence process is initiated to try to determine the source of GM material and what GM material is specifically involved. To date at the Reading Standards Office, no legal action has been taken and it is viewed as a measure of last resort.

In the UK there is clearly a political issue and a labelling issue surrounding GM agrifood products. The FSA has been established to enforce the labelling issue. While the regulations that are spelled out by legislation are very clear, enforcing them is becoming very difficult. Due to the limited resources some boroughs have devoted to this area, it may take as long as 5 years to investigate complaints. Because each borough is required to fund the Standards Office, some have made this a very low priority.

From an economic perspective, the new rules have replaced one economically inefficient system with another economically inefficient system. Economic theory clearly shows that if consumers are not given a choice to express their preference, utility (consumer surplus) is reduced. So, from that perspective, the fact that labelling systems provide a choice to consumers who do not want to consume GM foods improves their welfare. In the UK system, however, the labelling rules have effectively driven out GM foods from the market. Given that 25% of UK consumers are indifferent to GM elements in their food, forcing them to consume (probably higher priced) non-GM foods similarly reduces their welfare.

Essentially, in spite of the best of intentions, the poorly constructed labelling rules in the UK have produced an expensive and suboptimal market situation, with virtually all GM foods excluded from the market.

Dr Losey's Bt maize pollen research

Dr Losey, a researcher in the Department of Entomology at Cornell University, New York, conducted a study in 1998 to determine the effects of Bt maize pollen on monarch butterfly larvae. Monarchs lay their eggs on milkweed leaves, which can grow near maize fields. The study comprised three larvae feeding scenarios: milkweed leaves with no pollen, milkweed leaves with conventional maize pollen and milkweed leaves with Bt maize pollen. A spatula was used to place the pollen on the leaves and they were dusted until they visually resembled pollen densities in the wild. The larvae were allowed to feed for 4 days at which time the survival rate and larva weights were recorded. After 4 days, the survival rate for the larvae on the Bt maize leaves was 56% compared with 100% for the other two groups. This led Dr Losey to state that '[b]ecause there was no mortality on leaves dusted with untransformed pollen, all of the mortality on leaves dusted with Bt pollen seems to be due to the effects of the Bt toxin' (Losey, 1999). Additionally, the surviving Bt larvae had weight gains that were less than half those of the larvae fed on the milkweed leaves with no pollen. Dr Losey concluded that:

> ...plants transformed with genetic material from...Bt are generally thought to have negligible impact on non-target organisms [monarch larve], Bt corn plants might represent a risk because most hybrids express the Bt toxin in pollen, and corn pollen is dispersed over at least 60 meters by wind. Corn pollen is deposited on other plants near corn fields and can be ingested by the non-target organisms that consume these plants. (Losey, 1999)

This research was published in the scientific journal *Nature*, on 20 May 1999. This article was not peer reviewed prior to being published, in contravention of the norm. At the time the article was publicly released, Dr Losey expressed caution about any definite conclusions resulting from the study. He believed that further studies needed to be conducted before any conclusive statements could be issued.

The scientific reaction

The scientific reaction to this research was swift and overwhelmingly negative. Other entomologists highlighted flaws and inconsistencies. What resulted were numerous studies in the summer of 1999 that set out to test the validity of Dr Losey's initial findings. The major critiques are summarized below.

1. Dr Losey stated that pollen could drift up to 60 m. This was based on research done in 1972. With new varieties being released for over 25 years, the weight of pollen has increased to the point that most pollen only drifts 2–3 m. One study documented that 90% of the pollen landed with 16 feet of the field perimeter (*Chicago Tribune*, 1999).
2. Mark Sears at the University of Guelph conducted lab experiments and determined there were measurable toxic effects on larvae at pollen densities greater than 100 grains of pollen cm^{-2}. When milkweed plants were examined near maize fields, those at a distance of 1 m from the field were found to have 28 grains of pollen cm^{-2} and at 5 m the measurable level of pollen dropped to zero (Sears and Stanley-Horn, 2000).
3. Farmers spray for weeds, including milkweed, before monarchs lay eggs. This would effectively control milkweed in the maize field and, depending on the drift of the spray, for a short distance around the edge of the field.[1]
4. Dr Losey used pollen from the Bt maize variety that happened to have the highest level of Bt in the pollen, which makes up a very small percentage of the maize acreage grown in North Dakota, which is largely outside the monarch range (Downey, 1999).
5. Dr Losey suggests that usually larvae hatch and pollen shed occur over coinciding dates. Researchers in Nebraska found that 95% of the pollen had spread before the commencement of the larvae hatch (*St Louis Post-Dispatch*, 1999).
6. An Iowa State University researcher noted that monarchs typically prefer to lay their eggs on milkweed plants that are in the open and on plants that range in size from 3 inches to 18 inches in height. Plants of this size would rarely be found growing within maize fields, rather they would typically grow in ditches and fencelines (Rice, 1999).

These follow-up studies found that there was an extremely small correlation between the mortality of monarch butterfly larvae and toxic effects from Bt maize pollen. Had Dr Losey allowed for a peer review of his 1998 findings, a great deal of incomplete information about the effects of Bt maize would not have been disseminated.

Resulting government regulations

Dr Losey's article in *Nature* touched off a firestorm of criticisms from opponents of biotechnology. News magazines such as *Science News* screamed the headline 'Bt-Corn pollen can kill monarchs' (22 May 1999). Jeremy Rifkin, head of the Washington based Foundation for Economic Trends, called for an outright ban on the release of any new GM crop varieties until the potential environmental impacts were more fully known. The National Corn Growers Association received letters from school children demanding that the mean farmers stop killing butterflies (*Scientific American*, 1999). Concern quickly developed among the American public, leading the Environmental Protection Agency (EPA) to issue new standards for producers intending to grow Bt maize.

The incomplete information generated by Dr Losey's research contributed to the EPA developing and implementing mandatory planting restrictions for Bt maize. The new measures for insect resistance management that applied to Bt maize growers for the 2000 crop year were:

1. Required to plant a minimum structured refuge of at least 20% non-Bt maize;
2. Required in areas where Bt maize and cotton are grown to ensure a structured refuge of 50% non-Bt maize;
3. Expected to closely monitor fields to detect any potential resistance;
4. Required to communicate voluntary measures that will protect non-target insects, particularly the monarch butterfly; and

[1] Letter from Jeffery Stein, Director of Regulatory and Government Affairs, Novartis Seeds, Inc. to Dr Losey, 3 May 1999.

5. Outright restriction in planting Bt maize in some geographic locations (EPA web site, 2000).

These new rules were initially viewed as burdensome by many in the industry and many in the maize industry speculated that the rules could cause some maize producers to reduce their acreage of Bt maize. Subsequent scientific research conducted in the summer of 1999 suggests that the risk to monarch butterflies could have been managed by Bt maize producers maintaining a 50-foot summerfallow strip around the edge of the field. This would ensure that no milkweed plants could grow close to the maize field and that virtually all of the pollen would not drift beyond this refuge distance.

Economic and financial cost of the regulations

The higher level of trust in the American institutions responsible for food and environmental safety meant that the concern that developed from Losey's research could be addressed through existing institutions. The EPA increased the standards pertaining to refuge requirement for producers wanting to plant Bt maize. When these increased requirements were announced in December 1999, there was concern expressed by the maize industry that there would be lower levels of Bt maize planted. Even by March 2000, the industry believed that Bt maize planting would be down by as much as 15% over the previous year. The most recent information shows that overall maize planting is up by 3% and Bt maize planting (Table 16.1) is down slightly, which is generally attributed to the natural cycle of the European corn borer. The fact that Bt maize production is relatively stable following these new standards would indicate that they closely mirror what the industry was actually practising.

Table 16.1. Percentage of all GM maize and Bt maize varieties.

Year	All biotech varieties	Insect resistant (Bt) only
1998	26	18
1999	30	25 (top 7 states only)
2000	25	18

Source: USDA: www.usda.mannlib.cornell.edu/reports/nassr/field/pcp-bba/acrg0600.txt; and Gianessi and Carpenter, 1999.

The question that arises from recently released research is, are the new standards an over-reaction? Shelton *et al.* (2000) examined the issue of managing insect resistance in Bt-engineered plants and found that using a 20% refuge, which is allowed under the rules to be sprayed with the usual array of chemicals, reduces the resistance capabilities of the pest. If producers do not want to sacrifice such a large amount of their field to pest damage they would be better to move to the option of having a 4% refuge that is not sprayed.

> Within an individual field or farm, treating the refuge with a highly effective insecticide may dilute the abundance of susceptible alleles to such an extent that the refuge is rendered ineffective unless there is substantial immigration of susceptible alleles from wild hosts or from surrounding non-Bt crops. On the other hand, growers may be reluctant to sacrifice a large number of refuge plants to insects just to maintain susceptible alleles. An alternative to the strategy of having a 20% refuge that can be sprayed (the requirement for cotton [and Bt corn] is the EPA-approved strategy ... of having a 4% refuge that remains unsprayed.
>
> (Shelton *et al.*, 2000)

It is possible to estimate the economic impact of the formal regulations and Shelton's alternative. Bt maize has been adopted because it reduces the number of insecticide applications and increases yields; restricting its use causes losses. Gianessi and Carpenter (1999) document that Pioneer data shows that Bt maize yields were 17 bushel acre^{-1} higher than non-Bt maize in 1997 (35,000 comparisons) and 7 bushel acre^{-1} higher in 1998 (64,713 comparisons). While Gianessi and Carpenter calculated that average yield increases were much lower (they averaged the Pioneer results with other studies, which had as few as 84 observations), we have used a 2-year average of the Pioneer data, which provides a 2-year average increase of 12 bushel acre^{-1}. Table 16.2 illustrates what the losses may be with a sprayable 20% and sprayable 50% refuge. The price of maize (US$1.83 bushel^{-1}) was selected on 11 July 2000. The table also includes losses at non-sprayable 4% refuge for illustration purposes.

When the costs are summed using the 50/20 refuge requirement, they reach over US$72 million. While that seems high, it is significantly less than the opportunity cost of one of the alternatives, which was to ban the crop (which was the effect of the labelling laws in the UK). Gianessi and Carpenter estimate that Bt maize generated US$230

Table 16.2. Comparison of refuge costs.

	50% Refuge	20% Refuge	4% Refuge
% total Bt maize crop	10%	90%	90%
Acres of non-Bt refuge	715,000	2,574,000	514,000
Yield loss (12 bushel $acre^{-1}$)	8.58M bushel	30.89M bushel	6.18M bushel
Value of loss at US$1.83 $bushel^{-1}$	US$15.7 million	US$56.53 million	US$11.31 million

million total gross benefits to farmers in 1997 and 1998. While the net benefit was only US$46 million (low incidence of European corn borer in 1998 resulted in a small net loss), some of the US$184 million of farmer costs was profit to the owners of the technology and should be counted as net social gains (Moschini and Lapan, 1997). That said, the new regulations are more restrictive than may be necessary to achieve their objectives. The 50/4 refuge costs are only US$27 million but under the current mandatory regulations, that option cannot be pursued.

Analysis

Food safety systems must be able to handle multiple risks and uncertainties that occur at multiple stages in the food system, including probabilistic, hypothetical and speculative risks. The food risk analysis system must handle those risks through assessment, management and communications. The national and international regulatory systems for biotechnology based agrifood products is incomplete (especially the risk management and communications elements) and therefore is unable to handle hypothetical and speculative risks.

Returning to the earlier model that presented a structure for disentangling the debate about safety of GM foods, we can use it to demonstrate what has occurred with the incomplete information resulting from Dr Pusztai's and Dr Losey's research (Table 16.3). These experiments can be said to have produced speculative and hypothetical risks. Dr Pusztai speculated that consumption of GM potatoes could have harmful health effects, and he suggested this claim be applied to all other varieties of GM agrifood products. There is no theory that allows for claims against one food variety to be applied to other varieties without supportive research; therefore, his claims are speculative. Dr Losey's findings were hypothetical as they were only laboratory experiments and field trials needed to be conducted to validate his findings. There certainly was theory that suggests that Bt maize could hurt non-target organisms, so his claims were hypothetical.

This model documents how the approach taken by British institutions differed from the approach of US institutions when faced with similar problems. Due to the complete lack of trust in previous British food safety institutions, an entirely new institution had to be created to establish trust with consumers. In the USA a high level of trust already existed in food and environmental safety agencies, and as a result existing institutions were able to satisfactorily deal with public concerns. The key when dealing with incomplete information is to have strong institutions with existing structures that have a proven capability of successfully addressing rapidly evolving crises.

Table 16.3. Institutional incorporation of speculative and hypothetical risks.

Level of risk	Speculative risks of incomplete information on GM potatoes	Hypothetical risks of incomplete information on Bt maize
Risk assessment	Research showed reduced immune systems and lower organ weights	Research showed that Bt maize pollen had toxic effects on monarch larvae
Risk management	European Union implements a 1% tolerance level for raw and processed foodstuffs	EPA increases the refuge requirements for spring 2000 planting of Bt maize
Risk communication	Newly created British Food Standards Agency responsible for labelling	Publication of findings from numerous field experiments conducted in 1999

In the UK where the food safety network collapsed due to the BSE problem, an entire new food safety institution has had to be created to establish credibility in the minds of consumers regarding GM agrifood products. So far consumers view this institution as credible, at least partly due to the labelling legislation that was implemented. However, while this legislation is capable of clearly defining the process that needs to be followed, it is difficult to enforce. As the costs of labelling and providing information to consumers are passed on to consumers, there will be a threshold in terms of rising food prices beyond which consumers may begin to react adversely. Whether this adverse reaction will be enough to threaten the continuation of the new food safety institution is yet to be seen.

By comparison, in the USA, where a strong environmental safety institution existed, the problems that developed due to incomplete information were more readily assimilated into the existing institutional framework. The average American consumer appears to have accepted the EPA's actions as proper and reassuring. Consumers have continued to purchase GM agrifood products and do not face new price increases because of the change in Bt maize planting standards.

In the UK consumers expressed outrage when the safety of their food was questioned by new research. Consumers in America did not express concerns at anywhere near the level of those expressed in the UK. This can be attributed at least partly to the high level of trust that is present in the US food safety system.

Conclusions

While the incomplete information that developed from these two experiments was not entirely responsible for the resulting regulatory changes, the studies did contribute to concern and acted as catalysts for action. Had the unexpected media sensation that was created by this research not developed, it is questionable whether the new standards and regulations would have been implemented at the levels or times chosen. The research created a global media phenomenom, which forced food and environmental safety agencies to develop and implement action plans that did not consider all of the potential negative economic impacts. This was clearly the case in the UK where the British Government forced its labelling legislation into law to appease concerned consumers.

The absence of strong institutions to support food safety systems can lead to consumer outrage. British and European consumers have lost confidence in their food safety systems in the wake of the BSE problem, with the result that the British Government over-reacted with labelling legislation and tolerance levels in order to re-establish confidence in the minds of consumers. Reassurances by government and scientists were largely ineffective and in fact may have been viewed as offensive by those concerned given the claims that government and scientists made during the BSE debate. The incomplete information that resulted from Dr Pusztai's potato research is an example of institutional failure that led to an adverse effect on consumers.

The American food and environmental safety system faced crisis situations in the past and, as a result of this, their risk communication process has created a high level of trust among consumers. When faced with the incomplete information that resulted from Dr Losey's research the EPA responded thoughtfully through a strong institution and as a result there were limited effects on consumers and the economy. The EPA had previously allowed refuge requirements to be voluntary, but due to Dr Losey's research and other research regarding insects developing resistance to Bt, the refuge requirements were increased and became mandatory. The trust in this institution by the American public meant that unnecessarily restrictive standards were not created thereby increasing food costs for consumers.

In short, institutions matter, especially when dealing with new products that exhibit experiential or credence features. The stronger the institution, the greater its capability to deal with incomplete information and the lower the cost of inappropriate consumer and regulatory responses.

References

Chicago Tribune (1999) Monarch butterfly so far not imperiled: gene-altered corn gets an early ok in studies. *Chicago Tribune* 2 November, p. 4.

Craven, B. and Johnson, C. (1999) Politics, policy, poisoning and food scares. In: Morris, J. and Bate, R. (eds) *Fearing Food: Risk, Health and Environment.* Butterworth-Heinemann, Oxford.

Downey, K. (1999) Comments on the Environmental Benefits and Sustainable Agriculture through Biotechnology Symposium. Agriculture Canada, Washington, DC, 10–11 November.

The Economist/Angus Reid World Poll (1999) In: *The Globe & Mail*, 'Canadians wary of genetically altered food', 15 January 2000.

EPA (Environmental Protection Agency) (2000) Press release, 14 January 2000, Bt Corn Insect Resistance Management Announced for 2000 Growing Season. www.epa.gov/pesticides/biopesticides

Ewens, S. and Pusztai, A. (1999) Health risks of genetically modified foods. *The Lancet* 354, 684.

Gianessi, L. and Carpenter, J. (1999) *Agricultural Biotechnology: Insect Control Benefits.* National Center for Food and Agricultural Policy, Washington, DC.

Independent (1999) Scientists revolt at publication of 'flawed' GM study. *The Independent*, 10 October.

Isaac, G. and Phillips, P. (1999) Market access and market acceptance for agricultural biotechnology products. Paper presented to the ICABR Conference on The Shape of the Coming Agricultural Biotechnology Transformation: Strategic Investment and Policy Approaches from an Economic Perspective, Rome, 17–19 June.

James, C. (1999) Transgenic crops worldwide: current situation and future outlook. Paper presented at the Conference *Agricultural Biotechnology in Developing Countries: Towards Optimizing the Benefits for the Poor*, organized by ZEF and ISAAA in collaboration with AgrEvo and DSE in Bonn, 15–16 November 1999. www.zef.de/zef_englisch/f_veranstalt_biotech.htm

Losey, J. (1999) Transgenic pollen harms monarch larvae. *Nature* 399(6733), 214. www.nature.com

Moschini, G. and Lapan, H. (1997) Intellectual property rights and the welfare effects of agricultural R&D. *American Journal of Agricultural Economics* 79, 1229–1242.

Nature (1999) UK debates public's role in science advice. *Nature* 399(6733), 188.

Nuffield Council on Bioethics (1999) *Genetically Modified Crops: The Ethical and Social Issues.* Latimer Trend & Company, London.

Phillips, P. and Foster, H. (2000) *Labeling for GM Foods: Theory and Practice.* Paper presented to the 4th ICABR Conference on the economics of agriculture biotechnology. Ravello, Italy, 24–28 August 2000.

Rice, M. (1999) Monarchs and Bt corn: Questions and answers. 14 June. www.ipm.iastate.edu/ipm/icm/1999/6-14-1999/monarchbt-html

Sandman, P.M. (1994) Mass media and environmental risk: seven principles. In: Morris, J. and Babe, R. (eds) *Fearing Food: Risk, Health and Environment.* Butterworth-Heinemann, Oxford.

Science News (1999) Bt-corn pollen can kill monarchs. 15(5), 324.

Scientific American (1999) The butterfly effect. 281(2), 28–29.

SOAEFD (Scottish Office of Agriculture, Environment and Fisheries Department) Flexible Fund Project (1998) Report of project coordinator on data produced at the Rowett Research Institute. www.rri.sari.ac.uk/gmo/ajp.htm

Sears, M. and Stanley-Horn, D. (2000) Impact of Bt corn pollen on monarch butterfly populations. In: Fairbairn, C., Scoles, G. and McHughen, A. (eds) *The Biosafety of Genetically Modified Organisms.* University Extension Press, Saskatoon, Saskatchewan.

Shelton, A., Tang, J., Roush, R., Metz, T. and Earle, E. (2000) Field tests on managing resistance to Bt-engineered plants. *Nature Biotechnology* 18(3), 339–342.

St Louis Post-Dispatch (1999) Scientists discount threat to butterfly from altered corn. *St Louis Post-Dispatch*, 2 November, p. A5.

Tirole, J. (1988) *The Theory of Industrial Organization.* MIT Press, Cambridge, Massachusetts, p. 106.

UK Ministry of Agriculture, Fisheries and Food (2000) *The Genetically Modified and Novel Foods (Labelling) (England) Regulations 2000.*

van den Daele, W., Puhler, A. and Sukopp, H. (1997) *Transgenic Herbicide-resistant Crops: A Participatory Technology Assessment.* Summary report for the Federal Ministry for Research and Technology. Wissenschaftszentrum Berlin fur Sozialforschung, Berlin.

17 Quantifying Scientific Risk Communications of Agrobiotechnology

Leonie A. Marks,[1] Siân Mooney[2] and Nicholas Kalaitzandonakes[1]
[1] Economics and Management of Agrobiotechnology Center, Department of Agricultural Economics, University of Missouri-Columbia, Columbia, Missouri, USA;
[2] Department of Agricultural Economics, Montana State University, Bozeman, Montana, USA

Problem Setting

Changing demographics in Western societies have led to an increasingly urbanized population far removed from food production and distribution realities. The agrifood industry has continued to improve the overall quality and safety of food at both the production and the processing levels (e.g. pasteurization of milk, juices, cheeses and so on) though risks still exist from pathogens, toxins and other contaminants in the food supply. Despite improvements, and a food supply that is safer than ever, consumers are increasingly aware of food safety as an issue. A 1999 survey of five countries – Canada, the USA, Japan, China and Germany – based on a sample size of 1000 respondents in each country, found that the number one food issue for consumers was food safety (Environics International, 2000). On average (across all five countries) 46% of respondents were most concerned about food safety, followed by nutrition (22%), food quality (12%), price (8%) and shortages (7%). In terms of specific food safety concerns, on average 88% of consumers were very or somewhat concerned about pollution (where food is produced) and chemical pesticides, followed by bacterial contamination (84%). Overall, the least concern (69%) was expressed about genetically modified (GM) foods. Recent surveys of European and US consumers by the Food Marketing Institute (FMI) (Bruhn, 1997) and Hoban (unpublished) indicate that the most important food safety concern for consumers in both regions is food borne pathogens.[1] Among American respondents, 77% are concerned about microbial contamination; 66% about pesticide residues; 66% about product tampering; and 42% about antibiotic and hormone residues. Only 15% of US respondents are concerned about foods derived from biotechnology (see Fig. 17.1). European consumers are more concerned about food safety issues in general: 85% of respondents were concerned about microbial contamination. A greater percentage of respondents are concerned about biotechnology products as well (see Fig. 17.1).

Increased attention on microbial contamination is well placed. Each year, food borne pathogens

[1] Survey instruments vary by type of question asked, time period and geographical location. Hence, the difference (as a percentage of respondents) between the Environics International (2000), FMI (Bruhn, 1997) and Hoban (unpublished) studies. Nevertheless, these surveys indicate that microbial contamination and pesticide residues are of most concern to consumers in each country and region, and that GMFs have been found to be of least concern over time.

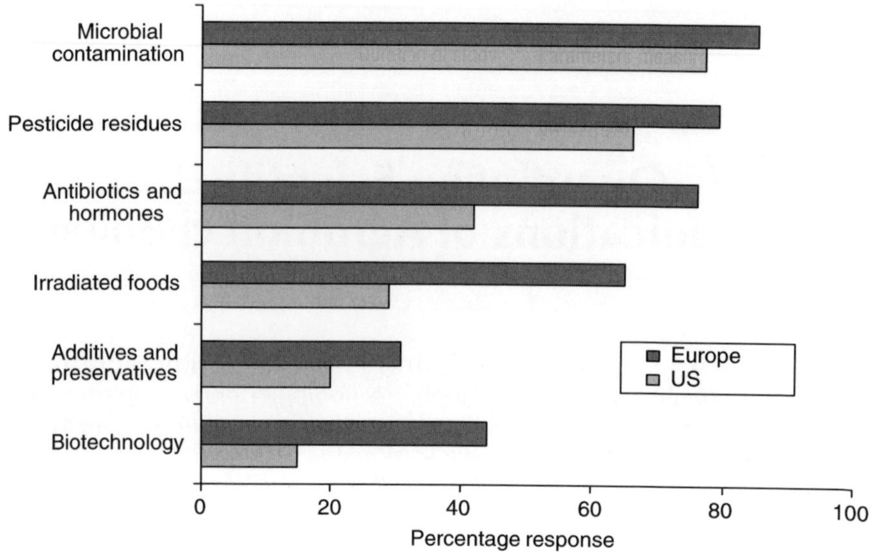

Fig. 17.1. Food safety concerns of US and European consumers. (Source: Bruhn, 1997; Hoban, unpublished.)

cause approximately 76 million illnesses, over 320,000 hospitalizations and over 5000 deaths in the US alone. Of these deaths, 1800 can be attributed to known pathogens (*Salmonella, Listeria* and *Toxoplasma*, for example) and the remainder to unknown pathogens (Mead *et al.*, 1999). However, focusing all attention on the agrifood industry may be misleading. Collins (1997) and Bruhn (1997) indicate that food handling at home is a serious food safety concern due to poor consumer knowledge regarding food risks and safe practices. Between 1987 and 1992, 79% of food illness outbreaks were bacterial. The majority of these cases were attributed to improper holding temperature and poor personal hygiene of food handlers (Bruhn, 1997; Collins, 1997).

As food borne contaminants can often be traced to specific manufacturers, 'food scares' are common and highly reported in the media. Recent US and European food scares include salmonella in eggs, *Escherichia coli* (*E. coli*) in hamburgers, listeria in soft cheeses, dioxin residues in animal feed in Belgium and the 'mad cow' or bovine spongiform encephalopathy (BSE) crisis in the UK (and more recently throughout Europe). By November 2000, some 87 people in the UK had contracted new variant Creutzfeldt-Jakob disease (nvCJD), the human variant of BSE (Brown *et al.*, 2001). Unlike its bovine counterpart, the nvCJD outbreak has been small with only a modest increase during its first 6 years (Brown *et al.*, 2001: 8). However, uncertainty about the extent of the nvCJD outbreak still exists, with potentially many more people infected, from fewer than a hundred to hundreds of thousands of people (Reuters, 2000; Brown *et al.*, 2001). The relative risk of having contracted nvCJD from contaminated food is still small relative to the associated everyday risk from food borne pathogens. Yet, media attention and uncertainty surrounding the causes and infection rate of BSE, have heightened public concern over the safety of the food supply in general.

Heightened public concerns about food safety have been frequently presumed to lie behind the negative stance of European consumers towards agrobiotechnology products as well. One hypothesis is that sensational coverage by the mass media has emphasized the potential risks of agrobiotechnology over its benefits against the backdrop of BSE and other food scares. Indeed, there is some empirical evidence suggesting that signal events, such as the BSE crisis, have affected communication on potential risks and benefits of agrobiotechnology by mass media (Marks *et al.*, 1999; Marks *et al.*, Chapter 18, this volume).

It has also been frequently hypothesized that the mass media has amplified perceived risks associated with biotechnology by unevenly reporting on a few scientific studies that have pointed to risks and hazards. An often cited example is the attention of mass media on the public statements of Dr Arpad Pusztai during 1998, who warned of potential serious health risks from products of agrobiotechnology. The firestorm of controversy surrounding Pusztai's study was heightened by the fact that, at the time of his comments, the research had not been subjected to peer review. Subsequent to such media attention, the study was published as a 'Research Letter' in *The Lancet* (Ewen and Pusztai, 1999). However, there has been much criticism of his methodology and results (FitzGerald *et al.*, 1999; Lachmann, 1999; Mowat, 1999) and many of his statements have been discredited by mainstream scientific societies across the globe (e.g. the Royal Society).

Implicit to all such hypotheses is the assumption that the scientists have been less concerned about possible risks of food safety (or other) hazards from agrobiotechnology, and that technical perceptions of risks diverge greatly from public perceptions of risks in agrobiotechnology. But is there really a gap between the way scientists and the general public look at potential risks associated with agrobiotechnology?

In this chapter, we quantify published knowledge of the associated food safety risks of GM foods. We use content analysis to systematically analyse the content of scientific journals with respect to food safety in general and the safety of biotechnology products in particular. We compare and contrast the content of these journals with key issues that have been raised in the media about the safety of GM foods: (i) possible long-term 'dreaded' health effects, such as cancer and birth defects; (ii) possible new allergens resulting from biotechnology; and (iii) possible side-effects of the technology, such as increased antibiotic resistance. Our goal is to quantify any possible gaps in technical risk assessment and risk communication in the case of food safety of agrobiotechnology.

Content Analysis: an Approach to Inventorying Scientific Knowledge

'Content analysis is a systematic method for analyzing and quantifying message content and message handling. It is a tool for observing and analyzing the overt communication behavior of selected communicators' (Budd *et al.*, 1967: 2). Instead of soliciting people's behaviour directly (through interviews) or measured responses to specific events or stimuli, content analysis may be used to analyse communications that people have produced as accounts of, or a framework for, behaviour (Kerlinger, 1964). The main advantage of the approach is that it allows investigation at any time and place of the investigator's choosing. In addition, with the increasing availability of electronic data sources and computing power, content analysis opens up possibilities to analyse research conducted over longer periods of time and with a larger geographical scope than before.

The literature on content analysis is vast and multidisciplinary. Journalists, psychologists, sociologists and linguists have carried out content analysis of a wide variety of issues including political campaigns, the environmental impacts of chemicals, the nuclear energy debate and so on. A comprehensive review of the literature is beyond the scope of this chapter. However, content analysis can be used to analyse any type of text including media reports, conference and regulatory proceedings, and journal articles. For example, Bengston and Xu (1995) have used content analysis to analyse changing national forest values in the USA over the period 1982 to 1993. They examined three types of textual information including: (i) newspaper articles; (ii) keynote and general session papers presented at national forestry conferences, and the complete text of articles from the *Journal of Forestry*; and (iii) articles published by forest-related environmental groups. Similarly, Hagedorn and Allender-Hagedorn (1997) compared public opinion surveys, the popular press and technical/regulatory sources for trends in issues related to agricultural biotechnology in the USA.

Methodological Approach

There are three steps in content analysis beyond the formulation of research questions and hypotheses: data collection, formulation of conceptual categories and coding of the data, and empirical analysis. Each step taken in this study is detailed next.

Data collection

There are several leading specialized journals that publish articles related to food safety; for example, *Food Technology*, *Food Biotechnology*, *Food Policy*, the *Journal of Food Protection*, *Applied and Environmental Microbiology* and the *Journal of Food*

Science among others. In addition, non-specialized journals, such as *Science* and *Nature* also address issues related to new technologies, such as biotechnology. We analyse research reported in two leading scientific journals; namely, the *Journal of Food Protection* and *Science.*

The *Journal of Food Protection* was analysed as it is the most widely read of the specialized food journals and considered a leading journal in its field (D. Holt, personal communication, 19 April 2000). The *Journal of Food Protection* is published monthly beginning with the January issue by the International Association of Milk, Food and Environmental Sanitarians.

Science was analysed as it is the world's largest circulation general scientific publication. In addition, the initial source for many science news stories is a breaking journal article published in an elite scientific journal, such as *Science.* Our analysis should therefore reveal the body of evidence about the safety of GM foods readily available to the media in *Science. Science* is published weekly by the American Association for the Advancement of Science. Two very different kinds of editorial material are published: the most important news of the week in science and in science policy, and a selection of scientific papers reporting the most significant breakthroughs in global research. The front of the magazine is written by a team of science journalists who translate the latest developments in science into clear, non-technical language for the general public, media and other scientists.

Our period of analysis is 1990 to 1999. This period coincides with many European food crises (e.g. BSE) and emerging food safety issues in the US, including detection of food borne pathogens and allergens, food irradiation, chemical residues and the emergence of antibiotic resistant bacteria in the food supply.

All articles in the *Journal of Food Protection* relate to food or food safety issues. Therefore, every article in a given issue was analysed. In the case of *Science,* however, a broad range of science-related topics are covered. Therefore, a list of keywords was used to identify articles related to key content categories. For example, words or word phrases such as biotechnology, food biotechnology, food safety, BSE, microbial contaminants and so on, were used to identify the subset of articles analysed.

Categorization and coding of data

The most important step in content analysis is the identification and categorization of the variables under study. Categories, such as subject matter or direction categories, serve the same function as variables in content analysis. As Budd *et al.* (1967) argue, 'no content is better than its categories'. Variables (categories) must be defined through an operational definition or set of definitions. In addition, these categories must be exhaustive and mutually exclusive.

In this study, several contextual categories were defined to mirror public food safety concerns. Articles were categorized if they related to contamination of foods from pathogens (including listeria, *E. coli,* salmonella and vomitoxin), chemical residues (as a separate sub-category of food contaminants), food allergens, irradiation, and antibiotic residues and hormones. In addition, articles were categorized on the basis of 'risk analysis' where risk analysis was broadly defined to include any mention of the words 'risk' or 'health hazard'. Hence, this category over-reports the actual level of risk analysis undertaken or discussed by scientists. A 'dreaded effects' category was also developed following Slovic's conception of such risks. This category included such words as 'deaths', 'mortality', 'long-term effects', 'birth defects' and so on. The purpose of this category was to quickly identify articles that might indicate possible dreaded effects of food hazards. This category, along with the 'risk analysis' category was particularly important for cross-checking of biotechnology-related articles.

A large sub-category of articles identified across both journals was biotechnology-related articles. Biotechnology tools and methods were broadly defined to include both basic (e.g. general assay techniques) and the more sophisticated techniques, such as genetic engineering, genetic modification, use and identification of DNA, ELISA, PCR, use of bacteriophages such as T4 and M13, and so on. This sub-category of articles was then cross-tabulated with the food safety categories using semantic-based software to identify how food safety issues in biotechnology are viewed and addressed by the scientific community.

Coding and context units

Several different types of coding unit can be used to conduct content analysis; for example, words, word assertions, themes and character units can all be counted. The coding unit used was 'word' or 'word phrases' related to the category being measured. The unit of analysis in this research is 'article'[2] as it relates to the conceptual category measured. Journal articles with content relating to several categories are counted more than once.

Empirical Results

Journal of Food Protection

Figure 17.2 categorizes research as a percentage of total articles in each year. It is clear that food scientists publishing in this journal have largely been concerned with investigating contaminants in the food supply: microbial pathogens such as *E. coli*, listeria and salmonella. Between 60 and 80% of the articles published dealt with food contaminants in

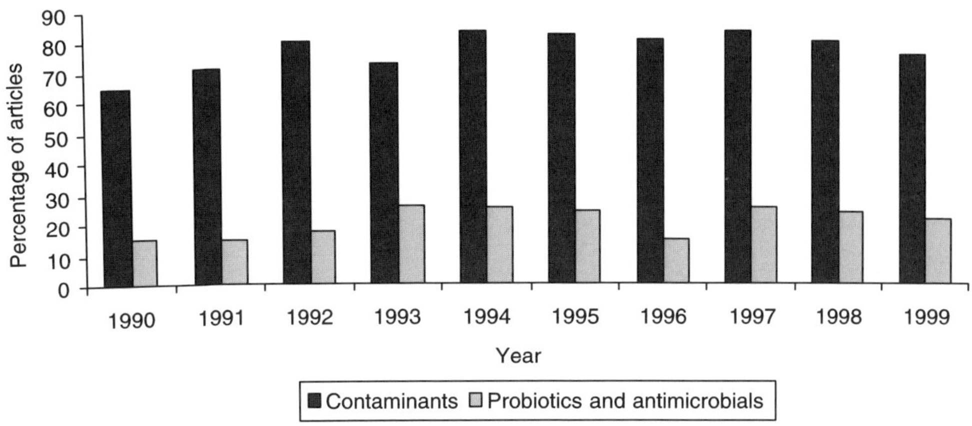

Fig. 17.2. Research on food contaminants and probiotics and antimicrobials in the *Journal of Food Protection*.

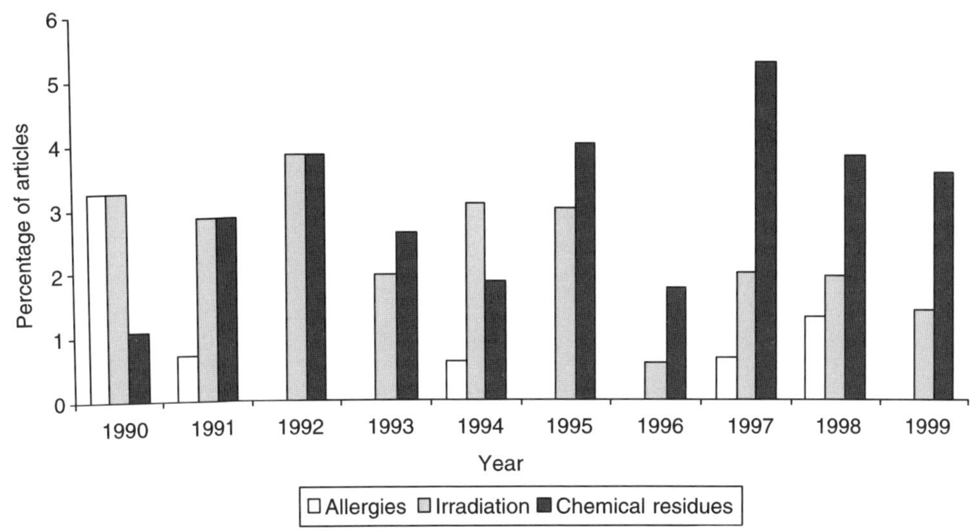

Fig. 17.3. Research on food allergies, irradiation and chemical residues in the *Journal of Food Protection*.

[2] Full text articles were not electronically available for the *Journal of Food Protection*, therefore, full abstracts were analysed instead. These abstracts, along with additional research descriptors, provided the basis for our analysis.

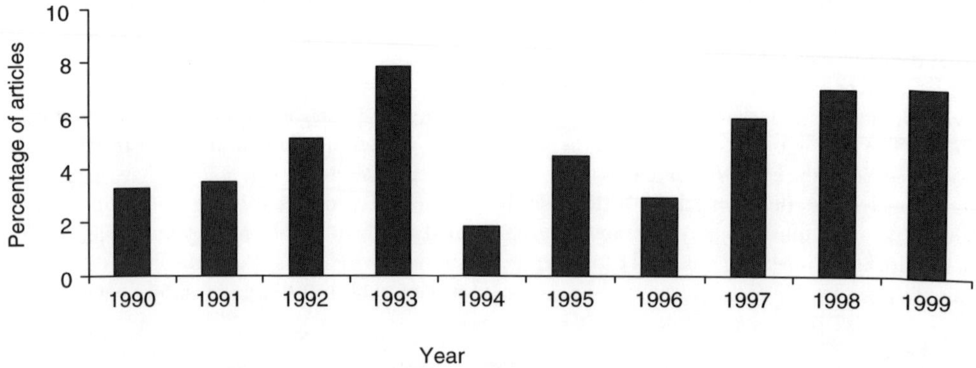

Fig. 17.4. Research on antibiotic residues and use of hormones in the *Journal of Food Protection.*

some form or other. Likewise, the next biggest area of research was in methods aimed at reducing or eliminating food borne pathogens from the food supply, that is, the use of probiotics (or 'friendly bacteria') such as *Lactobacillus* and other antimicrobial methods. Other food safety issues, such as chemical residues, food allergies and irradiation have been considerably less addressed by food scientists, with less than 4% of articles dealing with these concerns (Fig. 17.3). Development of methods for detecting and testing for antibiotic residues and antibiotic resistant bacteria in the food supply has been another area of research but again considerably less than the main food safety concern addressed (microbial contamination). Only a few articles dealt with the use of hormones in the food supply (Fig. 17.4). Anywhere from 4 to 13% of articles addressed possible risks (or lack of) and health hazards to consumers (Fig. 17.5). However, very few articles provided detailed risk assessment methodologies in relation to food safety risks.

Biotechnology-related research increased steadily over the time period; less than 10% of articles identified used biotechnology terms in 1990 compared with over 30% of total articles relating to biotechnology in 1999 (Fig. 17.6). It is striking, however, that biotechnology has typically been used in a positive way in relation to food safety, as a tool for detecting food contaminants and as a means for reducing or eliminating them in the food supply. Hence, as illustrated in Fig. 17.7, a large percentage of biotechnology-related articles are cross-tabulated with food contaminants and with the use of probiotics and anti-microbials.

Every article that was cross-tabulated with risk analysis, antibiotic resistance, allergen or dreaded effect was systematically checked across all 10 years. No article indicated or discussed any possible 'dreaded effects' from biotechnology food products. No tests were conducted into potential dreaded effects of GM foods. No articles discussed potential allergens from biotechnology products or conducted

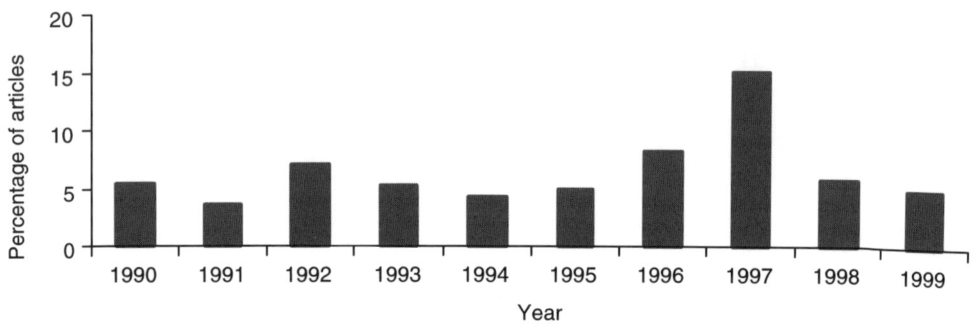

Fig. 17.5. Risk analysis in the *Journal of Food Protection.*

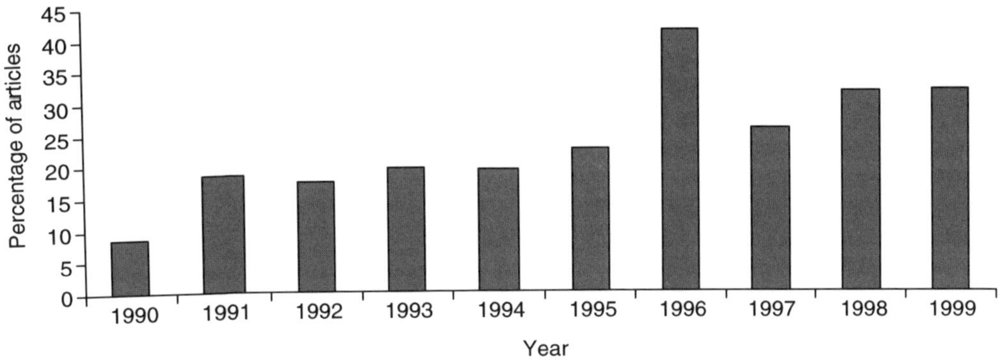

Fig. 17.6. Biotechnology-related research in the *Journal of Food Protection.*

any tests for potential allergens. The number of research articles on food allergens was extremely low and this related to known allergens, such as milk residues in non-related foods. One article used ELISA to detect whey proteins in food processing. In terms of antibiotic resistance, biotechnology was typically used as a tool or indicator for the presence of resistant bacteria. No article studied the use of antibiotic resistance marker genes as a food safety issue. Again, PCR and ELISA techniques were used to test for specific strains of resistant bacteria.

In terms of risks and risk assessment, no article dealt with food safety risks of GM foods. One article discussed the possibility of using GM bacteriocins (probiotics) to combat listeria contamination and suggested that, while promising, such GM organisms (GMOs) would have to clear regulatory hurdles because of possible risks. One article investigated enterococci isolated from dairy products as having useful biotechnological traits but that there was no conclusion on whether they posed a threat in themselves as food borne pathogens.

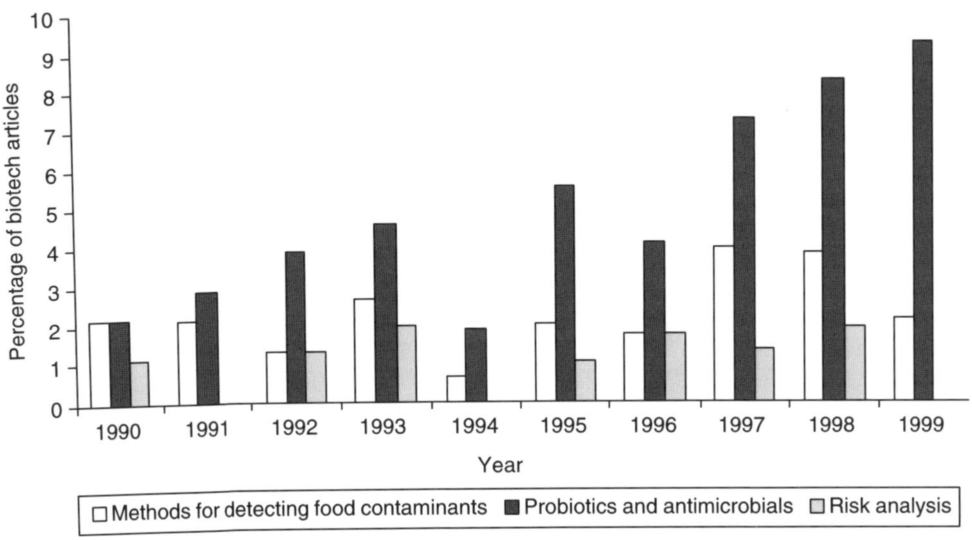

Fig. 17.7. Cross-tabulation of biotechnology-related articles in the *Journal of Food Protection.*

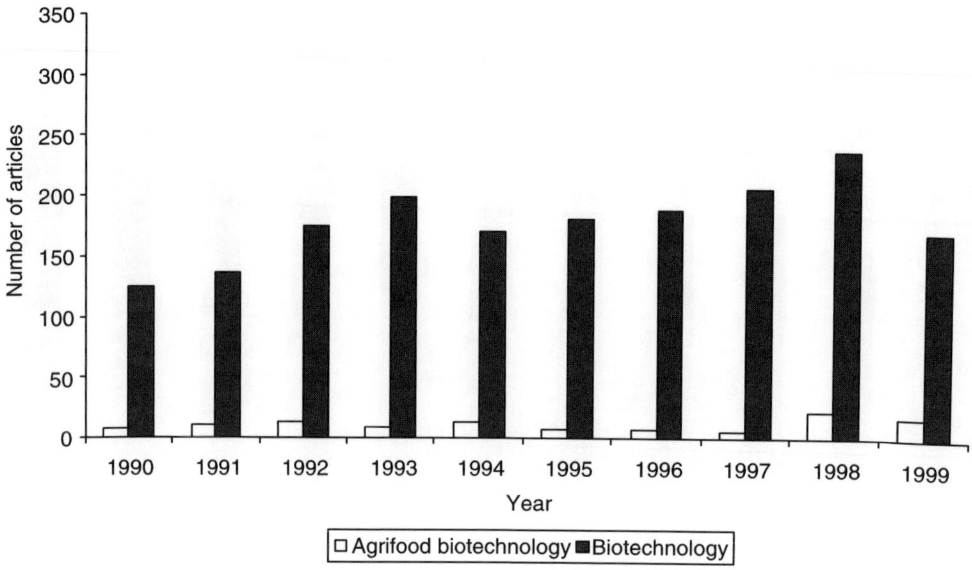

Fig. 17.8. Total articles related to biotechnology and food biotechnology in *Science*.

Science

Figure 17.8 details coverage in *Science* of biotechnology in general and agrifood biotechnology articles in particular. Coverage is expressed as a total number of articles, rather than as a percentage, as the total number of articles was very large in any given year.[3] Overall, less than 10% of articles dealt with biotechnology in general and less than 1% of articles related to agrifood biotechnology. Interestingly, risk analysis in relation to general biotechnology articles has received quite a lot of cov-

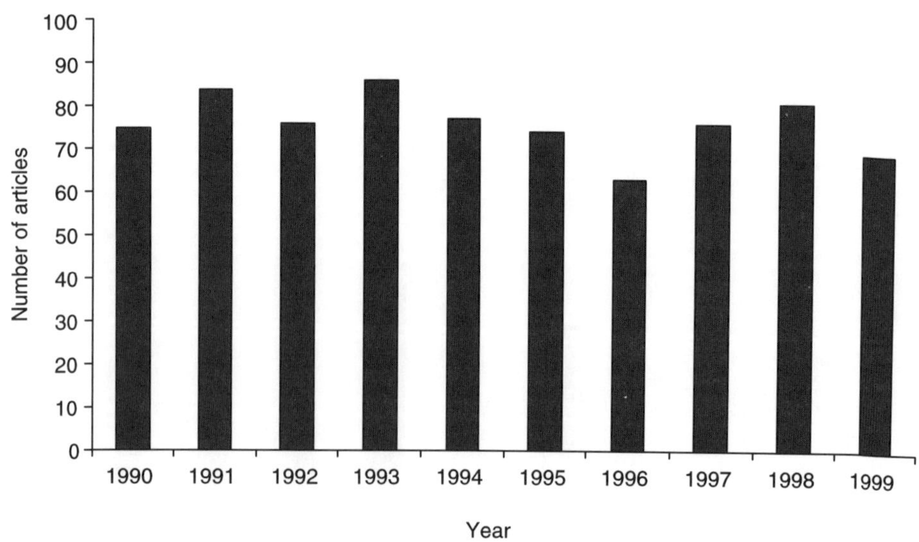

Fig. 17.9. Coverage of risk analysis in biotechnology-related articles in *Science*.

[3] As *Science* is published weekly, over 1664 articles were published during 1990. By 1999, the number of articles had steadily increased to 2467.

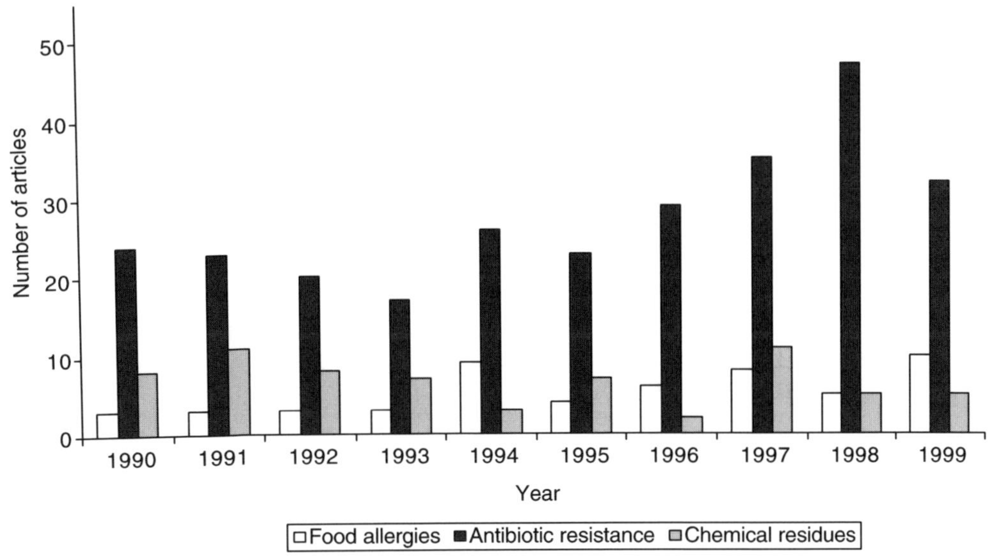

Fig. 17.10. Coverage of food allergies, antibiotic resistance and chemical residues in biotechnology-related articles in *Science*.

erage, along with antibiotic resistance (Figs 17.9 and 17.10). As already indicated, such risk analysis is broadly defined and is likely to overstate the actual level of risk assessment undertaken or discussed by scientists. In addition, such coverage includes biosafety as well as health safety issues.

In order to identify the body of research on the food safety of biotechnology, only research articles were analyzed on the basis of the food safety categories identified previously. Editorials, opinion pieces, book reviews and short articles by journalists were excluded from the analysis. Scientific commentary articles, that is articles discussing scientific research conducted by other researchers, were included. These articles could be literature reviews of scientific work or studies.

A small number of research articles per year cross-tabulated with agrifood biotechnology and food contaminants, chemical residues, risk analysis and antibiotic residues/hormones. Hence, a small fraction of the agrifood biotechnology coverage in *Science* provided scientific research articles that

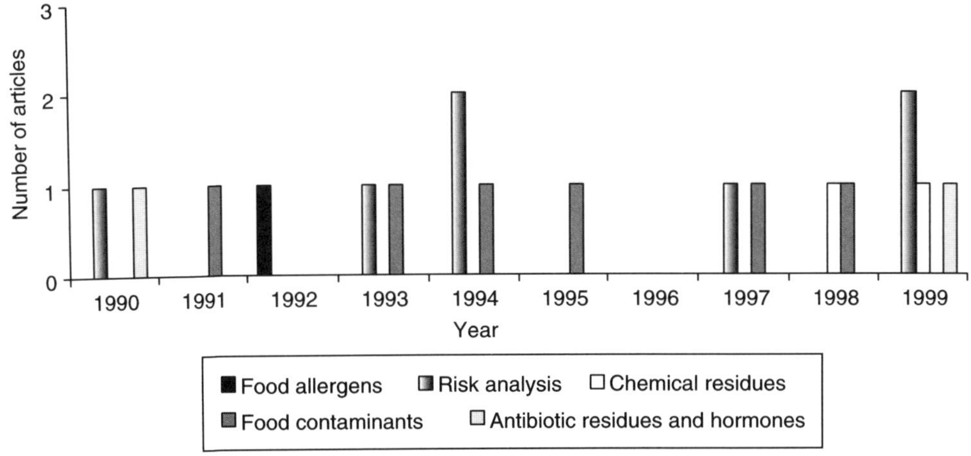

Fig. 17.11. Scientific research of food safety issues of biotechnology in *Science*.

addressed many of the food safety issues raised by the public. Figure 17.11 indicates the level of scientific coverage of each food safety category. For example, only one article addressed food allergies in relation to ag-biotechnology food products. This article discussed the possibility of allergenicity as novel proteins are inserted into food products. This article did not test specific proteins but suggested which proteins might need testing and which might not.

Several research articles did cross-tabulate with agrifood biotechnology on the basis of risk analysis over the 10 year period; however, most of these articles were indirectly related to food safety. For example, one article referred to consumer perception of risk in opinion polls; another article related to the risk of inheritance of prion diseases; and biotechnology was used as a method to detect prions (with transgenic mice). No scientists, based on our analyses, reported testing GM foods for possible health risks or finding potential health risks.

Concluding Comments and Implications

From our initial results, there appears to be a gap in the perceptions of food safety risks and benefits of agrobiotechnology between food scientists and reporters. Typical public concerns about the food safety of products of agrobiotechnology have not been a primary focus of scientific studies. Similarly, biotechnology has often been considered by scientists as a means of reducing other food risks (e.g. microbial contamination). These benefits are generally under-reported through mass media.

Lack of studies testing the food safety of agrobiotechnology products does not imply that food safety is not an issue. It is possible that scientists consider testing for food safety 'mundane' work that is best left to organizations pursuing commercialization and regulators. Our results indicate that food (and other) scientists have not put forward hypotheses of major food safety risks in the case of agrifood biotechnology. And, they have not seemed overly concerned with testing the risks for which hypotheses have been advanced.

These results suggest a possible gap between the scientific perception of risk of agrifood biotechnology and that of the general public. The emphasis of mass media on the potential food safety risks of agrobiotechnology suggests a communication problem for the scientific community and the biotechnology industry. Under-reporting of potential benefits of biotechnology in food safety hazards indicates a similar communication gap. Irrespectively, to ensure public confidence, studies testing the food safety of agrobiotechnology products must be subjected to a rigorous peer-reviewed process and must be placed in the public domain.

Recent regulatory developments seem promising in this respect. In May 2000, the Clinton Administration outlined several steps needed to provide the public with more information about the science-based regulation and food safety testing of GM foods (The White House, 2000). In particular, the US Food and Drug Administration (FDA) proposed a rule mandating that developers of bioengineered foods and animal feeds notify the agency when they intend to market such products (US Department of Health and Human Services [DHHS]/US Food and Drug Administration [FDA], 2000). Up until this point, the notification process had been voluntarily adhered to. The FDA has proposed that all submitted data or information, and the agency's conclusions, be made available to the public. This proposed rule has subsequently been published in the *Federal Register* (DHHS/FDA, 2001). Such data could be exempt from public disclosure at the request of the notifier under established exemptions as provided in the Freedom of Information Act (FOIA) (DHHS/FDA, 2001: 4714). However, in the view of the FDA, much of the data or information provided is unlikely to constitute a trade secret or be confidential. More publicly available information would go some way to providing food safety test data to the general public. However, this information is not publicly accessible at this time.

In addition, the previous administration proposed that the US Department of Agriculture (USDA), the Environmental Protection Agency (EPA) and the FDA support an expanded programme of competitively awarded, independent peer-reviewed research focusing on current and future safety issues related to GM foods (The White House, 2000). Based on our initial results, such research is needed to further place food safety studies within the public domain.

Finally, a word of caution. Our results should be considered suggestive and not conclusive. More journals must be analysed for content, as second-tier journals may have published research that is different from that analysed here. Likewise, medical journals such as *The Lancet* and the *Journal of the American Medical Association*, warrant further systematic investigation.

References

Bengston, D.N. and Xu, Z. (1995) *Changing National Forest Values: A Content Analysis,* Research Paper NC, 323. United States Department of Agriculture, Forest Service, North Central Forest Experiment Station, St Paul, Minnesota.

Brown, P., Will, R.G., Bradley, R., Asher, D.M. and Detwiler, L. (2001) Bovine spongiform encephalopathy and variant Creutzfeldt-Jakob disease: background, evolution, and current concerns. *Emerging Infectious Disease* 7(1), 1–17. www.cdc.gov/ncidod/EID/vol7no1/brown.htm

Bruhn, C.M. (1997) Consumer concerns: motivating to action. *Emerging Infectious Diseases* 3(4), 511–515. www.cdc.gov/ncidod/eid/vol3no4/bruhn.htm

Budd, R.W., Thorp, R.K. and Donohew, L. (1967) *Content Analysis of Communications.* Macmillan Company, New York.

Collins, J.E. (1997) Impact of changing consumer lifestyles on the emergence/reemergence of food-borne pathogens. *Emerging Infectious Diseases* 3(4), 471–479.

Environics International (2000) Food issues monitor. www.environicsinternational.com

Ewen, S.W.B. and Pusztai, A. (1999) Effects of diets containing genetically modified potatoes expressing *Galanthus nivalis* lectin on rat small intestine. *The Lancet* 354(9187), 1353–1354.

Fitzgerald, A.J., Goodlad, R.A. and Wright, N.A. (1999) Correspondence: GM food debate. *The Lancet* 354, 1725.

Hagedorn, C. and Allender-Hagedorn, S. (1997) Issues in agricultural and environmental biotechnology: identifying and comparing biotechnology issues from public opinion surveys, the popular press and technical/regulatory sources. *Public Understanding of Science* 6, 233–245.

Kerlinger, F.N. (1964) *Foundations of Behavioral Research.* Holt, Rinehart and Winston, New York.

Lachmann, P. (1999) Correspondence: GM food debate. *The Lancet* 354, 1725.

Marks, L.A., Kalaitzandonakes, N. and Zakharova, L. (1999) Public opinion of agbiotech in the US and UK: a content analysis approach. Paper presented at the American Agricultural Economics Association (AAEA) National Meetings, Nashville, Tennessee. www.agecon.lib.umn.edu/cgi-bin/pdf_view.pl?paperid=1495

Mead, P.S., Slutaker, L., Dietz, V., McCraig, L.F., Bresee, J.S., Shapiro, C., Griffin, P.M. and Tauxe, R.V. (1999) Food related illness and death in the United States. *Emerging Infectious Diseases* 5(5), 607–625. www.cdc.gov/ncidod/eid/vol5no5/mead.htm

Mowat, A. (1999) Correspondence: GM food debate. *The Lancet* 354, 1725.

Reuters (2000) Human mad cow cases rising, Reuters Report. www.msnbc.com/news/441480.asp

United States Department of Health and Human Services (DHHS)/Food and Drug Administration (FDA) (2000) *FDA to strengthen pre-market review of bioengineered foods,* press release No. P00-10. DHHS/FDA, Washington, DC. www.cfsan.fda.gov/~lrd/hhbioen2.html

United States Department of Health and Human Services (DHHS)/Food and Drug Administration (FDA) (2001) Premarket notice concerning bioengineered foods. *Federal Register* 66(12), 4706–4738.

The White House (2000) Clinton administration agencies announce food and agricultural biotechnology initiatives: strengthening science-based regulation and consumer access to information, press release, 3 May. www.cfsan.fda.gov/~lrd/whbio53.html

18 Time Series Analysis of Risk Frames in Media Communication of Agrobiotechnology

Leonie A. Marks,[1] Nicholas Kalaitzandonakes,[1] Kevin Allison[2] and Ludmila Zakharova[1]

[1] *Economics and Management of Agrobiotechnology Center, Department of Agricultural Economics, University of Missouri-Columbia, Columbia, Missouri, USA;*
[2] *John F. Kennedy School of Government, Harvard Business School, Harvard University, Cambridge, Massachusetts, USA*

Introduction

The story is well known. After decades of research and development (R&D) agricultural biotechnology products hit the US market in 1995 and 1996. Commercialization and adoption of transgenic crops in the USA was rapid, yet equally swift was rejection of the technology by many consumers in Europe and elsewhere. Many factors have been hypothesized to drive this consumer response:

- A refusal of consumers to accept any risk in the face of no perceived direct benefit.
- An alleged lack of trust in food regulatory agencies in Europe and elsewhere.
- The unfortunate coincidence of the commercialization of genetically modified (GM) foods with the bovine spongiform encephalopathy (BSE) ('mad cow') crisis.
- Protectionist interests on the part of European governments to prohibit trade in biotech crops.
- An unyielding attitude by the USA and corporate entities championing the technology towards labelling and the consumers' 'right to know'.
- Anti-American sentiment.
- Sensational coverage by the mass media.

In this chapter, we focus on the last hypothesis and we analyse media coverage of agrifood biotechnology. A vast amount of research has investigated the role of the media in amplifying risks well beyond what is implied by 'actual or objective' scientific risk. In the case of agrobiotechnology, it has been shown that key events do indeed have a significant impact on the level of coverage and the way risks and benefits are communicated (Marks *et al.*, 1999).

Media represent fora where public debates on agrobiotechnology have taken place for many years. They, of course, have not only been reflective of the ongoing public debates but they have also had a role in shaping them. While they are not a singular influence, they have been shown to play a role in the risk (and benefit) perception of biotechnology that the public holds (Bauer *et al.*, 1998). A recent survey of US consumers (Schulz *et al.*, 2000) indicates that US consumer awareness of GM foods has come from traditional media such as television, magazines and newspapers.

In this chapter, we analyse mass media coverage of agrobiotechnology in a risk communication framework using content analysis (see Wimmer and Dominick, 1987). We analyse coverage in US and UK newspapers over the period 1990 to 1999. US papers include the *Wall Street Journal*, *Washington Post* and *USA Today*. UK papers include the *Daily Telegraph* and *The Times*. In particular, we examine how environmental and food safety risks and benefits have been communicated through the media on the two sides of the Atlantic.

Risk communication and risk perception are key elements of the ongoing public debate on agrobiotechnology and allow insights into attitudes and behaviour of the general public. If policy makers, educators and scientists are to effectively communicate the risks and benefits of agrobiotechnology to the public, an understanding of how different media and reporters handle risk frames is important. This chapter contributes in that direction.

Risk Perception

Like all technologies and, indeed, all human action, biotechnology is associated with risks, both known and unknown. The study of risk, encompassing the fields of risk assessment, perception and communication, has developed over the past several decades in response to the challenges posed by an increasingly technologically oriented society (Covello, 1983; Kates and Kasperson, 1983; Slovic, 1987). Technologies such as nuclear weapons and chemicals are capable of affecting the Earth and its inhabitants on a global scale, resulting in a public interest in accurate assessment and communication of the level of risk posed by such technologies. Such hazards have the potential to cause catastrophic and long-term damage to human health and to the environment, yet possess difficult-to-assess degrees of risk. Their most harmful consequences, such as the health risks of nuclear fallout or the greenhouse effect, are rare and most often delayed in their manifestation, making them difficult to assess statistically. Moreover, the tremendous impact of such consequences leaves no possibility for a trial-and-error approach to risk management (Slovic, 1987).

The difficulties of statistical risk assessment are confounded by the existence of a substantial gap between the expert and public perception of risks. How the public forms perceptions of technological risk has been an area of considerable investigation (Slovic 1987, 1993, 1997; Krimsky and Plough, 1988; National Research Council, 1989). One conclusion is that public perceptions can be influenced by both expert opinion leaders and the mass media among other factors.

A substantial body of work has been dedicated to the psychological aspects of risk perception, uncovering a set of mental strategies or rules that all people use to simplify risk problems. Two of these rules, known as heuristics, are particularly important (Covello, 1983). The first heuristic is that of 'information availability' or the tendency to believe that an event or action occurs more frequently if instances are easy to imagine or recall. If the overall level of media coverage of the risks of a technology increases, then, this may increase the perceived likelihood of those risks becoming manifest. The second important heuristic is 'representativeness', or the tendency to assign comparable risk characteristics and degrees to activities or events that are roughly similar, such as nuclear power and nuclear war. Representativeness is the basis for the assertion that exposure to information about the risks of technologies will increase the perceived riskiness of the technology itself.

Studies by Slovic (1987) and Slovic *et al.* (1985) have shown that perceived risk is both quantifiable and predictable. Fitting different types of risks into a quantitative framework can reveal which risks are perceived in similar ways to one another. Most risks can be grouped in terms of two main factors. First, the degree to which the risk is a 'dread' risk: its consequences are catastrophic, uncontrollable, potentially fatal, not equitable in their distribution, pose high risk to future generations, are not easily reduced and are involuntarily imposed. Second, the degree to which the risk is an unknown risk: it is not observable, not evident to those exposed, its effects are delayed and not definitively known to science. These two broad categories provide the framework for a quantitative model of risk perception (see Fig. 18.1).

In a study based on this model, both nuclear technologies and DNA technologies, including agrobiotechnology, score high in both the dread risk and unknown risk factors, indicating that the risks associated with these technologies are perceived in similar ways (Slovic *et al.*, 1985; Slovic, 1987). According to the heuristic principle of representativeness mentioned above, exposure to information about risks associated with nuclear power will influence the perceived risks of agrobiotechnology, and vice versa. It has been shown that unfortunate events or accidents involving technologies that score high in both the dread and the unknown categories are particularly likely to produce broad social, political and policy consequences. That is, people are more likely to consider higher order impacts when dealing with these high-scoring risks, such as ethical and moral or environmental factors or threats to future generations that are not usually accounted for by

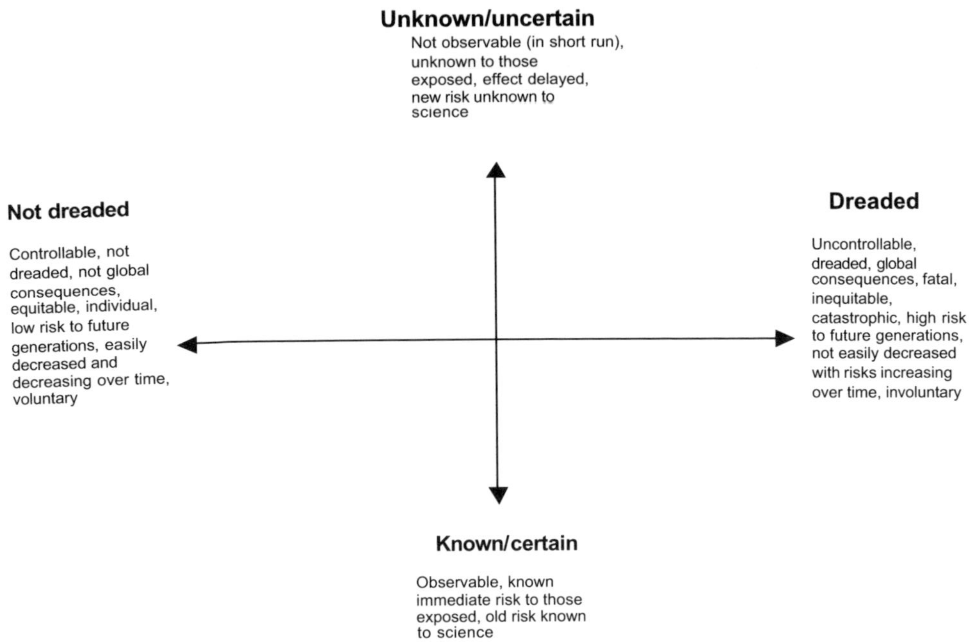

Fig. 18.1. Conceptualization of risk. (Source: adapted from Slovic, 1987.)

traditional statistical risk assessment (e.g. expected annual mortality rates (Slovic, 1987)). This tendency is certainly evident in the ongoing debates regarding nuclear technologies and agrobiotechnology.

Risk Communication and Frame Theory

Somewhere between its assessment and perception, risk must naturally be communicated. In addition to the two heuristics that individuals often use to make judgements about risks, sociologist Erving Goffman (1974) and others (Payne *et al.*, 1992; Irwin and Davis, 1995) have identified the importance of how information is 'framed' in risk judgements. Frames, according to Goffman (1974), 'allow people to locate, perceive, and label' social events. Frames provide meaning, and a way of thinking about our lives, events and the world in general. Payne *et al.* (1992) and Irwin and Davis (1995) argue that frames effectively frame the way that information is presented and, depending on the frame, judgements about the perceived risks versus benefits of a technology might be quite different. For example, simple wording changes in decision contexts, such as evaluating road safety in terms of lives saved rather than lives lost, can have significant impacts on decision-making processes. Through frames, communicators organize information by selecting certain aspects of a perceived reality and transforming them into an easily communicable context, thereby emphasizing a particular problem definition, causal interpretation or other feature of the concept at hand (Entman, 1993). Risk communication can take place in the context of any number of frames. While some attempts at risk communication might frame risk in terms of expected annual mortality rates, others might frame risk of the same technology in terms of less tangible elements, such as moral and ethical risk.

Several empirical studies have shown the importance of framing in the decision-making process when dealing with risky outcomes. Kahneman and Tversky (1984) found that when presented with an identically probable outcome, people will select one equivalent outcome over another depending on whether the outcome is framed in terms of a risk of loss or chance of benefit or gain. Though initially performed to evaluate decision-making behaviour when playing games of chance, and the tenet of rationality, subsequent studies have repeated these results in several different situations, from AIDS treatment to personal relationships (Levin and Chapman, 1990, as cited in

Boon and Griffin, 1996). These studies imply that the framing of risks over time may affect opinion formation, to the extent that opinion is formed by a series of decision-making events over an extended period of time. As Slovic (1997) writes, 'We now know that every form of presenting risk information is a frame that has a strong influence on the decision maker.'

The Media's Role in Framing Risk

The fact that risk frames can have such a pronounced impact on the decision-making process means that those responsible for providing information may influence perceptions and behaviour (Slovic *et al.*, 1984). In some countries (the US for example) the news media can serve as a primary source of risk information to the general public (Schulz *et al.*, 2000) in addition to the more trusted sources (i.e. the US Food and Drug Administration (FDA), the Environmental Protection Agency (EPA), American Medical Association). Even in countries such as the UK, where the media takes a lesser role to other more trusted sources of information (doctors, nutritionists, consumer advocacy groups), the media is still found to play a role.

As frames, news stories offer the public definitions of social reality. In framing risk, the news media may define the agenda of public concern about a given technology, although there has been considerable debate about the role of the media in the formation of public opinion. Lang and Lang (1966: 468) observed that 'the mass media force attention to certain issues ... They are constantly presenting objects suggesting what individuals in the mass should think about, know about, have feelings about.' The agenda-setting function of the media has been succinctly summed up by Cohen (1963: 13) who noted the press 'may not be successful ... in telling people what to think, but is stunningly successful in telling people what to think about'.

Tuchman (1978) has pioneered the concept of a 'story frame', the application of Goffman's frame theory to the mass media. In reporting a story, journalists turn an occurrence into a newsworthy event, a newsworthy event into a story, which is then communicated to the public. Journalists and editors adjust frames according to their own understanding, their ideologies, styles and practical limitations such as deadlines and space (Best, 1990, 1991). Writing on science and technology can thus emphasize scientific facts, their sociopolitical implications, environmental risks, human health concerns and so on (Hornig, 1990).

News, like all public documents, is a constructed reality, assembling facts and information within a narrative structure, or frame, that serves to communicate an event or story to the reader (Tuchman, 1976). Through frames, media highlight certain points of view and marginalize or ignore others, defining occurrences and explaining how they are to be understood (Hornig, 1993). Science stories, for example, can emphasize scientific facts and discoveries, or their sociopolitical implications (Horning, 1990). A technology's potential risk to the environment can be highlighted, while its potential economic benefits are ignored, or vice versa, depending on the story frame. Analysis of how the media has framed the different risks and benefits of biotechnology can be systematically conducted using media content analysis. What follows is a brief outline of content analysis before a discussion of our methodological approach and results.

Media Content Analysis

'Content analysis is a systematic method for analyzing and quantifying message content and message handling. It is a tool for observing and analyzing the overt communication behaviour of selected communicators' (Budd *et al.*, 1967: 2). Instead of soliciting people's behaviour directly (through interviews), or measuring response to specific events or stimuli, content analysis may be used to analyse communications that people have produced as accounts of behaviour (Kerlinger, 1964).

The main advantage of the approach is that it allows the investigator to observe public messages at any time and place of the investigator's choosing. In addition, with the increasing availability of electronic data sources and computing power, content analysis opens up possibilities to analyse trends over longer periods of time and with a larger geographical scope than before.

The main disadvantage to content analysis is the possibility of inter-coder bias if text is manually coded. Second, data collection, coding and analysis of the data are very tedious and time-consuming. Inter-coder reliability and manual coding can be reduced or eliminated by computer processing of text. If manual coding is needed, inter-coder bias can be reduced or eliminated through careful construction of the coding categories and coder training (Budd *et al.*, 1967; Risse *et al.*, 1998). However, data collection is still time-consuming, even with new electronic databases.

Methodological Approach

We use Slovic's model of risk perception to analyse biosafety and food safety frames in reporting of agricultural biotechnology in UK and US newspapers. The period of coverage is from 1990 to 1999.

Sampling technique

A comprehensive database of all articles related to agricultural and food applications of biotechnology, published in the selected media, was developed based on an exhaustive list of keywords. Both animal and plant biotechnology applications are included in the population of articles. Electronic data sources were searched, resulting in a comprehensive electronic database of articles.

Figure 18.2 details coverage of agrobiotechnology by newspaper for the period of coverage 1990–1999. All newspapers have trended upwards over time; however, coverage of agricultural biotechnology issues has increased dramatically in UK newspapers since 1998, reflecting the intense debate that has taken place about the technology in the UK.

Categorization and coding of data

The most important step in content analysis is the identification and categorization of the variables under study. Categories, such as subject matter or direction categories, serve the same function as variables in content analysis. As Budd *et al.* (1967) argue, 'no content analysis is better than its categories'. Variables (categories) must be defined through an operational definition or set of definitions. These definitions should allow for systematic observation that implies reliability and repeatability. In addition, these categories must be exhaustive and mutually exclusive.

In this research, two coding categories were developed across the food safety and biosafety frames; namely, associated 'benefits' and 'catastrophic and memorable events' in reporting of the technology. It is well documented in the decision and psychometric literature that individuals are willing to trade-off potential benefits versus risks of differing technologies. Tolerance of even very minor risks may be small if individuals perceive no benefits associated with the risk (Frewer, 1999). On the other hand, very high benefits accruing to the risk-bearers can mitigate relatively high associated risks (Slovic, 1987).

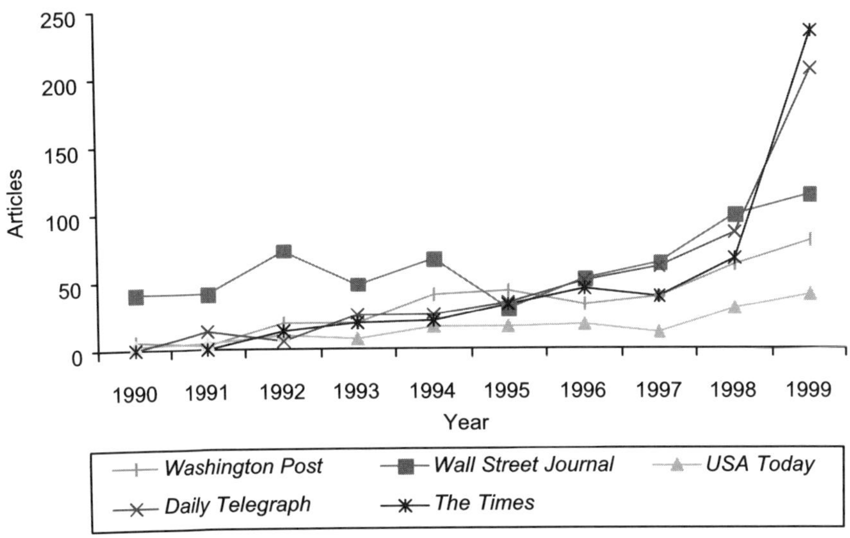

Fig. 18.2. Coverage of agrobiotechnology by paper, 1990–1999.

Coding and Context Units

Several different types of coding unit can be used to conduct content analysis, for example, words, assertions, themes and character units can all be counted. Words and phrases were used in this study because coding can be done electronically and word counting is objective. The contextual unit was words 'before' and 'after' the word or phrase included in each category. Hence, key-word-in-context (KWIC) analysis was used. Words or word phrases were included in categorical dictionaries that served to measure the degree of reporting on a specific benefit/risk frame. The dictionary capturing health and related benefits of GM foods included phrases such as 'enhanced flavour', 'food-fortification', 'abundant food supply', 'cheaper food', 'cleaner food' and so on. Catastrophic and memorable events related to health safety included the 'Alar scare', the 'mad cow disaster', the 'BGH scare', the 'dioxin scare', the 'Jack-in-the-Box scare' and so on. Biosafety benefits included phrases such as 'chemical free pesticides', 'cut pesticide use', 'eliminate the need for chemicals', 'reduce cultivation of the land', 'reduce soil erosion' and so on. Catastrophic and memorable events linked to biosafety included 'Chernobyl', 'DDT', 'Bhopal', the 'Exxon Valdez', 'Three Mile Island', 'Times Beach', the 'Monarch Butterfly', the extinction of the 'Dodo' and so on.

Table 18.1. Frequency of words by frame (1990–1999).

National daily newspaper	GM-food benefits	Catastrophic and memorable events
The Times	312	210
Daily Telegraph	161	263
Washington Post	181	45
USA Today	89	18
Wall Street Journal	233	43

Empirical Results

Food benefits and safety risks

Table 18.1 shows frequencies of words related to food safety by category and paper for the entire time period 1990–1999. Some general observations can be made about the frequencies in Table 18.1. First, the overall frequency of words is small as the objective but restraining coding dictionary limits the number of countable words. Furthermore, food safety is just one of several frames reported by the media. When a much broader categorization of content is used (such as unfavourable or favourable content) then frequencies are much higher (see Marks *et al.*, 1999). Despite the low frequency of reporting, the categorical variables provide a good indicator of overall coverage. Second, overall reporting of food benefits from biotechnology exceeds reporting of food risks for all papers except the *Daily Telegraph*. Reporting in *The Times* and the *Daily Telegraph* has regularly linked agrobiotechnology to memorable food safety lapses. Hence, this is quantitative evidence that the UK media has indeed reported more heavily on the food safety risks associated with biotechnology over its potential benefits.

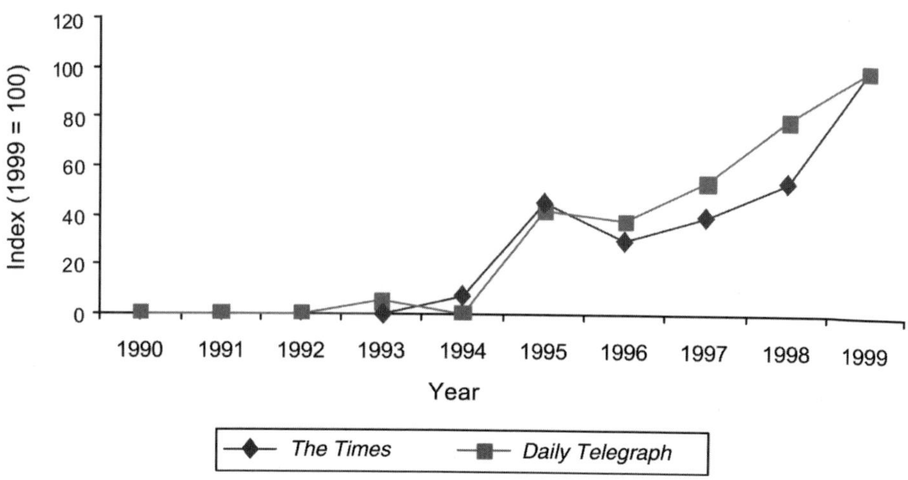

Fig. 18.3. Coverage of food safety risks (catastrophic and memorable events) in the UK.

One interesting result is the coverage of catastrophic and memorable events in the UK. Clearly, from 1995 onwards (1 year preceding the BSE food crisis) UK media set the agenda about biotechnology linking it explicitly with previous and ongoing food crises (see Fig. 18.3).

Results have varied by paper and by media frame. Formal chi-square tests for the food safety frame indicate that reporting of the perceived risks of GM food has been proportionately higher in the UK than in the USA across all papers. Specifically, the hypothesis tested is:

H_0: The proportion of reporting is the same between papers.
H_1: The proportion of reporting is different between papers.

In pair-wise comparison of the coverage between the *Daily Telegraph* and the *Washington Post*, $\chi^2 = 104.89$ (1 d.f., $P > 0.001$). The null is rejected indicating that the proportion of benefits to risks is statistically different. While there was no statistical difference among the US papers in their coverage of food safety, the UK papers were statistically different ($\chi^2 = 44.46$ (1 d.f., $P > 0.001$). The *Daily Telegraph* highlighted the risks significantly more than *The Times.*

Environmental benefits and risks

For the biosafety frame the results are quite different (see Table 18.2). Overall the reporting of environmental benefits is lower than food related benefits across all papers, with associated risks exceeding benefits for all papers except the *Wall Street Journal.*

Table 18.2. Frequency of words by frame (1990–1999).

National daily newspaper	Biosafety benefits	Catastrophic and memorable events
The Times	82	140
Daily Telegraph	64	167
Washington Post	53	153
USA Today	35	48
Wall Street Journal	142	110

The coverage of the *Daily Telegraph* and the *Washington Post* were significantly more focused on catastrophic and memorable environmental risks instead of benefits relative to other US and UK newspapers in pair-wise comparisons of frequencies. In pair-wise comparison of the coverage between the *Daily Telegraph* and *USA Today*, for example, $\chi^2 = 5.917$ (1 d.f., $P > 0.015$), and in comparison to the *Wall Street Journal*, $\chi^2 = 40.427$ (1 d.f., $P > 0.001$). Likewise, for the *Washington Post* versus *USA Today*, $\chi^2 = 7.551$ (1 d.f., $P > 0.006$), and versus the *Wall Street Journal*, $\chi^2 = 43.4$ (1 d.f., $P > 0.001$). Coverage between the *Daily Telegraph* and the *Washington Post* was not significantly different. Results for *The Times* and *USA Today* were more mixed. *The Times* was significantly more negative in its coverage of biosafety than the *Wall Street Journal*, $\chi^2 = 17.84$ (1 d.f., $P > 0.001$), but not significantly different from *USA Today*. The UK papers were statistically different from each other, $\chi^2 = 4.417$ (1 d.f., $P > 0.036$). The *Daily Telegraph* highlighted the risks significantly more than *The Times.* The *Wall*

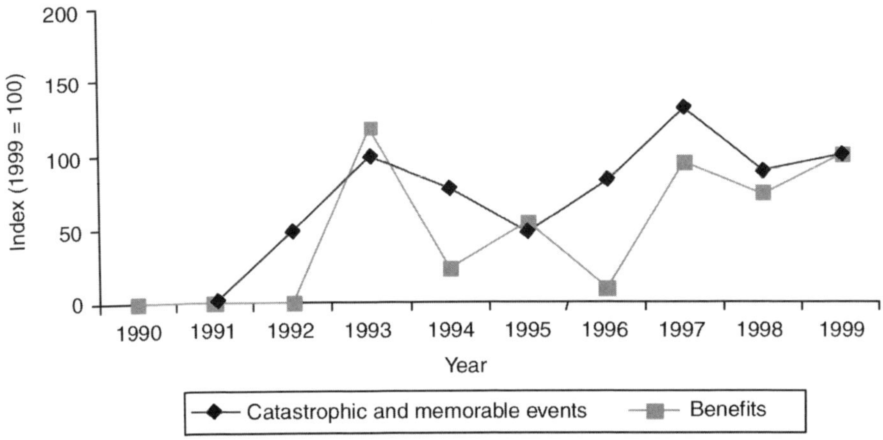

Fig. 18.4. Coverage of biosafety issues in *The Times*.

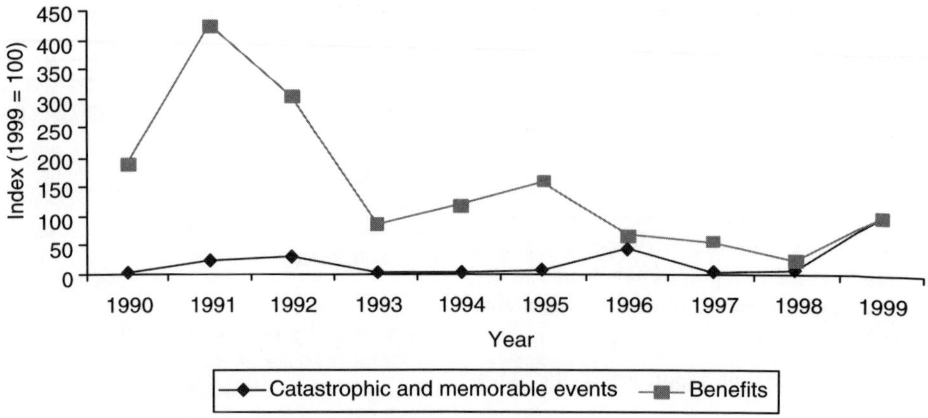

Fig. 18.5. Coverage of biosafety issues in the *Wall Street Journal*.

Street Journal was significantly more positive about biosafety than all the papers analysed (both US and UK).

These results indicate that on both sides of the Atlantic, environmental risks (e.g. uncontrollable transgenes) rather than benefits (e.g. lower pesticide use and associated benefits to water quality, land savings and lower impact on wildlife) have been the focus of newspaper reporting. Analogues to memorable and catastrophic environmental events (e.g. nuclear accident at Chernobyl) have been used in such coverage. On balance, UK newspaper coverage of such biosafety risks has been at least as negative, if not more so, than in the USA, and has had a modest upward trend in recent years (e.g. see Fig. 18.4). In the USA, on the other hand, the biosafety benefits of GM foods have been less emphasized over the time period (e.g. see Fig. 18.5).

Some Concluding Comments

In recent months, reports from various scientific societies (e.g. the US National Academy of Science, the Royal Society, the Chinese Academy of Science the Indian Academy of Science and others) have reported on agrifood biotechnology emphasizing the potential environmental and food benefits while acknowledging possible (but manageable) risks. Such treatises are in contrast with the broad media coverage of benefits and risks of biotechnology.

It is unclear whether negative media coverage of biotechnology emphasizing potential catastrophic and dread risks instead of possible benefits has played a significant role in shaping public opinion. What is clear from our initial results, however, is that those from the technical and scientific quarters supporting biotechnology have much work to do in effectively communicating their perspectives to broad public media.

References

Bauer, M.W., Durant, J. and Gaskell, G. (1998) *Biotechnology in the Public Sphere.* Science Museum, London.

Best, J. (1990) *Threatened Children.* University Press of Chicago, Chicago, Illinois.

Best, J. (1991) 'Road warriors' on 'hair trigger highways': cultural resources and the media's construction of the 1987 freeway shootings problem. *Sociological Inquiry* 61, 327–345.

Boon, S. and Griffin, D.W. (1996) The construction of risk in relationships: the role of framing in decisions about intimate relationships. *Personal Relationships* 3, 293–306.

Budd, R.W., Thorp, R.K. and Donohew, L. (1967) *Content Analysis of Communications.* Macmillan Company, New York.

Cohen, B.C. (1963) *The Press and Foreign Policy.* Princeton University Press, Princeton, New Jersey.

Covello, V. (1983) The perception of technological risks: a literature review. *Technological Forecasting and Social Change* 23, 285–297.

Entman, R. (1993) Framing: toward a clarification of a fractured paradigm. *Journal of Communication* 43(4), 51–58.

Frewer, L. (1999) Risk perception, social trust, and public participation in strategic decision making: implications for emerging technologies. *Ambio* 28(6), 569–574.

Goffman, E. (1974) *Frame Analysis.* Harper & Row, New York.

Hornig, S. (1990) Science stories: risk, power, and perceived emphasis. *Journalism Quarterly* 67(4), 767–776.

Hornig, S. (1993) Reading risk: public response to print media accounts of technological risk. *Public Understanding of Science* 2, 95–109.

Irwin, J.R. and Davis, J.H. (1995) Choice/matching preference reversals in groups: consensus processes and justification-based reasoning. *Organizational Behaviour and Human Decision Processes* 64(3), 325–339.

Kahneman, D. and Tversky, A. (1984) Choices, values, and frames. *American Psychologist* 39(4), 341–350.

Kates, R. and Kasperson, J. (1983) Comparative risk analysis of technological hazards (A Review). *Proceedings of the National Academy of Science, USA* 80, 7027–7038.

Kerlinger, F.N. (1964) *Foundations of Behavioral Research.* Holt, Rinehart and Winston, New York.

Krimsky, S. and Plough, A. (1988) The release of genetically engineered organisms into the environment: the case of ice minus. In: *Environmental Hazards: Communicating Risks as a Social Process.* Auburn House Publishing Company, Dover, Massachusetts.

Lang, K. and Lang, G.E. (1966) The mass media and voting. In: Berelson, B. and Janowitz, M. (eds), *Reader in Public Opinion and Communication,* 2nd edn. Free Press, New York.

Levin, I. and Chapman, D. (1990) Risk taking, frame of reference, and characterization of victim groups in AIDS treatment decisions. *Journal of Experimental Social Psychology* 26, 421–434.

Marks, L.A., Kalaitzandonakes, N. and Zakharova, L. (1999) Public opinion of agrobiotechnology in the US and Europe: a content analysis approach. Paper presented at the 1999 American Agricultural Economics Association (AAEA) Annual Meetings, Nashville, Tennessee. www.agecon.lib.umn.edu/cgi-bin/pdf_view.pl?paperid=1495

National Research Council (1989) *Improving Risk Communication from the National Research Council.* National Academy Press, Washington, DC.

Payne, J.W., Bettman, J.R. and Johnson, E.J. (1992) Behavorial decision research: a constructive processing perspective. *Annual Review of Psychology* 43, 87–131.

Risse, D., Lacy, F. and Fico, F. (1998) *Analyzing Media Messages: Using Quantitative Content Analysis in Research.* Erlbaum, Mahawaha, New Jersey.

Schulz, S., Burkink, T. and Marquardt, R. (2000) Consumer opinions toward the use of food made with genetically modified (GM) ingredients. Paper presented at the International Food and Agribusiness Management Association (IAMA) 2000 Agribusiness Forum, Chicago, Illinois. www.agecon.tamu.edu/iama/2000Congress/Forum%20%20Final%20PAPERS/Area%2011/Schulz_Steven.PDF

Slovic, P. (1987) Perceptions of risk. *Science* 236, 280–285.

Slovic, P. (1993) Perceived risk, trust, and democracy. *Risk Analysis* 13(6), 675–682.

Slovic, P. (1997) Trust, emotions, sex, politics, and science. In: Bazerman, M., Messick, D., Tenbrunsel, A. and Wade-Benzoni, K. (eds) *Environment, Ethics, and Behavior.* The New Lexington Press, San Francisco, California.

Slovic, P., Fischoff, B. and Lichtenstein, S. (1984) Behavioral decision theory perspectives on risk and safety. *Acta Psychologica* 56, 183–203.

Slovic, P., Fischoff, B. and Lichtenstein, S. (1985) Characterizing perceived risk. In: Kates, R., Hohenernser, C. and Kasperson, J. (eds) *Perilous Progress: Managing the Hazards of Technology.* Westview Press, Boulder, Colorado, pp. 91–125.

Tuchman, G. (1976) Telling stories. *Journal of Communication* 26(4), 93–97.

Tuchman, G. (1978) *Making News: a Study in the Construction of Reality.* Collier MacMillan, London.

Wimmer, R.D. and Dominick, J.R. (1987) *Mass Media Research: an Introduction,* 2nd edn. Wadsworth Publishing Company, Belmont, California.

19 Case Study in Benefits and Risks of Agricultural Biotechnology: Roundup Ready Soybeans

Janet E. Carpenter and Leonard P. Gianessi
National Center for Food and Agricultural Policy, Washington, DC, USA

Introduction

The development of crops tolerant to the herbicide glyphosate (Roundup) began in the early 1980s. The first generation of glyphosate tolerant soybeans was grown in a greenhouse during the winter of 1990–1991, the seeds of which were then planted in field tests during the summer of 1991 (Padgette *et al.*, 1996a). Approvals for commercialization of glyphosate tolerant soybeans were granted by FDA and USDA in 1994 and by EPA in 1995. Glyphosate tolerant soybeans, commonly known as 'Roundup Ready' soybeans, were first made available for planting by US farmers in 1996.

Glyphosate tolerant soybean varieties have been widely adopted by US growers. Figure 19.1 shows adoption of glyphosate tolerant soybeans since 1996 in the USA. By 2000, growers planted 54% of US soybean acreage to glyphosate tolerant soybeans (USDA NASS, 2000a).

Glyphosate controls weeds by inhibiting the enzyme 5-enolpyruvylshikimate-3-phosphate synthase (EPSPS), which catalyses the synthesis of amino acids essential for survival of plants and bacteria. EPSPS is present in plants, bacteria and fungi, but not animals, as animals do not make their own aromatic amino acids but rather receive them from plant, microbial or animal-derived foods.

Several bacteria exhibit tolerance to glyphosate. A glyphosate tolerant EPSPS from the soil bacterium *Agrobacterium* sp. strain CP4 was isolated and introduced into the genome of a soybean cultivar using the particle acceleration method. DNA is coated on to microscopic gold particles, which are then accelerated and penetrate target plant cells. Resulting cells are then incubated to produce shoots, which will eventually grow into mature plants. Successfully transformed plants are selected that exhibit unaltered agronomic traits from the parent line.

Soybean is the second largest crop in the USA after maize, planted on 30 million hectares in 2000 (USDA NASS, 2000a). Area planted to soybean has expanded in recent years owing to several factors. Improved yields through variety improvements, adoption of moisture-saving no-till practices, strong soybean prices relative to other crops, and elimination of acreage reduction programmes are all factors that have contributed to expanded plantings. Total soybean crop value in 1999 was US$13,000 million (USDA ERS, 1999).

Soybean acreage is centred on the Midwestern states, though 30 states have significant acreage planted to soybeans each year. Illinois and Iowa each plant over 4 million hectares of soybeans. Other major soybean states include Minnesota, Indiana, Missouri and Ohio.

The USA is the largest producer of soybeans in the world, growing nearly half of the total world soybean crop. Other major producing countries include Brazil, China and Argentina. The USA

(eds V. Santaniello, R.E. Evenson and D. Zilberman)

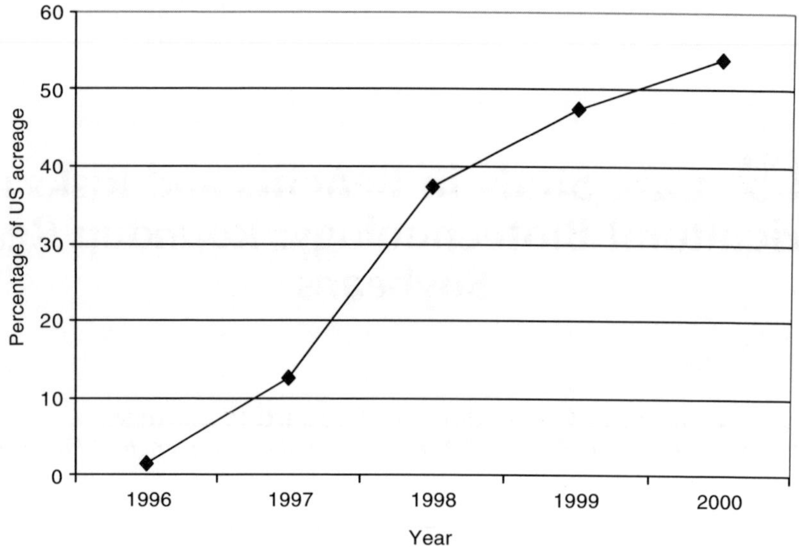

Fig. 19.1. Glyphosate tolerant soybean adoption. (Sources: K. Marshall, 2000, Monsanto, personal communication; USDA NASS, 2000a.)

exports approximately one-third of its soybean production, primarily to Asia and Europe, which together account for over 70% of total exports. Competition in export markets comes from Brazil and Argentina, as China is a net importer of soybeans (USDA NASS, 2000b).

Risks

Concern has been raised about the risks associated with GM crop varieties. Potential human health risks include allergenicity, toxicity and development of resistance to orally administered antibiotics. Environmental risks include potential for increased weediness of the crop plant, out-crossing of GM plants with closely related wild plant species, non-target effects and the development of pesticide resistance. In response to these concerns, US regulatory agencies routinely assess the risks involved with the introductions of GM plant varieties. The approval of crop varieties developed through biotechnology falls under the jurisdiction of three agencies: the US Department of Agriculture (USDA), the Food and Drug Administration (FDA) and the Environmental Protection Agency (EPA). The risks of agricultural biotechnology are examined in the context of the regulatory framework that governs the introduction of GM crops in the USA. The majority of risk studies that have been conducted on these new crop varieties has been conducted by the developers of the technology in order to meet the requirements of the regulatory agencies.

US regulatory framework

The US regulatory framework for agricultural biotechnology has evolved over time as new technologies emerged that allowed the manipulation of genetic material, beginning in the early 1970s. Over the past 25 years, policy has developed to address potential risks in a process open to public review and comment.

Initially, responsibility for oversight of the technology rested with the National Institutes of Health (NIH), but as applications of the technology changed, involvement of other agencies was deemed appropriate. As diverse products were presented for field-testing and commercialization (e.g. human insulin, ice minus bacterium, insect resistant tobacco, chymosin, rbST) involvement of the various agencies was required, and the system developed accordingly. Following is a brief overview of the development of regulations for agricultural biotechnology in the USA.

Concerns about the potential dangers arising from new recombinant DNA (rDNA) techniques first arose in the early 1970s. In 1973, scientists gathered at an annual conference on nucleic acids, known as the Gordon Conference, heard descriptions of experiments where DNA molecules from diverse sources were joined. By the end of the conference, many attendees had voiced reservations about the ethical and moral problems as well as the safety issues that might arise from the technology. The conference attendees voted that a letter should be sent to the National Academy of Sciences pointing out that a problem had been raised meriting investigation (Goodfield, 1977). It was also decided to go public with the issue, by publishing their letter in *Science* on 21 September 1973 (Singer and Soll, 1973). The letter noted that although no hazards had yet been established, 'prudence suggests that the potential hazards be seriously considered', and suggested that the Academies establish a study committee on the subject to recommend specific actions or guidelines as appropriate.

The National Academy of Sciences quickly convened a committee in 1974, publishing the recommendations in *Science* in July of that year (Berg *et al.*, 1974). The committee recommended that three types of experiments be deferred until the potential hazards were better evaluated or until adequate methods were developed for preventing the spread of biologically active recombinant DNA molecules: constructing replicating plasmids that would introduce either antibiotic resistance or bacterial poisons into bacterial strains; linking DNA from likely cancer-causing viruses to bacterial plasmids; and the linking of fragments of animal DNA to bacterial plasmid DNA or bacteriophage DNA. The committee also suggested that the director of NIH establish an advisory committee to develop an experimental programme to evaluate the hazards, develop procedures that would minimize the spread of such molecules within populations and devise guidelines to be followed by investigators. They also called for an international meeting of scientists to further discuss appropriate ways to deal with the potential biohazards of recombinant DNA molecules.

The Asilomar Conference was held in 1975, convening nearly 140 international scientists to 'review scientific progress in research on recombinant DNA molecules and to discuss appropriate ways to deal with the potential biohazards involved' (Berg *et al.*, 1975). The recommendations of the conference consolidated and extended those of the National Academy of Sciences Committee.

The Recombinant Molecules Advisory Committee (RAC) of the NIH began meeting as soon as the Asilomar Conference ended, working on research safety issues of experimental facilities and personnel, as well as of the proposed experiments themselves. In February 1976, the director of NIH called a public hearing in response to increased public interest in the subject. Four months later, in June 1976, NIH published its final guidelines for laboratories conducting recombinant experiments under federal grants (Goodfield, 1977).

As a standing committee, the RAC meets periodically to address and incorporate emerging scientific understanding of the potential risks involved with rDNA technologies. By 1983, experience with rDNA had allayed many fears, and NIH guidelines had been successively weakened to allow experiments that had been delayed awaiting better understanding of the associated risks. NIH had become comfortable with the vast majority of ongoing basic and biomedical research (Thompson, 1987). Risk assessment work helped to assure the scientific community and the public that many rDNA experiments were not as hazardous as originally believed (Korwek, 1997).

Although the NIH Guidelines govern only federally funded research, private industry and trade associations generally abide by the Guidelines as well. Through institutional biosafety committees, private industry reviews risk and ethical concerns of prospective research areas, referring any questions to the RAC for advice and consultation. It is believed that individuals and institutions that are not required to follow the NIH Guidelines do so for legal liability concerns. A 1987 General Accounting Office report found that private companies appeared to follow the Guidelines more closely than public sector organizations (Korwek, 1997).

The landscape of risk issues changed in the early 1980s as genetic engineering was to move out of the laboratory and into agricultural fields with the development of 'ice minus', a genetically altered bacterium intended for use on a variety of crops to reduce the risk of freezing. The regulation of a product that was to be purposely introduced into the environment presented quite a different set of issues from those involved with laboratory experimentation, the risks of which were controlled primarily by containing engineered materials and insuring against introduction into the environment.

Originally proposed in 1983, field testing of 'ice minus' was delayed through a series of legal challenges for 4 years. During this time, the authority of NIH over field tests was questioned, and EPA, USDA and FDA were proposed as the appropriate bodies for regulating in this area. The lack of coordination and uncertainty about oversight of biotechnology led to the formation of an interagency working group under the White House Cabinet Council on Natural Resources and the Environment. The working group was composed of approximately 13 member agencies, as an interagency effort to review regulatory requirements for conventional technologies, to clarify regulatory requirements for new products and to determine whether current regulatory requirements were adequate. Initial results of the working group were published for public comment in the Federal Register in 1984 (OSTP, 1984). The Office of Science and Technology Policy (OSTP) published its final notice of how each agency would regulate biotechnology applications in 1986, in the policy that would become commonly known as the 'Coordinated Framework' (OSTP, 1986). In this notice, existing laws were deemed adequate to oversee modern biotechnology applications. The notice also set forth which regulatory bodies were designated as the lead agency where the possibility of duplication of oversight existed (Korwek, 1997). USDA is the lead agency for plants grown to produce food or feed crops, while the food or feed itself is subject to regulation by FDA. EPA would primarily handle pesticide microorganisms. Notably, the initial policy of EPA addressed microbial pesticides, but did not address the regulation of pesticidal plants, which had not yet been developed at that point.

EPA, USDA and FDA each issued statements outlining their regulatory policy, which were incorporated into the Coordinated Framework. A common theme in the policies of all three agencies is the concept of product- not process-based risk assessment, based on the conclusion that the risks associated with the introduction of rDNA-engineered organisms are the same as those associated with introductions of unmodified organisms and those modified by other methods. This concept was supported by three reports, issued by the National Academy of Sciences and the National Research Council.

The first report was published in 1987, entitled *Introduction of Recombinant DNA-engineered Organisms into the Environment: Key Issues*, which concluded that the risks associated with the introduction of genetically engineered organisms were the same as those associated with introductions of unmodified organisms and those modified by other methods. In 1989, the National Research Council (NRC) issued *Field Testing Genetically Modified Organisms: Framework for Decisions*, which more specifically addressed the scientific foundation for regulatory decisions governing the release of genetically engineered microorganisms and plants into the environment. The 1989 report further supported the concept of the product not process-based standard for oversight put forth in the 1987 study. In 2000, the NRC released a report entitled *Genetically Modified Pest-protected Plants: Science and Regulation*, the purpose of which was primarily to evaluate the EPA's regulatory system for pesticidal plants. In the 2000 report, the committee was critical of EPA's policy of exemptions for plant varieties produced using particular methods.

The scope of regulation was the subject of a review prepared by the White House Council on Competitiveness, published in 1990 for public comment in the Federal Register. This document excluded from regulation organisms developed by traditional techniques, though the document did not propose any rules. The Council later published four principles of regulatory review for biotechnology and a report on national biotechnology policy. These publications, along with the final scope document, published in 1992, articulated a risk-based approach to regulation.

USDA

The USDA Animal and Plant Health Inspection Service (APHIS) is responsible for protecting US agriculture from pests and diseases. Under the Federal Plant Pest Act, USDA retains the authority to regulate plant pests and other articles to prevent direct or indirect injury, disease, or damage to plants, plant products and crops. In 1987, USDA published regulations that finalized the rule that was proposed under the Coordinated Framework (USDA APHIS, 1987). The requirements extended regulations imposed by APHIS for non-genetically engineered organisms or products which are plant pests or could harbour plant pests. APHIS promulgated these new regulations because it deemed that the existing regulations did not provide any way to determine whether or not a genetically engineered organism or product would fall under existing regulations of plant pests. The rule specifically notes that APHIS is not treating genetically engineered organisms and products differently from non-genetically engineered organisms. These regulations were amended in 1993 and 1997 (USDA APHIS, 1993a, 1997).

The regulations provide the rationale for determining whether a genetically engineered organism or product would be considered a 'regulated article', that is one with plant pest characteristics, and also calls for additional data for making a determination on the plant pest status of certain genetically engineered organisms or products. A genetically engineered organism is deemed a regulated article either if the donor organism, recipient organism or vector agent used in engineering the organism is listed in the regulation and is also a plant pest, is unclassified, or if APHIS has reason to believe that the genetically engineered organism presents a plant pest risk. This criterion for determining whether a particular modified plant is subject to regulation by APHIS was criticized in the recent National Research Council report, which noted that some pest-protected plant varieties did not fall under its scope given the definition of regulated article (NRC, 2000).

APHIS is responsible for approving introductions of GM crops at two stages: for field trials and for full market release. Prior to conducting field trials, it is necessary to either obtain a permit or notify APHIS. The notification option was established in 1993 for certain regulated articles with which the department is familiar, provided that the introduction is conducted in accordance with established requirements and standards (USDA APHIS, 1993a).

APHIS regulations also provide for a petition process for the determination of non-regulated status, which allows the unregulated movement and release of the product. In the petition for non-regulated status, applicants must 'describe known and potential differences from the unmodified recipient organism that would substantiate that the regulated article is unlikely to pose a greater plant pest risk than the unmodified organism from which it was derived' (USDA APHIS, 1993a).

Environmental assessments are prepared for field tests, and for petitions for non-regulated status. These assessments detail the nature of the genetic modification and assess the potential for environmental impacts from the introduction of the crop varieties into the environment. When a product is approved for full release, a Determination of Non-regulated Status is published in the Federal Register.

Lack of plant pest risk may be concluded when there is evidence that the plant under consideration: (i) exhibits no plant pathogenic properties; (ii) is no more likely to become a weed than its non-engineered parental varieties; (iii) is unlikely to increase the weediness potential for any other cultivated plant or native wild species with which the organism can interbreed; (iv) does not cause damage to processed agricultural commodities; and (v) is unlikely to harm other organisms, such as bees, that are beneficial to agriculture.

APHIS received a petition from Monsanto on 15 September 1993, seeking a determination from APHIS that glyphosate-tolerant soybean (GTS) line 40-3-2 and its progeny do not present a plant pest risk and are therefore not regulated articles (Re *et al.*, 1993). On 6 December 1993, APHIS announced receipt of the Monsanto petition in the Federal Register, stating that the petition was available for public review (USDA APHIS, 1993b).

APHIS received 33 comments on the Monsanto petition. With one exception, the comments were favourable to the petition. The one unfavourable comment stated that USDA should not approve the Monsanto petition or any other petition until the federal government has revised its oversight programme for transgenic crops at the commercialization stage, including establishment of standardized assessment and data collection schemes for consideration of risks of transgenic crops to ecosystems in the USA and worldwide, with particular attention to centres of diversity for food and fibre crops. The commenter also expressed the view that development of herbicide-tolerant crops should not be encouraged because they increase farmers' dependence on chemical herbicides (USDA APHIS, 1994b).

The Roundup Ready gene contained in GTS line 40-3-2 is a single insert of DNA comprising the enhanced 35S promoter derived from cauliflower mosaic virus, the chloroplast transit peptide coding sequence from *Petunia hybrida* fused to the 5-enolpyruvylshikimate-3-phosphate synthase (EPSPS) gene derived from *Agrobacterium* sp. strain CP4, and the nopaline synthase 3' terminator from *Agrobacterium tumefaciens*.

GTS line 40-3-2 has been considered a 'regulated article' because it contains components from organisms that are known plant pathogens: the bacterium *A. tumefaciens* and cauliflower mosaic virus. Field testing of GTS line 40-3-2 had been conducted with APHIS approval since 1991. Monsanto submitted its petition after the completion of field tests of GTS line 40-3-2 under nine APHIS permits. These permitted field tests took place at approximately 54 sites in 19 states and Puerto Rico. Additional trials were conducted in the USA and Puerto Rico under permit and notification during the 1993 growing season. All field trials were performed under conditions of physical and reproductive confinement.

The Monsanto petition describes the genetically engineered soybean plants and provides information relevant to determining whether glyphosate tolerant soybean plants are more likely than conventional varieties to become a plant pest. The petition addresses potential environmental consequences of unregulated release of glyphosate tolerant soybean varieties, including the development of glyphosate tolerant weeds, enhanced weediness, effects on non-target organisms, impacts of human and animal exposure, indirect effects on other agricultural products and the potential for outcrossing. Reports from field trials are included with observations of yields, plant growth, outcrossing, survival and gene expression gathered during field tests of glyphosate tolerant varieties compared with conventional varieties. Examples of the monitoring forms used by investigators who conducted the experiments are also included in the petition. In addition, letters from six land grant university weed scientists are included addressing the potential for development of weed resistance to glyphosate, weed population shifts and the overwintering of glyphosate tolerant soybeans, which were issues of concern to USDA.

APHIS granted the petition in May 1994, issuing a Finding of No Significant Impact. This conclusion was based on the nature of the genetic modification, the fact that soybean has no weedy relatives with which it can interbreed in the USA and its territories and the fact that this modification will not increase the weediness of the soybeans or negatively effect any non-target organisms, including beneficials (USDA APHIS, 1994b).

Weediness

Soybean (*Glycine max*) possesses few of the characteristics of plants that are notably successful weeds. *G. max* cv. 5403, the cultivar which was GM, is not considered to be a weed, and glyphosate tolerance is not expected to confer any additional weedy characteristics. Standard texts and lists of weeds give no indication that cultivated soybean is regarded as a weed anywhere (USDA APHIS, 1994a). Overwintering of soybeans is rare due a lack of innate dormancy. A lack of dormancy is selected for in commercial soybean seeds, so soybean seeds germinate quickly. Any seed that might remain in a field after harvest is likely to germinate, emerge and be killed by frost or field preparation for the following crop. Very few volunteers were observed in field testing. The number of seeds produced, germination characteristics, final stands, overwintering capability and disease or insect susceptibility were all found to be similar for the tested glyphosate tolerant line compared with conventional varieties. These findings were based on yield data and observations of germination, stand counts and disease or insect susceptibility. Further, increased weediness of the glyphosate tolerant soybean plant compared with conventional varieties would have to be due to selection pressure in association with glyphosate use. This was judged not to be an issue since glyphosate is not applied to the soybean for control of the soybean itself, but rather for controlling weeds in the field (USDA APHIS, 1994a).

Outcrossing

The genus *Glycine* is divided into two subgenera, *Glycine* and *Soja.* The first consists of 12 wild perennial species that are primarily distributed in Australia, South Pacific Islands, Philippines and Taiwan. The subgenus *Soja* consists of three annual species from Asia, *Glycine max, Glycine soja* and *Glycine gracilis.* The first species is the cultivated soybean, the second species is the wild form of the soybean and the third species is referred to as the 'weedy' form of the soybean.

Cultivated soybean is sexually compatible only with members of the genus *Glycine.* Cultivated soybean is the only member of the genus *Glycine* that grows in the USA and its territories and is sexually compatible with cultivated soybean, with the exception of specialized research collections. However, some members of the wild perennial species of subgenus *Glycine* may be found in US territories in the Pacific. There are no known reports of successful natural hybridization between the cultivated soybean and the wild perennial species.

The wild annual species, *G. soja,* is found in China, Taiwan, Japan, Korea and the former USSR. Natural hybridizations between *G. soja* and cultivated soybean occurs. *G. soja* is not native to North America and occurs only in research plots. There are no reports of its escape or dispersal from research plots. *G. soja* has never been found as a weed or naturalized in the USA. Thus, the possibility of gene transfer is very low within the USA.

Even if non-agricultural land containing any wild *Glycine* populations were near sites of commercial soybean production, it is highly unlikely that pollen from GTS line 40-3-2 would fertilize the wild relative, because soybeans are almost completely self-pollinated. The anthers mature in the bud and shed their pollen directly on to the stigma of the same flower, thus ensuring a high degree of self-pollination. Cross-pollination is generally very low and various studies have shown it to be from 0.03 to 3.62%. Honeybees are responsible for the occasional cross-pollination.

The limited potential for cross-pollination is evident in certified seed regulations for Foundation seeds, the most stringent category in the Certified Seed Regulations, which permit zero distance between different soybean cultivars in the field.

Non-target effects

Glyphosate tolerant soybeans were judged to have no detrimental effects on non-target organisms. EPSPS enzymes are present in plants and microorganisms and are therefore normally found in food and feed. No effects on non-target organisms were expected. The glyphosate tolerant EPSPS that was introduced into soybeans is not known to have any toxic properties. Field observations revealed no negative effects on non-target organisms including insects, birds or other species that frequent soybean fields (USDA APHIS, 1994a).

Weed resistance

Although the development of herbicide resistant weeds is not specifically considered by USDA in the approval process, Monsanto's petition to USDA provided information addressing this possibility. Glyphosate is considered to be a herbicide with a low risk for the development of weed resistance. Major factors which can contribute to the development of resistant weeds include: a single target site and a specific mode of action, broad spectrum of activity, long residual activity, and frequent applications without rotation to other herbicides or cultural control practices. Glyphosate essentially has no residual activity in the soil and is relatively quickly broken down by microorganisms in the soil. Also, there is no other herbicide on the market today that has the same mode of action as glyphosate. Glyphosate has been widely used for over 20 years, as a pre-plant burndown, directed, spot or postharvest treatments. However, some have questioned the impression of 'invincibility' of glyphosate to the development of resistance (Gressel, 1996). Resistant weed populations have been reported in Malaysia and Australia (Sindel, 1996; Doll, 1999).

Plant pest risk

APHIS also assessed the possibility that glyphosate tolerant soybeans would pose a plant pest risk due to the presence of pathogen-derived sequences. Neither of the gene sequences from *A. tumefaciens* or the cauliflower mosaic virus cause any plant or animal disease, is the source of pathogenicity in its host or encodes any polypeptide. No crown gall, the disease caused by *A. tumefaciens*, or cauliflower mosaic virus disease were observed in any glyphosate tolerant soybean plants during greenhouse or field studies.

Yields

Further information was submitted by Monsanto on 19 November 1993, to address a slight yield reduction observed at three of seven sites in initial yield trials.

EPA

The EPA assesses the safety of pesticides, both chemical and those that are produced biologically. Under the authority of the Federal Insecticide, Fungicide and Rodenticide Act (FIFRA), EPA regulates the distribution, sale, use and testing of plants and microbes producing pesticidal substances. Under the Federal Food, Drug and Cosmetics Act (FFDCA), EPA sets tolerance limits, maximum allowable residue concentrations, for substances used as pesticides on and in food and feed, or establishes an exemption from the requirement of a tolerance. The EPA also establishes tolerances for residues of herbicides used on novel herbicide-tolerant crops.

The goal of FIFRA is to register pesticides that do not have unreasonable adverse effects on human health or the environment and have benefits outweighing risks. Unreasonable adverse effects on the environment are defined as any unreasonable risk to 'man or the environment, taking into account the economic, social, and environmental costs and benefits of the use of any pesticide' (Korwek, 1997).

Any substance that is considered a pesticide under FIFRA is automatically subject to regulation under FFDCA if used on a food or feed crop (Nelson and Abramson, 1999). Until recently, EPA's decision making under FFDCA also involved a balancing of risks and benefits; however, only dietary risks to humans and other animals were considered, as opposed to FIFRA which also takes into account environmental risks (US EPA, 1994). Since passage of the Food Quality Protection Act in 1996, Congress has required EPA to apply a safety-only standard when examining the potential dietary risks associated with pesticide residues that may be found in food (Nelson and Abramson, 1999).

Early policy statements of EPA were focused on the regulation of GM microbial pesticides. A 1984 statement of interim policy required notification prior to small-scale field tests involving certain microbial pesticides, including those that had been genetically altered, in order to determine whether an experimental use permit (EUP) would be required for testing (US EPA, 1984). In 1986, as part of the Coordinated Framework, EPA published its statement of policy pertaining to regulating microbial pesticides under FIFRA, which sought to define which microbial products would be subject to review under FIFRA as well as the nature of the review (OSTP, 1986). In a 1989 Notice, EPA requested comments on the regulatory approach to microbial pesticides articulated in the 1986 policy statement (US EPA, 1989). A proposed rule was published by EPA in 1993, based on the 1984 interim policy and the 1986 proposed policy, addressing the requirements for small-scale field testing of microbial pesticides, as regards notification and EUPs (US EPA, 1993).

It was not until 1994 that the agency began to publish policy applicable to GM organisms other than microbial pesticides and products. That year, EPA published a proposed policy for 'plant pesticides' to be regulated under FIFRA and FFDCA. The 1994 proposed policy annunciated the agency's intent to regulate the pesticidal substances in plants, but not the plants themselves, leaving the regulation of the plants to USDA. This stance followed from an earlier policy by EPA to exempt from regulation under FIFRA all biological control agents, except for certain microorganisms, which has been interpreted to include plants (US EPA, 1994).

Several exemptions were proposed in the 1994 statement. First, plant pesticides derived through conventional breeding methods were granted a generic exemption from registration under FIFRA. Further, EPA proposed to exempt from regulation under FIFRA plant pesticides that are derived from sexually compatible plants. Viral coat proteins were also proposed to be exempt under FIFRA. Three categories of exemptions from tolerance setting under FFDCA were also proposed: plant pesticides that would not result in new dietary exposures, nucleic acids in plants and coat proteins from plant viruses. With these exemptions, the agency intended to regulate those plant pesticides that have the greatest potential for adverse effects, on the environment or on health (US EPA, 1994).

A recent report by the National Academy of Sciences addressed the issue of the exemptions proposed by EPA. Though the committee agreed that conventionally bred plants should be exempt for practical reasons based on historical safe use and benefits of these crops, the committee questioned the scientific basis used by EPA for this exemption. Regarding the exemption for plant pesticides derived from sexually compatible plants, the committee questioned the categorical nature of the exemption, while noting that exemptions for certain sexually compatible transgenic plant pesticides would be appropriate. The committee agreed that viral coat proteins should be exempt from regulation under FFDCA, but questioned the exemption under FIFRA due to concerns about potential outcrossing with weedy relatives (NRC, 2000).

The 1994 proposed policy also describes the risk issues with which the regulations are concerned. The following environmental risk issues are considered for both field testing and sale or distribution of a plant pesticide: increasing the ability of the modified plant to survive outside cultivation through the introduction of a specific trait; gene capture and expression of the introduced trait by a wild or weedy relative; potential for a trait conferring a selective advantage to a plant in a natural plant community with the result of increasing the 'weediness' of that species; environmental fate of the pesticidal substance, the dosage to soils after plant senescence and incorporation into the soil, rate of degradation or dissipation and transport in the environment. A further issue is whether or not the pesticidal substance is either exuded or volatilized from the plant during the growing season, resulting in a continuous application to the environment (US EPA, 1994).

Under FFDCA, EPA maintains jurisdiction over food safety issues related to the plant pesticide. Food safety issues related to compositional changes

Table 19.1. Glyphosate mammalian toxicology test results submitted to support revised glyphosate tolerances.

Subject animal	Type of study	Dosages	Results
Dogs	1-year feeding	0, 20, 100 and 500 mg kg^{-1} day^{-1}	NOEL 500 mg kg^{-1} day^{-1}
Mice	2-year carcinogenicity	0, 150, 750, 4500 mg kg^{-1} day^{-1}	No carcinogenic effects at 4500 mg kg^{-1} day^{-1}
Rats	Chronic feeding/ carcinogenicity	0, 3, 10 and 31 mg kg^{-1} day^{-1} (males) 0, 3, 11 and 34 mg kg^{-1} day^{-1} (females)	No carcinogenic effects at any dose level; Systemic NOEL of 31 mg kg^{-1} day^{-1} (males); Systemic NOEL of 34 mg kg^{-1} day^{-1} (females)
Rats	Chronic feeding/ carcinogenicity	0, 89, 362 and 940 mg kg^{-1} day^{-1} (males) 0, 113, 457 and 1183 mg kg^{-1} day^{-1} (females)	No carcinogenic effects at any dose level; Systemic NOEL of 362 mg kg^{-1} day^{-1} (males) based on increased incidence of cataracts and lens abnormalities, decreased urinary pH, increased liver weight and increased liver weight/brain ratio at 940 mg kg^{-1} day^{-1} (males); Systemic NOEL of 457 mg kg^{-1} day^{-1} (females) based on decreased body weight gain at 1183 mg kg^{-1} day^{-1}
Rats	Developmental	0, 300, 1000 and 3500 mg kg^{-1} day^{-1}	Developmental NOEL of 1000 mg kg^{-1} day^{-1} based on an increase in number of litters and fetuses with unossified sternebrae, and decrease in fetal body weight at 3500 mg kg^{-1} day^{-1}; Maternal NOEL of 1000 mg kg^{-1} day^{-1} based on decrease in body weight gain, diarrhoea, soft stools, breathing rattles, inactivity, red matter in the region of nose, mouth, forelimbs, or dorsal head and deaths at 3500 mg kg^{-1} day^{-1}
Rabbits	Developmental	0, 75, 175 and 350 mg kg^{-1} day^{-1}	Developmental NOEL of 350 mg kg^{-1} day^{-1}; Maternal NOEL of 175 mg kg^{-1} day^{-1} based on increased incidence of soft stool, diarrhoea, nasal discharge and deaths at 350 mg kg^{-1} day^{-1}
Rats	Multigenerational reproduction	0, 3, 10 and 30 mg kg^{-1} day^{-1}	Developmental NOEL of 10 mg kg^{-1} day^{-1} based on increased incidence of focal tubular dilation of the kidney of F3b pups
Rats	Two generation reproduction	0, 100, 500 and 1500 mg kg^{-1} day^{-1}	Developmental NOEL of 500 mg kg^{-1} day^{-1} based on decreased pup body weight and body weight gain on lactation days 14 and 21 at 1500 mg kg^{-1} day^{-1}; Systemic NOEL of 500 mg kg^{-1} day^{-1} based on soft stools in F0 and F1 males and females at 1500 mg kg^{-1} day^{-1}; Reproductive NOEL of 1500 mg kg^{-1} day^{-1}

NOEL, no observable effects level.

in the plant itself are under FDA jurisdiction. Environmental issues related to the plant itself are regulated by USDA APHIS, as mentioned above.

Crops that have been genetically modified to be herbicide tolerant do not face regulation under FIFRA, as the plants contain no pesticidal substance. EPA must grant any changes in tolerances for residues that might be needed to accommodate altered use patterns for in-season applications of the herbicide. Further, EPA must approve the modification of the label for the herbicide to allow for in-season use of the herbicide over the growing crops, which would not have been allowed previously.

In April 1996, EPA established new tolerances and feed additive regulations for the residues of glyphosate on several commodities for several end uses, in response to a number of petitions submitted by Monsanto. The revised tolerances were based on data submitted from several toxicological studies, as summarized in Table 19.1. In addition to those studies listed in Table 19.1, several acute toxicology studies were submitted that placed technical grade glyphosate in Toxicity Categories III and IV. All mutagenicity tests were negative. The carcinogenic potential of glyphosate has been judged to belong in Group E, evidence of non-carcinogenicity for humans, based on the lack of convincing carcinogenicity evidence in adequate studies in two animal species (US EPA, 1996). Revised tolerances for glyphosate are provided in Table 19.2. EPA approved a change in the label for Roundup to allow use of Roundup over the top of growing soybean plants in 1995. Since this change did not affect the registration of Roundup, this approval was not published in the Federal Register (Korwek, 1997).

Table 19.2. Glyphosate tolerances in soybeans (p.p.m.). (Source: EPA, 1996.)

	Revised tolerance
Soybeans	20
Soybeans, grain	20
Soybeans, aspirated grain fractions	50
Soybeans, forage	100
Soybeans, hay	200

FDA

FDA regulates foods and food ingredients, including animal feed and feed additives, under the FFDCA. The agency's authority to regulate the safety of food is generally exercised under two sections of the Act. Section 402(a)(1) applies to unintended occurrences of unsafe levels of toxicants in food. This section is the agency's primary legal tool for regulating the safety of whole foods, placing liability for food safety on the producer of a new food, and it is under this section that new plant varieties, including those produced using conventional techniques, have historically been regulated. Under this section, the agency retains the authority to remove a food from commercialization if it is found to be unsafe. However, under this section, there is no requirement for safety testing prior to commercialization. Section 409 of the Act applies to food additives, or intentional changes in the composition of foods. Under this section, premarket approval is required unless the food additive is generally recognized as safe (GRAS), or is a pesticide and therefore regulated by the EPA. The GRAS exception allows many ingredients derived from natural sources (e.g. salt, pepper, spices) and some chemical additives (some sweeteners, preservatives, artificial flavours) to be marketed without having been formally reviewed by FDA (US FDA, 1992).

In its 1986 statement, as part of the Coordinated Framework, FDA announced its intention to apply the existing regulatory framework to genetically engineered plant varieties. In that statement, FDA clearly states its intention to base its regulation of food on rational and scientific evaluation of the product, not on the process used to develop the product (US OSTP, 1986).

Further refinements to FDA policy were made in 1992 as the agency issued its policy statement establishing the regulatory framework under which FDA currently operates with regard to foods developed using biotechnology (US FDA, 1992). Under the 1992 policy, regulation of genetically engineered varieties under the food additive provisions of FFDCA which would require premarket review are interpreted to apply to the transferred genetic material and the intended expression product. The introduced genetic material itself is considered to be GRAS, as nucleic acids are present in the cells of every living organism. Expression products, such as proteins, carbohydrates, fat or oil, would only require premarket review if they differ significantly in structure, function or composition from a substance found currently in food, or sufficient safety issues are raised.

Several scientific issues are highlighted in the 1992 statement, including unintended effects, known toxicants, nutrients, new substances, allergenicity and antibiotic resistance selectable markers. These issues are the focus of FDA regulation of new plant varieties.

FDA has been particularly attuned to the potential of new plant varieties to cause allergies. The agency's principal concern is the possibility that an allergy-causing protein would be transferred from one food plant to another, making the recipient plant cause an allergic response in those allergic to the donor plant. In the case where a protein is derived from a commonly allergenic source, it is possible to test the new variety for allergenic responses in individuals known to be sensitive to the donor plant. For proteins that are derived from non-food sources, testing for potential allergenicity is less straightforward.

In April 1994, FDA, EPA and USDA hosted a scientific conference on allergenicity in transgenic food crops. Attendees concluded that methods are available to assess allergenic potential for proteins that are derived from sources to which consumers have reacted and for which serum is available, but it may be useful to establish a serum bank. There are no direct methods to assess potential allergenicity of proteins from sources that are not known to produce food allergy. Some assurance can be provided to minimize the possibility that a new protein will cause an allergic reaction by evaluating its similarity with characteristics of known food allergens. However, this is an area where more research has been called for. The National Academy of Sciences recommended that priority be given to developing improved methods for identifying potential allergens (NRC, 2000).

FDA is also concerned with the use of antibiotic resistance marker genes in transgenic plants and the risk of reducing the effectiveness of antibiotics in humans and animals (FDA, 1998). The kanamycin resistance marker gene is commonly used in transgenic plants. Calgene, the developer of the FlavrSavr® tomato, the first transgenic crop to be approved by FDA, requested that FDA subject the kanamycin resistance gene to evaluation under food additive regulations. At the time, FDA convened a Food Advisory Committee to consider Calgene's petition. The committee considered both direct risks of allergenicity and toxicity and the effects on the efficacy of antibiotics.

The 1992 policy statement includes a section on guidance to the industry for foods derived from new plant varieties, which describes scientific considerations for the evaluation of the safety and nutritional aspects of new plant varieties. The guidance section of the statement includes decision trees to assist developers in determining whether their product would be subject to regulation as a food additive or if consultation with FDA is necessary to determine the regulatory status of the product. Informal consultation with the agency has been standard practice for the food industry, and FDA expects that developers of genetically engineered varieties would continue this practice (US FDA, 1992).

One controversial aspect of the FDA policy is that no premarket review has been required for these crops. Consultations have been technically voluntary, though the agency knows of no product that has been commercialized without prior consultation with the agency. However, in 2000, consultations with the agency became mandatory (US FDA, 2000).

The most controversial aspect of FDA's policy has been the decision that foods developed using rDNA technology would not require labelling (FDA, 1993). This decision was based on the judgement that these products do not differ in any significant way from their conventional counterparts solely due to the process through which they were developed. It should be noted that labelling is required for genetically engineered foods that contain genetic material from foods that are commonly allergenic, unless it can be demonstrated that the allergenic property has not been transferred to the new plant variety. Further, plant varieties that have altered nutritional characteristics, such as modified oil content, would also require labelling.

Monsanto began the consultation process with FDA in June 1993. In accordance with the consultation guidelines, data describing the crop that was being transformed, the introduced genetic material, the identity and function of the expression product, comparison of composition of GM and conventional soybeans was included, as well as data and information addressing potential allergenicity and toxicity issues.

The safety evaluation can be broken down into two categories, unintended effects and intended effects, in accordance with the statutory structure, which regulates these effects differently, requiring premarket review only for intended effects under section 409.

Unintended effects

In order to address the possibility of the genetic modification having unintended effects on the crop,

Table 19.3. Animal studies submitted to FDA on glyphosate tolerant soybeans.

Animal	Feed	Duration of study	Parameters measured
Rats	Processed soybean meal	4 weeks	Mortality; body weight; cumulative body weight gain; organ weight; food consumption
Rats	Unprocessed soybean meal	4 weeks	Mortality; body weight; cumulative body weight gain; organ weight; food consumption
Broiler chickens	Processed soybean meal	6 weeks	Body weight; body weight gain; feed intake; feed/gain; liveability
Dairy cows	Raw soybeans	4 weeks	Milk production; fat-corrected milk*; milk composition; dry matter; net energy intakes; body weight changes; dry matter digestibility; nitrogen balance; volatile fatty acids in rumen; rumen nitrogen
Catfish	Processed soybeans	10 weeks	Feed efficiency; percentage weight gain; survival; food consumption*; body composition; moisture, protein, fat and ash in fillets
Bobwhite quail	Raw soybean meal	5 days	Mortality; body weight gain; food consumption

* Statistical differences found between animals fed conventional soybean product and GM soybean product.

studies were performed to assess the composition of the GM soybeans compared with conventional soybeans. In addition, wholesomeness studies were performed to evaluate any differences in feeding characteristics of GM soybean feed and conventional feed.

In the compositional analysis, evaluations were performed on seed, toasted meal, defatted meal (flour), protein isolate, protein concentrate, crude lecithin and refined, bleached, deodorized oil. Differences in seed composition were seen to be an indication that differences in other products would be found. Other products were chosen as they represent the various uses of soybeans. Toasted meal is widely used in animal feed. Defatted meal, protein isolate and protein concentrate are commonly used in food, as are lecithin and soybean oil. For seeds, the parameters compared were: protein, fat, fibre, ash, carbohydrate, amino acids, fatty acids, soybean seed proteins, trypsin inhibitor, lectin and isoflavones. Significant differences in fat, ash and carbohydrate were observed in one study, while no significant differences in these parameters were observed in similar studies conducted the next year. Protein, fat, fibre, ash and carbohydrate content were measured for defatted toasted meal, defatted non-toasted meal, protein isolate and protein concentrate, and no significant differences were found for these values between GM soybeans and conventional soybeans. Antinutrient content (trypsin inhibitor and urease, phytate, stachyose, raffinose, lectins, isoflavones) was measured in toasted meal and, apart from lectin concentrations which were below detection limits, no significant differences were found. Fatty acid composition was measured for soybean oil and no significant differences were found. The composition of crude lecithin was also compared, with no significant differences found.

Animal feeding studies were performed using rats, broiler chickens, dairy cattle, catfish and bobwhite quail. Two separate rat studies were performed, using processed and unprocessed meal. The study using processed meal was intended to address mammalian health issues, while the unprocessed meal was intended to address risks to wild animals that might feed on unharvested beans in the field. Broilers were included due to the prevalent use of processed soybeans in broiler operations. Similarly, dairy cattle were fed raw soybeans to reflect the widespread use of soybeans in cattle feed. Catfish were fed processed meal as this composes a great portion of feed used in aquaculture. Finally, bobwhite quail were fed unprocessed soybeans in order to address potential risks to birds that might feed on soybeans in the field. These studies were not designed as toxicology tests, but rather were undertaken to determine whether there were any differences in wholesomeness, or the ability to support growth and well-being. Table 19.3 summarizes the setup of the animal studies that were performed by Monsanto. It was concluded that no material differences were found in the wholesomeness of soybean products in any of the animal studies.

Intended effects

In the evaluation of intended effects, several aspects of the GM crop are considered: expression level of introduced protein, similarity of introduced protein to those already common in food and feed, allergenic potential, toxicity, prevalence of protein in food and feed, and changes in carbohydrate, fat or oil composition, structure or levels.

The expression level of CP4 EPSPS in soybean seed and processed soybean products was evaluated. In whole seed, the concentration of CP4 EPSPS was found to be 0.3 μg mg^{-1} fresh weight. Concentrations in toasted meal, defatted meal, protein isolate and protein concentrate were measured and found to be less than 0.1% of total protein. No enzymatic activity was found in any of the processing fractions.

The introduced protein, CP4 EPSPS was found to be similar to EPSPS already commonly present in food due to similarity in the reaction catalysed, amino acid sequence, homology of active site residues and three-dimensional structure.

Soybeans are known to cause allergies to some sensitive individuals. The allergenicity of GM soybeans was assessed in relation to conventional varieties. Known allergenic proteins of soybeans were found to be unchanged, based on an evaluation of protein extracts from non-toasted, defatted soy flour. Assessing the allergenicity of proteins that are not derived from allergenic sources is more problematic, as discussed above. CP4 EPSPS fits one of the criteria common to allergenic proteins, that of molecular weight, but does not share any of the other characteristics. Table 19.4 shows the characteristics common to allergenic proteins.

Table 19.4. Characteristics of known allergenic proteins (source: Monsanto, data submitted to FDA, obtained by FOIA request from FDA).

Characteristic	Allergens	CP4 EPSPS
Molecular weight 10–70 kDa	yes	yes
Glycosylated	yes[a]	no
Stable to digestion	yes	no
Stable to processing	yes	no
Similar to known allergens	—[b]	no
Similar to soybean proteins	—	yes
Prevalent protein in food	yes	no

[a] Typically but not absolutely.
[b] Implicit for allergenic proteins from soybeans.
FOIA, Freedom of Information Act.

Potential toxicity was assessed by considering the similarity of CP4 EPSPS to known protein toxins, an acute mouse gavage study and the study of the stability of CP4 EPSPS to digestion. First, CP4 EPSPS was not found to show any meaningful amino acid sequence homology when compared with known protein toxins in available databases. Next an acute mouse gavage study was performed, which resulted in no adverse effects (body weight, cumulative body weight and food consumption) at a dose representative of a 1300-fold safety margin relative to the highest potential human consumption of the protein in a diet including GM soybeans, maize, tomatoes and potatoes (assuming no loss in processing). An acute study was judged to be adequate in the toxicity assessment as proteins act as toxins by acute mechanisms. Finally, CP4 EPSPS was found to have a short half-life in simulated digestive fluids. The half-life was measured as less than 15 s in gastric fluids and less than 10 min in intestinal fluids. The relatively short digestion time of the protein indicates a reduced likelihood that the protein would be toxic.

Finally, the prevalence of CP4 EPSPS in the diet was considered. As CP4 EPSPS was found to represent 0.025% of the extractable protein in soybean seed tissue, it was not expected to become a macroconstituent of the human or animal diet. The addition of the CP4 EPSPS gene was also not found to alter the carbohydrate, fat or oil composition, structure or levels of the soybean compared with conventional varieties, as described in the compositional analysis above.

Monsanto has published the research results that were submitted to FDA on the composition of glyphosate tolerant soybeans, toxicity and feeding studies in a series of peer-reviewed articles in the *Journal of Nutrition* (Hammond *et al.*, 1996; Harrison *et al.*, 1996; Padgette *et al.*, 1996b). In addition, research results on the composition of glyphosate-tolerant soybeans treated with glyphosate, which were not submitted to FDA, were published in the *Journal of Agriculture and Food Chemistry* (Taylor *et al.*, 1999).

Benefits

The primary reason why growers have adopted Roundup Ready weed control programmes is the simplicity of a weed control programme that relies on one herbicide to control a broad spectrum of weeds without crop injury or crop rotation restrictions. Before the introduction of Roundup Ready soybean varieties, growers would choose between many herbicides, often applying three or more active ingredients, some of which would cause damage to the growing soybean plants, or cause harm to maize crops that commonly follow soybeans (Gianessi and Carpenter, 2000).

Roundup is a highly effective broad spectrum herbicide that controls both broadleaf and grass weeds. Each year, state extension services release weed control guides for field crops including soybeans. The guides provide information on the efficacy of available herbicide treatments on specific weed species, as well as ratings of crop safety. In the Michigan State University weed control guide, in which 182 treatments are rated on 24 different weed species, Roundup used over Roundup Ready soybeans received 23 good or excellent ratings. In addition, the Roundup treatment is rated with a minimal risk of crop injury. The next best available treatment with similar crop safety received 16 good or excellent ratings (Kells and Renner, 1999).

Growers also have more flexibility in timing herbicide treatments with the Roundup Ready system. Maximum weed heights at which Roundup is effective on most weed species are higher than other available herbicides. This allows growers to treat later if needed and still get effective weed control. Further, some commonly used soybean herbicides may cause injury to rotation crops. Because of this potential for injury to crops following soybeans, rotation restrictions are specified on the labels of these herbicides. For instance, sugarbeets may not be planted for 40 months after a field is treated with imazethapyr, a commonly used soybean herbicide.

Potential impacts of adopting Roundup Ready weed control programmes include changes in costs, yields and pesticide use. Roundup Ready programmes were introduced to be price competitive with existing conventional programmes. The introduction of competitively priced Roundup Ready programmes resulted in manufacturers of other products dropping their prices, in some cases by 40%. This resulted in an estimated US$216 million cost savings for soybean growers in 1999 compared with 1995, the year before Roundup Ready varieties were introduced, including the technology fee paid by growers who planted Roundup Ready varieties. Table 19.5 shows estimated soybean weed control programme costs for 1995, 1998 and 1999.

Table 19.5. Soybean weed control costs (US$ millions).

	1995	1998	1999
Herbicide expenditures	1865	1482	1441
Technology fee	0	160	208
Net weed control costs	1865	1642	1649

Calculated assuming herbicide expenditures in 13 states represent 80% of US total.

The impact on yields is less clear (Carpenter and Gianessi, 1999). Survey data on which to base a comparison of yields from Roundup Ready fields with conventional fields are scarce. Two areas of research assist in understanding the consequences of the adoption of Roundup Ready varieties on yield. The first is weed control research, comparing weed control strategies. The second type of research is variety trials, where the yield potential of conventional and Roundup Ready varieties have been compared.

In weed control trials, weed control programmes are compared in terms of efficacy against particular weed species and resulting yields. The purpose of these types of studies is to determine optimal herbicide application rates and timing to achieve control of various weeds. In general, these tests are conducted using a single variety. Recently, researchers have chosen to use Roundup Ready varieties in order to include Roundup treatments in their studies. Yield differences in these studies are due to more effective weed control and from avoiding crop injury. However, since only one variety is used in each study, the yield potential of the variety is not directly considered. It is difficult to generalize about the results from the weed control studies, although there seems to be no resounding yield advantage or disadvantage in the Roundup Ready systems compared with conventional programmes (Breitenbach and Hoverstad, 1998).

Variety trials are conducted by state universities to assess the characteristics of the varieties that will be available to growers the following year. The trials assess yield, maturity, lodging, protein and oil content, and resistance to pathogens and soybean cyst nematode and are generally maintained weed free, in order to eliminate competition from weeds as a factor influencing yield. Based on a compilation of variety trials from several states, it appears that available Roundup Ready varieties generally yield lower than conventional varieties (Minor, 1998; Oplinger *et al.*, 2000).

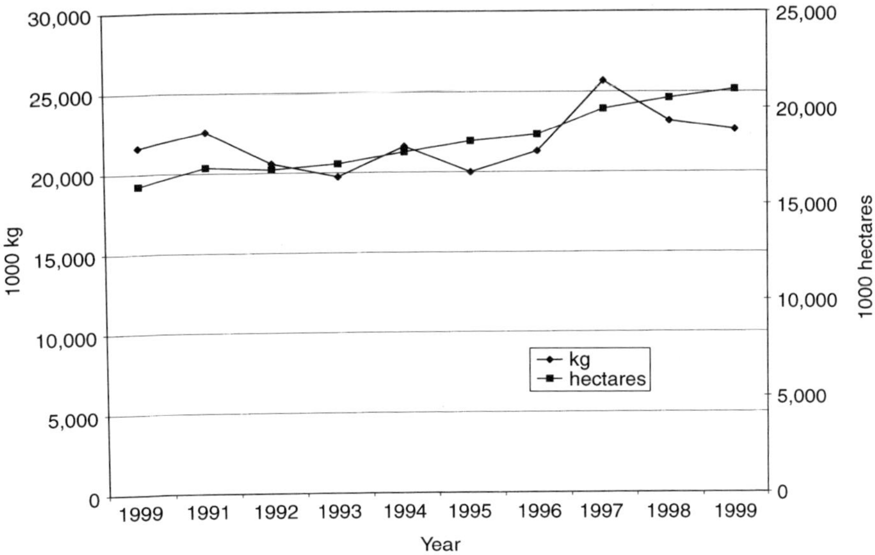

Fig. 19.2. Herbicide use in soybeans (AR, IA, IL, IN, MN, MO, NE, OH). (Source: USDA NASS, 1991–2000.)

Based on weed control and variety trials, it appears that Roundup Ready varieties do not have a yield advantage over conventional varieties.

Herbicide use in soybeans has been affected dramatically by the introduction of Roundup Ready soybean varieties. The USDA estimates the total number of acres treated and number of treatments by herbicide each year. The total number of pounds of herbicides used per soybean acre has remained unchanged since the introduction of Roundup Ready soybeans. The mix of herbicides being used in soybeans has changed. As one would expect, the use of glyphosate has increased, from being used on 20% of acreage in 1995 as a burndown or spot treatment, to being used on 62% of acres in 1999. The use of other herbicides has decreased. Imazethapyr, the most widely used soybean herbicide in 1995, was used on 44% of soybean acres in 1995, compared with 16% in 1999. Figure 19.2 shows trends in herbicide use and land area for 1990–1999 for eight states. Growers have also reduced the number of herbicide applications. Comparing 1995, the year before Roundup Ready varieties were introduced, and 1999, the last year for which data are available, the number of herbicide application-acres has decreased by 19 million, or 12%.[1] These changes in herbicide use occurred even though the total number of soybean acres increased by 18% between 1995 and 1999. The decrease in herbicide applications demonstrates that growers are using fewer active ingredients and making fewer trips over the field, which translates into ease of management.

Summary

Roundup Ready soybeans have been rapidly adopted by US farmers, yet their approval for commercialization is under scrutiny. This case study provides a description of the regulatory process governing agricultural biotechnology and traces the approval of Roundup Ready soybeans, summarizing the information that was submitted to US regulatory agencies by Monsanto. Estimates of the impact that the adoption of Roundup Ready soybeans has had on US agriculture are also provided.

The regulatory structure for agricultural biotechnology has evolved over the past 25 years, as

[1] An application-acre is the number of different active ingredients applied per acre multiplied by the number of repeat applications, and differs from the number of trips over the field, as one trip across the field to apply two active ingredients is treated as two applications, as is two treatments each containing a single ingredient.

technology allowing for genetic modification developed. The system continues to evolve as new and different applications of the technology emerge. Indeed, the most recent report of the National Research Council recommended that any new rules for regulating GM plants be flexible to reflect improvements in scientific understanding (NRC, 2000).

In reviewing the studies that were conducted on the safety of Roundup Ready soybeans, no indication of greater health or environmental risks were found compared with conventional varieties. The benefits of the introduction of Roundup Ready soybeans include a cost savings of US$216 million in annual weed control and 19 million fewer soybean herbicide applications per year.

References

Berg, P., Baltimore, D., Boyer, H.W., Cohen, S.N., Davis, R.W., Hogness, D.S., Nathans, D., Roblin, R., Watson, J.D., Weissman, S. and Zinder, N.D. (1974) Potential biohazards of recombinant DNA molecules. *Science* 185, 303.

Berg, P., Baltimore, D., Brenner, S., Roblin, R.O. III and Singer, M.F. (1975) Asilomar conference on recombinant DNA molecules. *Science* 188, 991.

Breitenbach, F. and Hoverstad, T. (1998) Roundup Ready soybeans. *Crop News, University of Minnesota* 4(29), 162–163.

Carpenter, J. and Gianessi, L. (1999) Herbicide tolerant soybeans: why growers are adopting Roundup Ready varieties. *AgBioForum*, 2(2). www.agbioforum.org

Doll, J. (1999) Glyphosate resistance in another plant. University of Wisconsin, *Wisconsin Crop Manager Newsletter* December.

Gianessi, L.P. and Carpenter, J.E. (2000) *Agricultural Biotechnology: Weed Control Benefits of Herbicide Tolerant Transgenic Soybeans.* National Center for Food and Agricultural Policy, Washington, DC.

Goodfield, J. (1977) *Playing God: Genetic Engineering and the Manipulation of Life.* Random House, New York.

Gressel, J. (1996) Fewer constraints than proclaimed to the evolution of glyphosate-resistant weeds. *Resistant Pest Management* 8(2), 2–5.

Hammond, B.G., Vicini, J.O., Hartnell, G.F., Naylor, M.W., Knight, C.D., Robinson, E.H., Fuchs, R.L. and Padgette, S.R. (1996) The feeding value of soybeans fed to rats, chickens, catfish and dairy cattle is not altered by genetic incorporation of glyphosate tolerance. *Journal of Nutrition* 126, 717–727.

Harrison, L.A., Bailey, M.R., Naylor, M.W., Ream, J.E., Hammond, B.G., Nida, D.L., Burnette, B.L., Nickson, T.E., Mitsky, T.A., Taylor, M.L., Fuchs, R.L. and Padgette, S.R. (1996) The expressed protein in glyphosate-tolerant soybean, 5-enolypyruvylshikimate-3-phosphate synthase from *Agrobacterium* sp. strain CP4, is rapidly digested *in vitro* and is not toxic to acutely gavaged mice. *Journal of Nutrition* 126, 728–740.

Kells, J.J. and Renner, K.A. (1999) *Weed Control Guide for Field Crops.* Extension Bulletin E-434, Michigan State University, East Lansing, Michigan.

Korwek, E.L. (1997) *1997 United States Biotechnology Regulations Handbook,* Volume 1. FDLI, Washington, DC.

Minor, H. (1998) Performance of GMO's vs. traditional varieties: a southern perspective. *28th Soybean and 53rd Corn/Sorghum Seed Research Conferences,* Chicago, Illinois, 9–10 December.

National Research Council (2000) *Genetically Modified Pest-protected Plants: Science and Regulation.* National Academy Press, Washington, DC.

Nelson, D.T. and Abramson, S.H. (1999) High-tech crops: emerging legal issues. *ACCA Docket,* no. 6, 34–56.

OSTP (Office of Science and Technology Policy) (1984) *Proposal for a Coordinated Framework for Regulation of Biotechnology.* 49 Fed. Reg. 50,856.

OSTP (Office of Science and Technology Policy) (1986) *Coordinated Framework for Regulation of Biotechnology; Announcement of Policy and Notice for Public Comment.* 51 Fed. Reg. 23,302.

Oplinger, E.S., Martinka, M.J. and Schmitz, K.A. (2000) *Performance of Transgenetic Soybeans – Northern U.S.,* Proceedings of the 28th Soybean Seed Research Conference, American Seed Trade Association, Washington, DC.

Padgette, S.R., Re, D.B., Barry, G.F., Eichholtz, D.E., Delannay, X., Fuchs, R.L., Kishore, G.M. and Fraley, R.T. (1996a) New weed control opportunities: development of soybeans with a Roundup Ready gene. In: Duke, S.O. (ed.) *Herbicide Resistant Crops,* CRC Press, Boca Raton, Florida, pp. 53–84.

Padgette, S.R., Taylor, N.B., Nida, D.L., Bailey, M.R., MacDonald, J., Holden, L.R. and Fuchs, R.L. (1996b) The composition of glyphosate-tolerant soybean seeds is equivalent to that of conventional soybeans. *Journal of Nutrition* 126, 702–716.

Re, D.B., Padgette, S.R., Palmer, R., Delannay, X., Kolacz, K., Nida, D., Peschke, V., Derting, C., Rogers, S., Edwards, J., Barry, G. and Biest, N. (1993) Petition for determination of nonregulated status: soybeans with a Roundup Ready gene. Monsanto, St Louis, Missouri.

Sindel, B. (1996) Glyphosate resistance discovered in annual ryegrass. *Resistant Pest Management,* Winter, 702–716.

Singer, M. and Soll, D. (1973) Guidelines for DNA hybrid molecules. *Science* 181, 1114.

Taylor, N.B., Fuchs, R.L., MacDonald, J., Shariff, A.R. and Padgette, S.R. (1999) Compositional analysis of glyphosate-tolerant soybeans treated with glyphosate.

Journal of Agriculture and Food Chemistry 47, 4469–4473.

Thompson, P.B. (1987) *Agricultural Biotechnology and the Rhetoric of Risk: Some Conceptual Issues,* Discussion Paper FAP87-01. National Center for Food and Agricultural Policy, Resources for the Future, Washington, DC.

USDA Animal and Plant Health Inspection Service (1987) *Introduction of Organisms and Products Altered or Produced Through Genetic Engineering Which are Plant Pests or Which There is Reason to Believe Are Plant Pests.* 52 Fed. Reg. 22,892.

USDA Animal and Plant Health Inspection Service (1993a) *Genetically Engineered Organisms and Products; Notification Procedures for the Introduction of Certain Regulated Articles; and Petition for Nonregulated Status.* 58 Fed. Reg. 17,044.

USDA Animal and Plant Health Inspection Service (1993b) *Monsanto Co.: Receipt of Petition for Determination of Nonregulated Status of Genetically Engineered Soybean Line.* 58 Fed. Reg. 64,287.

USDA Animal and Plant Health Inspection Service (1994a) *APHIS-USDA Petition 93-258-01 for Determination of Nonregulated Status for Glyphosate-Tolerant Soybean Line 40-3-2: Environmental Assessment and Finding of No Significant Impact.*

USDA Animal and Plant Health Inspection Service (1994b) *Response to Monsanto Petition P93-258-01 for Determination of Nonregulated Status for Glyphosate Tolerant Soybean Line 40-3-2.*

USDA Animal and Plant Health Inspection Service (1997) *Genetically Engineered Organisms and Products; Simplification of Requirements and Procedures for Genetically Engineered Organisms.* 62 Fed. Reg. 23,945.

USDA Economic Research Service (1999) *Oil Crops Situation and Outlook.*

USDA National Agricultural Statistics Service (2000a) *Acreage.*

USDA National Agricultural Statistics Service (2000b) *Agricultural Statistics.*

USDA National Agricultural Statistics Service (1991–2000) *Agricultural Chemical Usage: Field Crops Summary.*

US Environmental Protection Agency (1984) *Microbial Pesticides; Interim Policy on Small Scale Field Testing.* 49 Fed. Reg. 40,659.

US Environmental Protection Agency (1989) *Microbial Pesticides; Request for Comment on Regulatory Approach.* 54 Fed. Reg. 7,026.

US Environmental Protection Agency (1993) *Microbial Pesticides; Experimental Use Permits and Notifications.* 58 Fed. Reg. 5,878.

US Environmental Protection Agency (1994) *Proposed Policy; Plant-Pesticides Subject to the Federal Insecticide, Fungicide, and Rodenticide Act and the Federal Food, Drug, and Cosmetic Act.* 59 Fed. Reg. 60,496.

US Environmental Protection Agency (1996) *Pesticide Tolerances for Glyphosate.* 61 Fed. Reg. 15,192.

US Food and Drug Administration (1992) *Statement of Policy: Foods Derived From New Plant Varieties.* 57 Fed. Reg. 22,984.

US Food and Drug Administration (1993) *Food Labeling; Foods Derived From New Plant Varieties.* 58 Fed. Reg. 25,837.

US Food and Drug Administration (1998) *Guidance for Industry: Use of Antibiotic Resistance Marker Genes in Transgenic Plants: Draft Guidance.* 4 September.

US Food and Drug Administration (2000) *FDA to Strengthen Pre-market Review of Bioengineered Foods.* May 3.

20 Labelling for GM Foods: Theory and Practice

Peter W.B. Phillips and Heather McNeill
Department of Agricultural Economics, University of Saskatchewan, Saskatoon, Saskatchewan, Canada

Introduction

Genetically modified foodstuffs entered the global food system in the early 1990s and are now in a wide selection of raw and processed foods. The Canadian Food Inspection Agency (CFIA) has indicated that processed foods could contain up to 70% genetically modified (GM) ingredients if GM oils are measured. By the end of 1999 more than 40 genetic modifications related to 13 different crops were approved. Twelve countries produced at least one GM crop and, to varying degrees, these crops were available to other countries through international trade (James, 1999). A number of countries have also approved release of one or more varieties of GM fish (e.g. salmon), active ingredients (e.g. brewers' yeast and chymosin), drugs (e.g. recombinant bovine somatotropin) and vaccines for animals.

Since 1998, consumers have become increasingly concerned about the safety of GM foods. While extensive risk assessments have been undertaken and most countries and firms have developed some level of risk management systems, communication of the risks has lagged so much that many consumers are demanding mandatory labelling systems. Consumers argue that labelling will allow them to make their own choices about GM foods. The regulators, as well as distributors and other representatives of the supply chain recognize their role in developing efficient policies to deal with consumer concern about GM foods.

Our survey shows that 18 countries plus the European Union (EU), 29 manufacturers, 21 retailers and six restaurant chains around the world had signalled intentions to adopt voluntary or mandatory labels for GM foods or to eradicate GM ingredients by February 2000. The survey also illustrates that to date there has not been any convergence towards common standards. Rather, coverage, tolerances, conformity measures and implementation dates vary among firms and states.

In this chapter we examine the theory, practice and implications of GM food labelling systems that have emerged to date. The next section offers a short outline of the background of GM foods to the recent rise in demand for labelling. There follows a summary of the theory and literature on labelling for GM foods and a description of the survey methods used to identify labelling policies and practices. The following section sets out the results of the survey and a summary of private and public labelling policies. The final section discusses possible effects of the proposed labelling requirements on international food trade.

Background and Circumstances

Genetically modified foods entered the food chain in the early 1990s, with the commercialization of a variety of modified industrial ingredients (e.g. chymosin and brewers' yeast). Beginning in 1995 a

(eds V. Santaniello, R.E. Evenson and D. Zilberman)

wide variety of modified plants was introduced in a selection of countries and then exported to many consumer markets around the world.

Also in 1995, the UK beef industry and UK regulators learned that there could be a causal link between a new variant Creutzfeldt-Jakob disease (nvCJD) in humans and bovine spongiform encephalopathy (BSE). BSE had been endemic in UK cattle herds since the mid-1980s. Although the UK and European Union (EU) food safety regulators responded by tightening rules for producing and exporting UK cattle, their response often lagged behind private policies. The regulatory system appeared unprepared and therefore incompetent in developing public policy.

A number of events converged in 1998 which heightened awareness of GM foods and contributed to consumer concerns about food safety. Most notably, Arpad Pusztai of the Rowett Research Institute in Aberdeen reported on a BBC television show that research he had done with Stanley Ewen indicated that diets containing GM potatoes expressing the lectin Galanthus nivalis agglutinin (GNA) had variable effects on different parts of the rat gastrointestinal tract. This result was misleading because it was lab research; the GM potato had not been evaluated by the regulatory system and therefore had not been released into the food system. As a result, the public viewed the regulatory board as less credible. This was the first scientific result that questioned the safety of GM foods. UK and EU regulators were not prepared to confront public concerns.[1] The UK Government responded by developing national labelling which came into force in March 1999.

Also in 1998, Iceland Foods, a high-street food retailer, announced that it would remove GM ingredients from its own-label products. Greenpeace UK, Friends of the Earth and a wide range of public interest groups supported this move. The issue was picked up by British tabloids and became the subject of a vigorous circulation battle (Marks, 1999).

Although many of the triggers for a change in consumer attitudes occurred in the UK, the concerns quickly spread through Europe, into North America and to other importing countries. Bungling of the BSE problem by UK regulators was shortly followed by food safety scares across Europe (e.g. dioxin contaminated feed products, tainted cheeses, 'bad' CO_2 in Coca Cola) which regulators were either unwilling or unable to handle efficiently. These events caused many consumers to question the domestic and international regulations for GM foods. For many consumers, the regulators' answers were not satisfactory. During the late 1990s in Europe the regulations were more open to political influences. Consequently, the regulations could be affected by consumer groups and were usually tentative responses. In contrast, the North American system was science based and did not provide an opportunity for public influence. Outside Europe and North America, there was limited effective regulatory capacity (Isaac and Phillips, 1999).

Given the extensive trade in GM foods and food products, many consumers are uncertain about what they are consuming. Many consumers want to have a choice about what they consume. The Organization for Economic Cooperation and Development (OECD) (1999) recently completed a survey in key markets that asked consumers whether they wanted labels. UK consumers were 94% in favour, EU consumers were 74% in favour and Australian and New Zealand consumers responded 91% in favour of labels. Depending on how the question was asked, consumers in Canada were 83–99% in support and consumers in the USA were 45–93% in support of labels.

A review of recent customer surveys by Environics Research Group (2001) showed that between 1998 and 2000, anywhere from 72 to 98% of Canadians think that companies should be ordered to label their products.

A recent survey by Angus Reid and *The Economist*, shown in Table 20.1, reported that between 57 and 82% of consumers in the EU and North America would be less likely to buy GM labelled products if labels were used. In contrast, only 5–37% of consumers would be more likely to buy food labelled as 'genetically modified'. A further survey by Angus Reid (1999) in Canada reported that 66% of Canadians would pay more to have GM content labelled; 79% of those would be willing to pay 5% more, 48% would pay 10% more and 20% would pay 20% or more.

The consumer polls indicate that there is a variety of consumer concerns and awareness about

[1] When the authors subsequently submitted it for peer review, they chose to withdraw some of their conclusions, and the key scientific societies in the UK and elsewhere have challenged both the methodology and the results of the study. Nevertheless, the public largely believes that GM products are unsafe.

Table 20.1. Consumer attitudes to GM food, 1999 (source: Angus Reid/*The Economist* World Poll, 1999).

Country	Read or heard anything about GM foods (%)	Would less likely buy GM labelled products (%)	Would more likely buy GM labelled products (%)
Germany	95	82	5
UK	94	67	25
Japan	89	70	23
Australia	85	63	14
France	79	78	18
Canada	78	68	28
USA	66	57	37

GM foods. The demand for labelling indicates that there are commercial opportunities to segment the food market into GM and non-GM food products.

The Literature

Neoclassical economic theory suggests that two factors are necessary to create value in a product: consumer tastes and preferences and producer efforts to develop consistent, safe, affordable food products that meet consumer demand. Neoclassical theory further proposes that there are no costs associated with gathering information regarding product characteristics and that most, if not all, of these elements can and should be produced within minimally regulated markets.

Increasingly, however, the literature is suggesting that trust and confidence are vital in the creation and operation of markets (Fukuyama, 1995; Stiglitz, 1999). Markets for many products are often not able to create, by themselves, the conditions of trust that generate the socially optimal quantities of goods and services produced and consumed. If trust exists between two parties, there are fewer costs associated with gathering information, monitoring and enforcing a contract. In short, trust lowers the transaction costs and therefore creates more efficient markets.

In the food industry, the existence of a food product is often enough to signal that the product is safe for human consumption. Still, regulators can create trust through safety standards and labelling. Establishing a sense of trust in the market for GM agrifood products, where perceived risks and public uncertainties abound, is critical to the success of markets in which information costs are very high.

The characteristics of a good affect the type of regulation it requires. Tirole (1988) identifies three types of goods: search goods, where consumers can visually identify attributes before consumption; experience goods, which require consumption to determine the attributes; and credence goods, where the unaided consumer cannot know the full attributes of consuming a good, at least for some period after consumption. Markets for search goods are for the most part able to function efficiently based on simple transactions; primitive barter economies and street markets all thrive with little or no government intervention. Experience and credence goods, such as GM foods, require a greater element of trust, which must involve active communication about the product's attributes. Those product factors that involve probabilistic or hypothetical public health and safety risks are usually regulated by the state. Formal communication of this regulation is sometimes signalled through labels (e.g. Canada Choice meats) but more often is simply implied by the presence of the product in the food chain. GM foods, however, also involve a wide array of speculative risks that the state does not handle (Phillips and Isaac, 1998). Following on Ackerlof's work (1970) on the market for lemons, Bureau *et al.* (1997) and Giannakas (1999) suggest that in some instances where consumer fears are high enough, the absence of labels could result in global welfare losses. As long as public regulators are unable both to handle and to communicate their efforts in handling this risk, consumers will remain concerned.

Most food safety regulatory systems, both domestically and internationally, regulate risks based on the product attributes and not on the processes used to generate the product (Phillips and Buckingham, 2000). Process risks are usually managed by other legislation or regulation. All countries hold GM foods to at least the labelling requirements for other foods. In short, it is mandatory that GM foods in all markets must be labelled if they introduce a protein that commonly poses an allergy risk or the nutritional content differs from the norm for that food (Nenon, 1999). Most national regulatory systems assume that speculative risks are simply consumer preferences and not risks requiring mandated labelling. As with many rules, however, there is an exception, which makes the rule. In the 1970s all governments agreed to require labelling for irradiated food, and enshrined the agreement in the *Codex Alimentarius.* One result of that was that the supply chain did not adopt irradiation because of consumer fears. This experience at least partly explains producers' and exporters' concerns about mandatory labelling of GM elements in foods.

Caswell (1998) argues that labelling is an essential element in an efficient market because it offers consumers full information. Labels help to match producers that use different technologies with consumers who want to buy products with specific process attributes. Mandatory labelling would operate similarly to the irradiation labelling while voluntary labelling would probably operate in a similar way to other labelling systems that signal a wide variety of 'non-risky' processes that consumers prefer, such as 'green' status, organics, kosher, halal and ethically produced foods. She suggests that either voluntary or mandatory labelling will do the job, but that mandatory labelling would be more costly as the entire market would need to be segregated and labelled, rather than only the portion of production and consumption that valued specific process traits. Assuming that the minority preference would be segregated, Table 20.1 suggests that at most 43% of any single market (the US market that is indifferent to GM foods) and at a minimum 18% (the German market that is indifferent to GM foods) would need to be segregated. Caswell concludes 'governments are likely to prefer voluntary or mandatory approaches based on their perceptions of what proportion of their citizens want information about the technology'. Caswell suggests that public demand for information regarding new technology will influence a government's decision to pursue voluntary or mandatory labelling.

This chapter examines the practice of labelling as it is used both by governments and by private firms.

The Data

The authors undertook a search of various Internet sources between 20 December 1999 and 8 February 2000, in order to identify countries or firms that have adopted or have announced plans to adopt either voluntary or mandatory labelling for GM foods or food products. The search focused primarily on two main Internet sources: Agnet, an Internet archive of news and journal briefs produced by researchers at the Agri-Food Risk Management and Communications Project at the University of Guelph (www.plant.uoguelph.ca/ risk comm/newstoday/today.html#Agnet) and the *Financial Times* of London (www.ft.com/). Additional searches were undertaken of the OECD BioTrack Online (www.oecd.org/ehs/country.htm) and *The Economist* (www.economist.com/). The authors acknowledge that these sources may offer incomplete data.

The State of Labelling for GM Foods

The objective of the survey was to describe labelling regulations designed for food products containing (or perceived to contain) GM organisms. The survey summarized the labelling requirements of 19 countries plus the EU, 29 food manufacturers, 21 food distributors/retailers and six restaurant chains. The survey reveals that states and the food industry have adopted a variety of GM labelling regulations. The regulations differ in terms of administration of the regulation, the degree to which the regulation affects food production and the date that the regulation was or will be effective.

National legislation

At a national level, labelling regulations can be categorized as voluntary, mandatory or mandatory only above some tolerance level. Under a voluntary labelling scheme GM-free foods will be positively labelled. Table 20.2 presents the status of national rules for labelling GM foods.

Voluntary labelling has been adopted by Canada, the USA and Argentina. Canada and the USA have initiated efforts to assist industry to develop standards and implement consistent and credible voluntary labelling systems. In September 1999 the Canadian Government announced it would support efforts by the Canadian Council of Grocery Distributors (CCGD) and the Canadian General Standards Board (CGSB) to develop a Canadian standard for the voluntary labelling of GM foods. The CGSB set up a committee supported by food

Table 20.2. Status of national rules for labelling GM foods, February 2000.

States	Labels	Coverage	Effective date
Australia–New Zealand	M	0% GM content in processed foods, fruits, vegetables, take-aways, restaurants	n/a
Argentina	V	No details	n/a
Canada	V	Being developed	2000 or 2001
China	M	Considering a bill for mandatory labelling	n/a
Czech Republic	M	All products of GM origin or ingredient	n/a
Ethiopia	M	All products	n/a
EU	M	Dir 90/220: law requiring labelling of all foods and food products containing GMOs; no tolerances set	1990
	M	Reg 258/97: 1% tolerances; mandatory labelling of foods; no regulation for chymosin, additives or feeds	15 May 1997
	M	Reg 1139/98: specific rules for GM soybeans and maize	26 May 1998
UK	M	Includes both grocery store and restaurant foods; not for additives/flavourings/food on sale in UK before 1 September 1998	1 March 1999
Ireland, Spain, France	M	Want to label GM additives and preservatives	n/a
Austria	M	Opposed to labelling; want full ban on GM foods	n/a
The Netherlands	M	Propose mandatory labelling for animal feed	n/a
Hungary	M	Products containing/derived from GM material (excluding feed and novel food)	1 July 1999
Indonesia	M	Proposed regulations	n/a
Japan	M	Proposed MHW regulations could require all GM products to be labelled, regardless of the tolerances; proposed MAFF regulations exempt additives, animal feeds and any ingredient representing less than 5% of content	1 April 2001
Mexico	M	Senate has approved a bill for GM foods to be labelled as 'transgenic' or 'made with transgenic products'	n/a
Poland	M	Conform to EC 219/90 and 220/90	n/a
Russia	V	Plans to study/promote the use of GM crops for feed	n/a
Slovenia	M	Conform to EC 219/90 and 220/90	n/a
South Africa	M	A new law is proposed for 2002	2002
South Korea	M	GM maize, soybean and bean sprout	March 2001
Switzerland	M	Conforming to EC 219/90, 220/90 and 90/679	n/a
Thailand	M	No details	n/a
US	V	GM food must be 'substantially equivalent'; food exporters will meet EU standards	2000?

Notes: M, mandatory; V, voluntary; n/a, not available.

industry, producer and consumer interests, which was scheduled to report recommendations in 2000. Agreement proved difficult to attain, because the committee works on consensus, and is still some way away (*The Western Producer*, 2000b).

In the USA, the USDA announced in September 1999 an independent scientific review of its biotechnology regulatory system. Public meetings in December 1999 highlighted public concerns about the lack of labels. In response, the industry is working with the government to develop a voluntary labelling system. Meanwhile, in November 1999 the 'GE Right to Know Act' which would rule for mandatory labelling was introduced to Congress and by February 2000 had the support of 48 congressional members (equal to about 10% of the voting members). One recent development has been the increase in local efforts to impose new labelling rules. As of May 2000, altogether 16 US states had introduced bills that would require labelling for GM foods (Niiler, 2000).

At the other extreme, 15 countries, including the EU and its member states, have adopted or announced plans to implement mandatory labelling systems. As of May 2000, the EU and Japan had revealed the structure of their labelling rules. The UK was the only country that had implemented labelling rules (mainly Directive 258/97). A number of countries have proposed mandatory labelling. For example, Hungary, Poland and Switzerland have indicated that they are following or will adhere to the EU standards. Still, there is no available evidence that these countries have developed domestic systems to manage those regulations. China, the Czech Republic and Thailand have also announced deadlines for implementing mandatory labelling regulations for GM foods, but none of them has indicated when their system might be operating. At least three countries have announced definite dates to implement mandatory labelling systems. Japan and South Korea plan to implement labelling regulations by spring 2001 while South Africa announced it intends to implement its labelling regulations by 2002.

Although many countries have announced plans to implement mandatory labelling, only a few of them have developed their proposals sufficiently to be able to identify the tolerance level for GM content that will trigger mandatory labelling. Four merit some discussion.

In the EU, a number of directives set the framework for labelling systems in member states. Dir 90/220, the environmental regulation (in particular Annex III), sets the basic legal framework for labelling in the EU, mandating that products that contain GM elements should be labelled; the Directive did not, however, set tolerances. In the first instance, this Directive required labelling of GM varieties in the seed guides. The EU Novel Foods Regulation 258/97 then set the 1% tolerance level for foods. Because GM varieties of maize and soybeans were already released before Reg 258/97 was adopted, the EU passed Reg 1138/98 to explicitly require regulations for those two varieties. None of these directives or regulations requires labelling for GM additives, flavourings or active ingredients. Furthermore, although animal feed must be labelled, the meat produced using those feeds does not require labels under current rules. Finally, EU regulators have ruled that processed edible oils from GM maize, soybeans and canola will not require labels, as they do not contain any novel proteins. All of the novel traits are left in the meal, which if consumed by humans must be labelled. Given the practice of subsidiarity in the EU, Union Directives and Regulations do not come into effect until member states enact those provisions in their national laws. A number of EU member states have indicated intentions to go beyond the EU base, extending labelling requirements to additives and preservatives.

So far, the UK is the only EU member state to enact national legislation to activate the EU rules. As of March 1999 all foods, additives and flavourings that entered the market after 1 September 1998 that contain more than 1% GM content require labels. In April 2000 the new Food Safety Agency extended that provision to all GM foods, additives and flavourings, including those on the market before 1998. The UK also requires that all restaurant meals involving GM foods are labelled. In support of these rules, the UK has adopted a range of financial penalties for mislabelling of product. One feature that could complicate the UK regulatory system is the recent devolution of legislative authority to the Welsh Assembly and Scottish Parliament. Wales has already attempted to exert some influence over UK regulation of GM foods by proposing to reject approval of a GM maize variety that had been approved in England. In this case, if the Welsh decision had stood, the variety would not have been commercialized in the UK. By extension, the devolution of authority to the regional assemblies could lead to some inconsistencies across the UK.

Japan has proposed a set of mandatory

labelling regulations to be effective on 1 April 2001. Two agencies have offered rules that will affect how GM products are marketed. The Ministry of Health and Welfare (MHW) has developed regulations that will require all GM food products and food additives to be assessed for safety. To date, MHW has approved 29 GM crops and 6 GM food additives as safe to enter Japan. Furthermore, MHW requires that food products made from the safety-approved crops and additives must be labelled. As of April 2001, no GM food is authorized for marketing in Japan without having secured the required safety assessments by the MHW.

In contrast, the Ministry of Agriculture, Food and Fisheries (MAFF) has proposed regulations that will require labelling for 24 proscribed food products (no food additives are included). MAFF regulations would require labels for rDNA ingredients only if the ingredient is one of the top three food ingredients by weight and composes at least 5% of the total weight of the product. Labelling would not be required on packages less than 30 cm^2. In support of MAFF regulations, Japan will require importers to label as GM all bulk shipments with more than 5% GM content. Between 1 and 5% tolerances, products would have to be labelled 'may contain'. Shipments containing less than 1% GM material would not require any labelling.

It is not clear whether both the MHW and MAFF rules will be in effect. MHW requires all GM food products and additives to be labelled while MAFF requires the specified 24 food products (no food additives included) to be labelled. Under MAFF regulations, Japan could allow unlabelled products with up to 5% GM content. Alternatively, MHW would require labelling of products with even a trace of GM material. It is too early to tell how the system will actually operate; however, the MHW is expected to follow MAFF legislation.

Australia and New Zealand, via the joint Australia–New Zealand Food Standards Council, have announced plans to introduce mandatory labelling for any food, processed food, fruit, vegetable, take-away or restaurant meals that have any detectable amount of GM content. In effect, their tolerance is 0%.

Although these tolerance levels provide some potential for consumer choice, there is no consistency or consensus among national governments over the level or enforceability of tolerance levels. Furthermore, it is not clear exactly what the label will say. There are a variety of messages that could be used, offering a variety of information and connotations (e.g. 'GM ingredients', 'Warning: contains GM ingredients' or a symbol indicating the presence of GM products). Several EU and Japanese firms have taken the initiative to label products with less than 1% GM content as 'GM free' or 'non-GM'.

Labelling policy is being administered by a wide variety of governmental bodies. For example, in Argentina and Japan the Agriculture Ministry is in charge, while in South Korea it is the Ministry of Foreign Affairs. In Canada the Canadian Food Inspection Agency and Health Canada are both involved. In the USA, the FDA is responsible. In the UK, the Department of the Environment, Transport and the Regions enforces the labelling requirement for GM foods, while the FSA remains responsible for developing and communicating the rules and standards. These different locations for responsibility also affect the orientation and procedures used to evaluate options and manage the process.

Application of the rules also varies widely. Many of the proposed labelling rules tend to focus on consumption of soybeans and maize, which together accounted for more than 80% of the global GM acreage in 1999. The other two main GM crops are cotton, which is not ingested by humans, and canola, the oil of which is deemed GM free and the meal which is not ingested directly by humans. Meanwhile, only Australia, New Zealand and the UK have proposed labelling regulations for take-aways and restaurants. Furthermore, only the UK has rules in place to require labelling of GM material in food additives (e.g. chymosin in cheese, lecithin and artificial flavours). Most countries (e.g. Canada and the EU) already regulate additives through other laws and already have extensive labelling requirements for those elements. For example, in Canada the Food and Drug Act regulates the use of additives. However, the EU recently announced that it would pursue labelling requirements for additives with detectable GM protein. Finally, no country has yet required meats fed on GM feeds to be labelled, although the EU has mentioned the need for such regulations. Scientific evidence (*The Western Producer*, 2000a) so far does not show any transmission of GM elements between the feed and the animals. Furthermore, initial results from consumer surveys tend to suggest that so far consumers have few concerns about GM feeds.

Finally, a number of countries have expressed interest in or concern about GM foods but have yet to announce whether they will impose labelling rules. Brazil, Turkey, Ethiopia and Singapore are all in the undecided camp while Russia has recently started to study and promote GM crops for feed, but has not stated whether GM products for humans should be labelled or consumed.

Private industry labelling systems

Private companies in many cases have not waited for national legislation to clarify the requirements for GM food labels. Following Iceland Foods' decision in 1998 to make their own-label products GM free and to market them as such, a wide array of retailers, processors, restaurateurs and wholesalers have followed suit (Table 20.3).

Several firms in the EU, particularly in the UK, have initiated private labelling. Legislation at both the EU level and at the member state level has signalled to producers the direction the industry is going. Rather than wait for the final dust to settle, however, private firms have in many cases acted before the legislation has been promulgated. The main catalyst can probably be traced to the UK. All foods in the UK have had to be labelled since March 1999. Given that many European food processors have integrated their operations on a continental basis, it is next to impossible to produce GM-free foods for the UK and GM foods for the rest of Europe. Those companies with significant market shares in the UK have a clear incentive to make their entire EU production GM free (Bredahl, 1999). Several UK firms have endorsed more stringent labelling rules than required by national law. For example, Marks & Spencer has announced it will tolerate less than 0.1% GM content in own-label products; otherwise products will not be sold. Marks & Spencer also has a line of 'GM free' meat and eggs. The predominance of UK firms with strict private regulations is related to the extensive regulations the UK has adopted, including greater scope for labelling and hefty fines (£5000) if GM contents are found in an unlabelled product.

EU retailers have also responded aggressively. Even though EC Reg 258/97 does not operate yet in many of the EU member states, seven large European supermarket chains have joined forces to eliminate GM ingredients in their own-label products. The consortia initially included Carrefour (France), Delhaize (Belgium), Marks & Spencer (UK), Migros (Switzerland), Sainsbury's (UK), SuperQuinn (Ireland) and Tesco (UK). In addition, Effelunga in Italy, Edeka Association on the continent and UK retailers Somerfield, Tesco, Waitrose, Iceland Foods and Northern Foods, and many UK wholesalers claim that their own-label products do not contain GM materials. When Carrefour, France's largest supermarket and the world's third largest retailer, decided to remove GM ingredients from its own-brand products, it discovered that 516 of its 1783 own-label products contained GM ingredients. They replaced GM ingredients with non-GM substitutes in 286 of the products; for 221 products where alternative ingredients were not available, Carrefour offers a guarantee of origin and has demanded that its suppliers guarantee and prove their products do not contain GM ingredients. Nine product lines were discontinued because it was impossible to guarantee their GM-free status (*Ram's Horn*, 1999).

Although national legislation has induced many private labelling requirements, there are several food retailers that have pursued labelling policies without national encouragement. For example, Woolworth's (South Africa) and Park N'Shop (China and Hong Kong) have announced they will eliminate all own-brand products with GM ingredients (or find substitutes) from their shelves. While the Grocery Manufacturers of America strongly oppose mandatory labelling, three US-based retailers have announced intentions to go GM free. Two organic and natural-food stores – Whole Foods Mkt Inc. and Wild Oats Mkts Inc. – have been joined by Genuardi's Family Markets in Philadelphia. Three chains – Loblaws in Canada and Pick n Pay and Checkers (Shoprite) in South Africa – have indicated that they will not remove GM products from their shelves nor will they make their own label products GM free.

The action by large grocery chains and the pending labelling rules under EU legislation has prompted at least 23 manufacturers and food processors to remove all products with GM content or replace GM with non-GM ingredients in their product lines (Table 20.4). In addition, five processors have signalled that they do not want GM wheat or durum (which is not expected in the market until 2003) while only one processor (in Japan) has said it does not want to segregate and label GM products. While a few processors have announced that they will make all of their products worldwide GM free (e.g. Gerber, FritoLay, Seagram), most of the processors have targeted to remove GM traits from only

Table 20.3. Labelling systems planned or adopted for retail food chains as of February 2000.

	Target market	Non-GM product coverage
Carrefour	France	Own label food products
(Shoprite) Checkers	South Africa	Will not remove GM products
Delhaize	Belgium	All products
Edeka retail association	Germany, Czech Republic, Denmark, France, Poland	All products
Effelunga	Italy	All products
Iceland Group	UK	Own label products (as well as artificial colours/flavours)
Loblaws	Ontario, Canada	Does not want segregation of GM products
Marks & Spencer	UK	Own label goods (~ 1800 products), < 0.1% tolerance level
Migros	Switzerland	All products
Northern Foods	UK	All products (or label), except derivatives
Park N'Shop	Hong Kong	Own label products (~ 600 products)
Pick n Pay	South Africa	Will not remove GM products
Sainsbury's	UK	All products
Somerfield	UK	All products
Superquinn	Ireland	All products
Tesco	UK	All products; encouraging non-GM feed
UK Co-op	UK	Eggs to be produced from chickens fed on a GM-free diet
Waitrose	UK	Own label products
Walmart (Asda)	UK	Own label products
Whole Foods Mkt Inc.	US (Texas)	All products
Wild Oats Mkts Inc.	US (Colorado)	All products
Woolworths	South Africa	Will seek alternatives to GM products, or label

selected markets. Some have removed only selected GM inputs (e.g. potatoes, tomatoes, maize and soybean), some have targeted selected products (e.g. baby foods) while some have adjusted only for selected markets (especially EU and Japan). Kellogg's, for example, produces Corn Flakes with non-GM maize from Argentina for the European market. Meanwhile, Unilever and Nestlé in the UK, both global food processors, announced in 1999 that they would remove GM ingredients from their products going into the UK market (Bowditch, 1999, www.bowditchgroup.corn/index2.htm). International food processors such as Dannon (Danone), Gerber, Heinz, Kraft, M&M, Pillsbury, Seagram and Quaker have announced that they will eliminate GM content in their products on the European market. Only four processors have targeted other markets: Groupo Maseca, a Mexican taco manufacturer, is pursuing a GM-free supply of maize flour; Yan Wai Yun in Thailand claims its soy sauce is GM free; and McCain Foods and Midwest Foods in Canada announced recently that they will only buy GM-free potatoes for their French fries.

Japanese firms are the second largest group to have adopted strict labelling systems. Several Japanese firms removed or announced they intended to remove GM ingredients rather than labelling products above the 5% GM tolerance level even

Table 20.4. Labelling systems planned or adopted by food processors as of February 2000.

	Market	GM product coverage
Barilla	Italy	Signalled to Canadian Wheat Board does not want GM durum
Bestfoods Deutschland	Germany	Does not want any GM products
Danone	Europe	Signalled to Canadian Wheat Board does not want GM durum
FritoLay	World	Reported to be buying only GM-free maize
Gerber	World	Removed GM ingredients from own-label baby foods
Groupo Maseca	Mexico	Removed GM maize flour from tacos
Heinz	World	Removed GM ingredients from own-label baby foods
	EU	Removed GM tomatoes from own-label products
Japanese Nat. Tofu Manufacturer	Japan	Reports that it will remove GM soybeans from tofu products
Japanese Oilseed Processors	Japan	Reports that it does not support segregation of GM products
Kellogg's	Europe	Removed GM maize from Corn Flakes
Kibun Food Chemifa Co.	Japan	Replacing GM maize oil with isomeric sugar in products
Kikkoman	Japan	Switching to organic soybeans for own-label soy sauce
Kirin	Japan	Removing GM maize from own-label products
Kraft	Europe	Removing GM foods from all own-label products
McCain Foods	Canada	Will only buy non-GM potatoes for French fries; will continue to buy canola oil for cooking
Midwest Foods	Canada	Will only buy non-GM potatoes for French fries
M&M Mars	Europe	Removing all GM foods from own-label products
Nabisco	Europe	Signalled to Canadian Wheat Board does not want GM grains
Nestlé	Europe	Removing GM foods from own-label products where possible; labelling all own-label products with GM ingredients
Nippon	Japan	Replacing GM maize starch with wheat starch
Nisshin Flour Milling Co.	Japan	Replacing GM maize and soybean with wheat products
Perdue Chicken	USA	'Positioning' to use only GM-free feed
Pillsbury	Europe	Removing GM foods from all own-label products
Quaker	Europe	Removing GM foods from all own-label products
Sapporo (brewery)	Japan	Seeking GM-free maize
Seagram (distiller)	World	Seeking GM-free maize
Taihei Co.	Japan	Labels one soy sauce product as 'GM free'
Unilever	Europe	Signalled to Canadian Wheat Board does not want GM grains
Warburtons (bread maker)	UK	Signalled to Canadian Wheat Board does not want GM grains
Yan Wai Yun (soy sauce)	Thailand	Label all own-label products as 'GM free'

before the law came into effect on 1 April 2001. For example, the Japanese National Tofu Manufacturers Association and soy sauce producers Kikkoman and Taihei Co. have announced that they will use only non-GM soybeans in food production. Japanese breweries Kirin and Sapporo have stopped using transgenic maize in their products. Nippon (food processor), Kibun Food Chemifa Co. (food processor) and Nisshin Co. (flour mill) have announced that they will replace maize products with wheat products.

Several firms have made prospective statements, warning their suppliers that they will not accept GM inputs. For example, food manufacturers such as Danone, Unilever and Warburtons have announced that they will eliminate GM ingredients from their products and have warned the Canadian Wheat Board (CWB), in particular, that they will not accept GM wheat. Barilla, an Italian pasta maker, has also warned the CWB not to supply GM durum. These statements demonstrate the firms' intentions, as no wheat, barley or durum is yet produced by GM means. They have already had some effect, as Agriculture and Agri-food Canada decided in the face of rising consumer opposition to GM products to stop research efforts to develop transgenic, herbicide tolerant durum (Calamai, 2000).

So far there has been only limited interest by processors in extending GM labelling to animals. While there has been some debate in the EU and USA, so far there appears to be only one processor (Perdue Chicken, one of the largest poultry producers in the USA) and one retail chain (UK Co-op) that was 'positioning' to use only GMO-free feed. In France, Carrefour and Cana-Caval, an agricultural cooperative, have removed GM ingredients from the pet food they make and sell (Caplan, 2000).

Finally, a number of international fast-food chains, responding in the first instance to UK labelling laws that require GM foods in restaurant meals to be labelled, have acted to remove GM foods from their system. Burger King, Domino's Pizza, McDonald's, Wagamama and Wimpy in the UK claim to be producing food with GM content below the EU tolerance levels while Burger King in the USA announced late in 1999 that it would act to ensure its French fries were produced using GM-free potatoes (Table 20.5).

There appear to be two reasons for firms to adopt labelling policies or to eradicate products with GM ingredients. Some firms claim to have taken this route because of ethical views about GM ingredients or about the consumer's right to know. The majority of firms have adopted policies that will maximize their profits. Iceland Foods, for instance, gained market share in the UK retail industry and saw strong profit growth after it declared it was removing all products with GM content. It is likely that the consortium of EU retailers that responded to Iceland's policy and announced their own GM policies were trying to increase their market share. Other retailers followed with their own GM policies to remain competitive. Once part of the market had declared itself to be GM free, the risk of being isolated rose. The same could be said of many of the processors and food chains. Companies in those sectors have been slow, and somewhat hesitant, to adopt GM labels or GM-free status. In spite of strongly held private views, very few firms have publicly opposed labelling or removing GM elements from their products.

Early in 2000 a consortium of 30 US investment groups sent proxy resolutions to 19 well-known restaurant, food, grocery and seed companies. The consortium planned to use this proxy to have a debate about the use of GM foods (*USA Today*, 2000). Resolutions involving Quaker Oats, Coca-Cola Co. and PepsiCo were rejected (*The Western Producer*, 2000b). It is too early to say whether this shareholder activism will force US companies to engage more fully in the GM debate.

Table 20.5. Labelling systems implemented by restaurant chains.

	Target market	Non-GM product coverage
Burger King (USA)	USA	Potatoes
Burger King (UK)	UK	All ingredients and possibly derivatives
Domino's Pizza (UK)	UK	All ingredients and possibly derivatives
McDonald's (UK)	UK	All ingredients and possibly derivatives
Wagamama	UK	All products and additives
Wimpy (UK)	UK	All ingredients and possibly derivatives

Conformity measures and costs

While labels may improve consumer information, they can only do so if the seller actually delivers the product promised. Carrefour's experience shows that a wide range of products may actually contain GM ingredients. Once the ingredients have been identified, it has so far been relatively easy to secure supplies of GM-free ingredients from markets not yet using GM seeds. However, as GM technology spreads it will become more difficult to find supplies that one can confidently state are GM free. The tests to determine if products are GM free are costly and slow, with the result that few tests can realistically be undertaken.

Stave and Durandetta (2000) identify two general techniques of evaluating the GM content of crops or foods. One technique is based on a polymerase chain reaction (PCR) that detects the presence of transgenes or their proteins but they cannot establish their levels in samples. A second technique uses immunoassays ELISA to detect novel proteins in crops and processed foods. The ELISA technique can determine whether the novel protein is above or below a particular threshold. At a 2% threshold, one form of the ELISA test was identified as being 99% confident that samples identified as below the threshold were correct and that samples measured as being over the threshold had at least 0.85% GM content. The within laboratory repeatability was ± 7% and the reproducibility between laboratories was ± 10%.

The PCR test requires sophisticated laboratory equipment, costs anywhere from US$400 to 700 per sample, and routinely takes 3–10 days to complete. Current ELISA methods require only 1–4 hours and cost approximately US$40–70 per test. Work is underway on a new form of the immunoassay test that would use an immunochromatographic strip. This test could cost less than US$10 and be completed in the field in 5–10 min.

Because the current testing methods are unreliable and costly, several firms have established identity preserved production and marketing systems (with feedback loops and continuous improvement through Hazard Analysis Critical Control Points (HACCP)-style programmes) to secure GM-free varieties. Even with these systems, it will be difficult to deliver truly GM-free product. In Canada, for instance, pedigreed seed varietal purity standards, which are established by the Association of Official Seed Certifying Agencies (AOSCA) and referenced in the regulations of the Canadian Seed Growers' Association (CSGA), permit up to 0.5% of other varieties. Hence, a 1% threshold for mandatory labelling for GM foods in or from Canada could already have 0.5% of the GM level due to tolerances in the seed variety, if that seed was produced in Canada.

Some companies are relying on their suppliers to secure and provide adequate non-GM inputs. In 1999, for example, Archer Daniels Midland (ADM), a major grain wholesaler, informed US producers that it would be segregating GM from non-GM corn. ADM indicated that it might not buy GM varieties in the 2000 crop year. Early in 2000, ADM reported that only 5% of sales (by volume) from July to December 1999 demanded GM-free products. As a result, ADM softened its policy, indicating it would continue to buy both GM and GM-free crops.

Somewhat hesitantly the grain marketing system is adapting to demands from its export markets. There are a number of examples in Canada where producers in the canola industry have for limited periods adopted identity preserved production and marketing systems (IPPMs) for GM and non-GM novel trait varieties (Phillips and Smyth, 1999; Smyth and Phillips, 2002). There are several successful IPPMs in the USA as well. For example, Dupont's Optimum Quality Grain (OQG) is segregated through the supply chain by a series of contracts. ADM, ConAgra, and Consolidated Grain and Barge manage the logistics of the export marketing system for OQG. Similarly, several Japanese firms make private contracts with US producers for specific types of soybean. One example is the relationship between Clarkson Grain (USA) and Nisshin Shokai. Clarkson Grain has a contract to produce GM-free soybeans primarily for Japanese soy food manufacturers. The Illinois Crop Improvement Association tests the soybeans to meet a 99.5% GM-free standard.

The USA, the single largest grain and oilseed exporter in the world, also announced in January 2000 that it would establish a new export policy to ensure that exports from the USA meet the rules of import markets. The policy is yet to come into effect. At the same time, many of the biotechnology and seed companies (e.g. Monsanto and Aventis) have agreed to work with importing nations and the grain trade to ensure that their products conform to national regulations (e.g. EC 90/220 and Japanese labelling laws) and processor requirements. For example, in early 2000, Monsanto provided, free of charge, genetic information to facilitate tests to detect their NewLeaf™ GM potatoes. This assisted Canadian potato producers to sort their seed stock which allowed them to plant crops that will meet McCain Foods and Midwest Foods specifications for GM-free potatoes.

There are few estimates of what an effective labelling and segregation system might cost. One estimate suggests that a temporary, voluntary, IPPM system for GM canola in Canada during 1995–1996 cost an average 13–15% of farm gate price to establish and manage (Phillips and Smyth, 1999). In Australia and New Zealand, industry compliance with the proposed labelling law was estimated to cost A$3 billion (approximately 6% of turnover) in the first year of operation of the standard and around 3% of turnover in subsequent years. The study concluded that the food industry could not absorb these cost increases and that prices for major ingredients could rise by around 10–15%. The increase in the price of ingredients could translate into retail price rises of 0.5–15% for processed foods, depending on the proportion of the retail cost represented by the raw material (KPMG, 1999).

Implications

Labelling regulations could have significant impacts on production and trade flows. Table 20.6 shows the number of total importing countries and the percentage of them that either now have or have announced intentions to introduce mandatory labelling systems. For example, countries with mandatory labelling import 77% of the canola exports. Flax is exported to 74 countries, 87% of which is destined for countries that currently require labelling. This table assumes that Argentina, Canada, the USA and Russia require voluntary labelling.

The USA, Argentina and Canada are three key producers and exporters of GM foods. Together, these three countries produced 99% of total GM foods in 1999. Table 20.7 shows that in 1999, over 50% of Canadian exports of agricultural commodities went to countries that propose to set up voluntary or mandatory labelling laws. The only exception was durum exports, of which only 31% went to countries currently requiring or proposing labelling.

The quantity of imports to countries that have proposed mandatory labelling indicates that forgoing those export markets with labelling rules will not be a realistic option. It will be very difficult and expensive for exporters to conform to the labelling rules of each importer. Indeed, the variety of labelling for GM foods in the world marketplace poses serious threats to the international trade system. Inconsistent policies cause uncertainty and therefore threaten investment in agricultural biotechnology applications.

One way to manage the changes that are needed will be through international negotiation (see Phillips, 2002). The current option being pursued would be to negotiate a common set of labelling rules at *Codex Alimentarius* that could help to

Table 20.6. International trade related to countries with labelling policies (source: FAO, 2000).

Commodities	Total global imports (thousand tonnes)	Total number of countries importing	% of total imports into countries with mandatory labelling	% of total imports into countries with voluntary labelling
Canola	14,750	133	77	16
Flax	665	74	87	3
Maize	74,437	168	62	3
Melon	55	13	15	0
Papaya	114	56	22	45
Potato	10,706	177	70	12
Rice	6,832	95	32	2
Soybeans	81,752	173	75	2
Squash	525	32	56	41
Sugarbeet	169	26	80	0.3
Tobacco	3,019	196	51	19
Tomato	4,813	140	55	29

Table 20.7. Potential export exposure to foreign labelling rules (source: Industry Canada (2000) for US and Canada; Oil World (1998) for Argentina).

	USA	Canada	Argentina
% Total exports to countries with mandatory labelling rules			
Canola	37	64	
Soybeans	63	38	64
Maize	52	46	
Wheat	33	36	
Durum	46	16	
Barley	84	50	
Potatoes	56	6	
% Total exports to countries with voluntary labelling rules			
Canola	57	31	
Soybeans	5	23	10
Maize	4	54	
Wheat	4	15	
Durum	1	15	
Barley	4	36	
Potatoes	19	81	

resolve the conflicting requirements. Although this strategy has been proposed, there is no certainty that it will be successful. In the absence of any international agreement, exporters could challenge labelling laws as trade barriers under the Sanitary and Phytosanitary and the Technical Barriers to Trade Agreements of the World Trade Organization (WTO). However, importing countries could ignore the decision and continue their policies. Alternatively, producers in exporting countries may take the more costly route of segregating to meet the requirements of importing countries.

The Cartagena Protocol for Biosafety, which was concluded on 29 January 2000 but must be ratified by at least 50 countries to come into effect, requires that all international trade in commodities that are or could be co-mingled with GM commodities be labelled as 'may contain GMOs'. Importers will then be able to consult with the proposed Clearing House to determine which GM traits might be in the shipment. This identification system could provide the base for an extension of GM labelling rules to a much wider group of countries. The challenge, then, will be to have some form of international consensus, developed through Codex, the OECD, the WTO or some bilateral mechanism, to ensure that the labelling systems that evolve nurture consumer sovereignty and do not simply shut down trade. In other areas private companies have taken the lead to resolve consumer concerns either by developing industry based, quality assured standards or by private investors establishing proprietary, private brands to assure the public of their commitment to their product (see Phillips *et al.*, 2000).

Conclusions

Labelling regimes for GM foods are not homogeneous among countries at either private or public levels. While some markets have aggressively adopted the new technologies in the food chain, consumer concerns have led many national governments and private companies to adopt mandatory labelling. Regardless of the safety of GM foods, there appears to be a need to address consumer concerns.

In most cases, mandatory labelling has led to eradication of GM products in those markets. Eradication has not improved consumer choice. Given that even in the markets with the most consumers concerned about GM products (Table 20.1) there are some consumers that are indifferent or positively oriented to GM products, removing their choice is as undesirable as not allowing consumers to avoid GM foods. In the absence of labelling that allows consumers to choose between competing GM and GM-free food products, however, it is impossible to tell how consumer preferences will affect the demand for GM and GM-free foods. The difficulty is not whether to develop a labelling system for GM foods but rather how to develop a system that provides real consumer choice.

It is unclear from this survey whether private labelling systems will harmonize standards across products and markets (perhaps establishing and adopting International Standards Organization (ISO) or Good Manufacturing Practices (GMP) quality assurances) or whether member states, individually or collectively through one or more of the international institutions, will negotiate a settlement. There are precedents for both approaches in other areas. In fact, many of the current sanitary and phytosanitary standards that are enshrined in the Codex system or the international plant or animal protection conventions started as private industry standards and then were adopted as international standards through negotiation.

References

Ackerlof, G. (1970) The market for lemons: quality, uncertainty and the market mechanism. *Quarterly Journal of Economics* 84, 488–500.

Angus Reid Group Inc. (1999) *Biotechnology in Foods: Executive Highlights Report.* Prepared for Ontario Agri-Food Technologies, December. Angus Reid Group Inc., Ottowa.

Angus Reid Group Inc. and *The Economist* (1999) World Poll. www.ccu-cuc.ca/en/polls/data/angus.html

Bredahl, M. (1999) Supply, demand and greed. Presentation to the *Missouri Agricultural Extension Station Research Expo IX Panel Discussion on Biotechnology*, Columbia, Missouri, 20 October. University of Missouri, Columbia.

Bureau, J.-C., Marette, S. and Schiavina, A. (1997) Trade, labels and consumer information: the case of hormone-treated beef. Contributed Paper, *XXII Conference of IAAE*, Sacramento, August. International Association of Agricultural Economists, Sacramento, California.

Calamai, P. (2000) Ottawa gives up on splicing genes into pasta ingredient: durum decision is partly due to consumer outcry. *Toronto Star* 19 April, A6.

Caplan, A. (2000) *Are Genetically Modified Foods Fit for a Dog? Breaking Bioethics: Not in France.* www.med.upenn.edu/bioethics/breaking/25May00.html

Caswell, J. (1998) Should use of genetically modified organisms be labeled? *AgBioForum* 1(1). www.agbioforum.missouri.edu/

Environics Research Group (2001) Secondary Analysis of Public Opinion Research Regarding Genetically Modified Food and Related Biotechnology Issues. Report commissioned for the Canadian Biotechnology Advisory Committee, 6 June. www.cbac.gc.ca/english/

Food and Agriculture Organization (FAO) (2000) FAOSTAT: Agriculture data. www.apps.fao.org/cgi-bin/nph-db.pl?subset=agriculture

Fukuyama, F. (1995) *Trust: the Social Virtues and the Creation of Prosperity.* Penguin Books, London.

Giannakas, K. (1999) Enforcement issues in GMO segregation systems. Paper presented to the *Economics of Quality Control in Agriculture Conference*, Saskatoon, 3 December. University of Saskatchewan, Saskatoon.

Industry Canada (2000) Strategis: Imports and exports by product. www.strategis.ic.gc.ca/sc_mrkti/tdst/engdoc/tr_homep.html

Isaac, G. and Phillips, P.W.B. (1999) Market access and market acceptance for agricultural biotechnology products. Paper presented to the *ICABR Conference on The Shape of the Coming Agricultural Biotechnology Transformation: Strategic Investments and Policy Approaches from an Economic Perspective*, Rome, 17–19 June. International Consortium on Agricultural Biotechnology Research, University of Rome, Tor Vergata.

James, C. (1999) Transgenic crops worldwide: Current situation and future outlook. Paper presented at the *Conference on Agricultural Biotechnology in Developing Countries: Towards Optimizing the Benefits for the Poor* organised by ZEF and ISAAA in collaboration with AgrEvo and DSE, Bonn, 15–16 November 1999. ZEF, Bonn. www.zef.de/zef_englisch/f_veranstalt_biotech.htm

KPMG (1999) *Report on the Compliance Costs Facing Industry and Government Regulators in Relation to Labelling Genetically Modified Foods.* Report commissioned by the Australia New Zealand Food Standards Council, Canberra, October. www.anzfa.gov.au/documents/gen24_99.asp

Marks, L. (1999) Media content analysis for biotechnology. Presentation to the *Missouri Agricultural Extension Station Research Expo IX Panel Discussion on Biotechnology*, Columbia, Missouri, 20 October. University of Missouri, Columbia.

Nenon, S. (1999) The USDA perspective on labeling. In: McGaughey, B. and Redick, T. (eds) *Liability and Labeling of Genetically Modified Organisms.* Council for Agricultural Science and Technology, Ames, Iowa. www.cast-science.org/0002abab.htm

Niiler, E. (2000) Local efforts to enforce GM food labeling are thwarted. *Nature Biotechnology* 18(6), 590.

OECD (1999) *Modern Biotechnology and Agricultural Markets: a Discussion of Selected Issues and the Impact on Supply and Market.* AGR/CA/APM/CFS/MD*(2000)2. OECD, Paris.

Oil World (1998) *Oil World* 15 January 1999. ISTA Mielke GmbH, Germany.

Phillips, P. (2002) International trade in genetically-modified agri-food products. In: Moss, C. *et al.* (eds) *Agricultural Globalization, Trade and the Environment.* Kluwer, Dordrecht.

Phillips, P. and Buckingham, D. (2001) Agricultural biotechnology, the environment and international trade regulation: your place or mine. In: Michelmann, M. *et al.* (eds), *Globalization and New Agricultural Trade Rules in the 21st Century.* Lynne Rienner Publishers.

Phillips, P. and Isaac, G. (1998) GMO labeling: threat or opportunity. *AgBioForum* 1(1). www.agbioforum.missouri.edu/

Phillips, P.W.B. and Smyth, S. (2000) Managing the value of new-trait varieties in the canola supply chain in Canada. *Proceedings of the 4th Wageningen Supply Chain Conference,* May. Wageningen Agricultural University, The Netherlands.

Phillips, P. and Smyth, S. (2002) Competitors co-operating: establishing a supply chain to manage genetically modified canola. Special issue of the *International Food and Agribusiness Management Review.*

Phillips, P., Porter, T. and Henry, O. (2000) *Standards in Emerging Biodiverse Industries: Western Canadian Agri-food Interests in the Canadian Standards Strategy.* Western Economic Diversification, Saskatoon.

Ram's Horn (1999) GE-free in France. *Ram's Horn* 167, 7.

Stave, J. and Durandetta, D. (2000) GM crop testing grows amid controversy. *Today's Chemist at Work,* 9(6), 32–33, 37.

Stiglitz, J. (1999) *Wither reform? Ten years of the transition.* In: *Annual World Bank Conference on Development Economics,* 28–30 April. The World Bank, Washington, DC.

Tirole, J. (1988) *The Theory of Industrial Organization.* MIT Press, Cambridge, Massachusetts, p. 106.

USA Today (2000) Firms face battle over altered food. *USA Today,* 14 February.

The Western Producer (2000a) Tests show no GMO residue in animals. *The Western Producer,* 10 February, p. 25.

The Western Producer (2000b) Quaker rejects ban on GMOs. *The Western Producer,* 25 May, p. 60.

21 Estimating the Costs of Segregation for Non-biotech Maize and Soybeans

William W. Lin
ERS, USDA, Washington, DC, USA

Introduction

Adoption of biotechnology in the production of maize, soybeans and cotton by US producers has grown rapidly since its commercialization in 1996. In 1999, the proportion of total acreage in major producing states planted to biotech crops reached 33% for maize, 57% for soybeans and 55% for upland cotton (USDA, 2000).[1] In 2000, while biotech soybeans and cotton remained popular with farmers in major producing states, accounting for more than one-half of acreage for both crops, US farmers cut back the share of acreage planted to biotech maize to 25%. In 2001, herbicide-tolerant soybeans became even more popular with US farmers, reaching an adoption rate of 68% (USDA, 2001). Biotech cotton also gained popularity, reaching an adoption rate of almost 70% of all cotton areas, and the share of biotech maize in actual plantings increased slightly to 26%.

The adoption momentum for biotech maize has slackened since 2000. Factors that affect farmers' net returns – such as whether yield-increasing potential offsets the higher cost for biotech seed, and whether observed infestation levels of certain target pests indicate likely savings on pesticide costs – play a major role in producers' decisions regarding the adoption of biotech crops versus conventional varieties. Uncertain market prospects for biotech crops triggered by potentially widening interest in food labelling regulations in various countries, and possible shifts in consumer preferences towards non-biotech foods might also contribute to the cutback.

Although the decline in 2000 biotech maize plantings in part reflects an overall market uncertainty for biotech crops, market demand for non-biotech maize is currently very limited, accounting for only 1–2% of 1999 US maize production (Lin *et al.*, 2000). Similarly, market demand for non-biotech soybeans is also small, accounting for about 2–3% of US soybean production. Segregation for meeting these non-biotech demands is largely limited to Japan, the EU (where less than 1% of US maize exports is shipped, see Fig. 21.1) and a handful of domestic food manufacturers that recently decided to use only non-biotech ingredients.

The purpose of this chapter is to develop a scenario analysis under which the costs of segregation for non-biotech maize and soybeans are estimated for grain handlers, from country elevators to subterminals and export elevators. It is important to note that these cost estimates do not take into account any additional costs that could be associated with segregation at the farm level and shipment expense beyond export elevators.

[1] Herbicide-tolerant varieties include those developed using both biotechnology and conventional breeding techniques. The adoption rate is expressed in terms of the percentage of harvested acres.

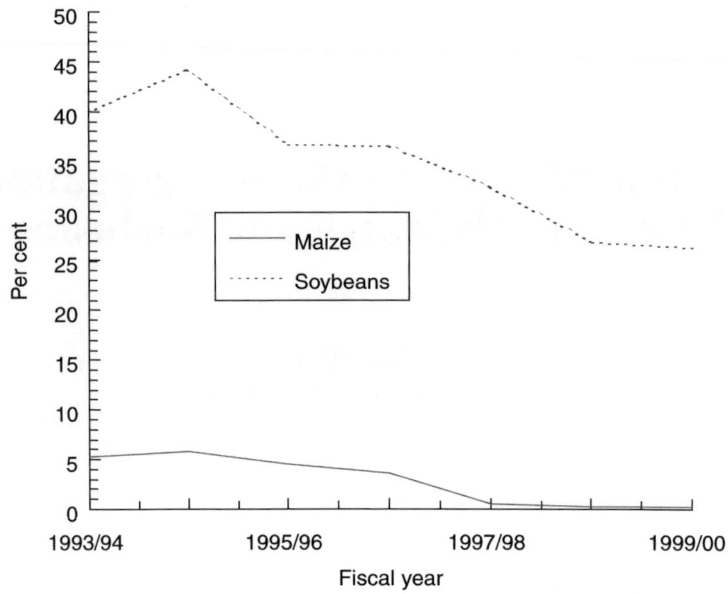

Fig. 21.1. EU share of US maize and soybean exports: 1993/94–1999/2000.

This chapter is organized into five sections. The first section discusses the impetus, concept and implications of segregation. The second section discusses the costs of segregation for speciality grains (e.g. high oil maize) and oilseeds (e.g. STS soybeans) based on a recent University of Illinois study by Bender *et al.* (1999). The third section discusses the procedures of estimating the costs of segregation for non-biotech maize while the fourth section discusses those for non-biotech soybeans. Finally, the last section discusses the issue of 'who bears the costs of segregation?'

Segregation: Impetus, Concept and Implications for Grain Handlers

Despite US farmers' rapid adoption of biotech crops, market prospects for genetically modified (GM) crops are tinged with uncertainty. Some consumers in the US and abroad – particularly the EU – remain wary of the use of this new technology in crop production and the use of these crops as ingredients in food production. These consumers hold this view despite reviews by the US Food and Drug Administration that have determined that biotech foods currently in the market are as safe for human consumption as their non-biotech counterparts. As a result, grain handlers, food manufacturers and others in the global marketing chain are attempting to balance the issue of divergent consumer demand with producers' desire to capture the cost-saving potential of biotech crops.

Although trade pattern changes arising from shifts in consumers' preferences have been quite modest so far, segregation of grain into biotech and non-biotech may increasingly become a consideration by producers and grain handlers. If buyers are willing to pay premiums for non-biotech crops, some US grain handlers can meet these demands, passing along additional premiums to farmers. However, demands for non-biotech maize and soybeans are likely to be limited if biotech labelling is confined to foods in some segments of US export markets.

Some large US grain processors (e.g. A.E. Staley and Archer Daniels Midland, ADM) announced in April 1999 that they would not accept EU-unapproved maize biotech varieties for processing for fear of jeopardizing their by-product exports to the EU. In summer 1999, ADM advised producers to segregate biotech crops from non-biotech crops, but reversed this decision in February 2000; ADM started to accept the delivery of biotech crops again so long as they are EU-approved varieties.

In addition, some countries have begun to require that foods containing biotech ingredients be labelled. The EU has implemented its labelling regulations for foods since mid-2000 and is currently drafting feed labelling regulations. Japan implemented its mandatory labelling policy for bioengineered foods in April 2001 and South Korea started implementing its own in mid-2001. Australia, New Zealand and China are among other countries adopting mandatory labelling policies for bioengineered foods. Potentially widening interest in food labelling regulation could be an impetus for more farmers and grain handlers to assess their ability to segregate or begin to take steps necessary to segregate.

Over the last few years, a few food manufacturers decided to end the use of biotech crops in their operations. In July 1999, the Gerber and Heinz companies announced that their baby food processing facilities would immediately stop using biotech ingredients. In January 2000, Bestfoods, Inc. decided to end its use of biotech ingredients in manufactured foods destined for the EU, in order to avoid the biotech labelling requirement. Recently, Frito-Lay Inc. announced that it would cease using biotech maize in its snack food manufacturing.

Segregation requires that crops be kept separate to avoid commingling during planting, harvesting, loading and unloading, storage and transport. This supply chain system thus requires cleaning of equipment such as combines and augers, as well as transport and storage facilities. Such a handling process may not involve containerized shipment, but testing to check for the presence of biotech content throughout the marketing system is critical.

Identity preservation (IP) is a more stringent (and expensive) handling process and requires that strict separation, typically involving containerized shipping, is maintained at all times. IP, which is not a focus of this chapter, is often used for marketing commodities like food-grade maize and soybeans. Testing for biotech vs. non-biotech status typically occurs just before containerization. IP lessens the need for additional testing as control of the commodity changes hands, and it lowers liability and risk of biotech and non-biotech commingling for growers and handlers. This handling process might be required to meet a stringent threshold level of biotech content, such as the 1% required under EU labelling regulations. However, no IP system can guarantee 100% purity.

Because of limited demand for non-biotech maize and soybeans and the expense of maintaining separate storage facilities, only a small to modest percentage of grain elevators have attempted to segregate and market non-biotech products. In September 1999, Sparks Companies conducted a survey of 100 Midwestern grain elevators and found that 11% were differentiating for non-biotech maize and 8% for non-biotech soybeans. Of the surveyed elevators, only 1% offered producer premiums for non-biotech maize and 3% offered producer premiums for non-biotech soybeans. The premium varied widely, depending on the elevator's location and the intended consumer market for the product. According to other industry sources, common non-biotech price premiums in 2000 ranged from US\$0.05 to 0.10 bushel^{-1} for non-biotech maize and US\$0.10 to 0.15 bushel^{-1} for non-biotech soybeans. The lower end of the premium range reflects less strict tolerance levels (i.e. more biotech content) and vice versa. In February 2000, the Farm Progress Company's survey of 1200 US elevators indicated that 24% planned to segregate maize and 20% planned to segregate soybeans in autumn 2000. Elevators were probably anticipating food labelling regulations in countries other than the EU. A more recent survey conducted by the American Corn Growers Foundation in autumn 2000 suggests a higher level of potential segregation. The survey polled 1107 grain elevators in nine states and found that 30.5% were either requiring or suggesting segregation at their elevator facilities.

Effective segregation, which begins at the farm level, is particularly difficult for maize because of cross-pollination. In addition, farmers are required to clean combines during harvesting and may find themselves in need of expanding on-farm storage facilities for segregating grains into biotech and non-biotech varieties. A straw poll of 400 US farmers conducted by Reuters in January 2000 found that 15% of farmers had made or were planning to make the necessary investments to handle or segregate non-biotech crops in the autumn.

Elevators must also develop stricter control over handling procedures in order to maintain segregation. A key problem is that segregation is likely to slow the rate of turnover in high-volume business and delays can be a serious problem during peak harvest periods (Lin *et al.*, 2000). A study of Canadian grain segregation suggested that storing segregated grains, in addition to the base-grade grain, leads to a reduction in capacity turnover at terminal elevators by more than 15% (McKeague *et al.*, 1987), thereby increasing handling costs. In addition, grain elevators will have to give up the possibility of blending to enhance value, and they will incur additional costs of clean-out.

The elevator's ability to segregate depends in large part on the size of the operation and the type of facilities at each location. Segregation may require new investments for some elevators. According to the National Grain and Feed Association, roughly 5% of the nation's elevators can achieve segregation without major new investments if the tolerance level for biotech content is set at a low level approaching 1% (Lin *et al.*, 2000).[2]

Segregation also poses logistical problems for grain transport. Currently, grains are commonly transported to port elevators in unit trains of up to 100 cars or by barge. If segregation makes it necessary to shift away from unit trains towards smaller units, transport costs, according to industry sources, could potentially double if the tolerance level is set at 1% (Lin *et al.*, 2000). In contrast, the increase in transport costs would be modest if it is set at 5% or higher.

Costs of Handling Speciality Maize and Soybeans

Segregation of non-biotech maize and soybeans is essentially an extension of the handling process for speciality grains and oilseeds, which has been in place for some time. Thus, it would be instructive to first examine the costs of handling speciality grains and oilseeds before estimating the costs of segregation for non-biotech maize and soybeans.

A recent University of Illinois study examines segregation costs based on a survey of US grain elevators and speciality grain firms (Bender *et al.*, 1999). In spring 1998, a mail survey was sent to over 200 US firms that were identified as possible handlers of speciality grains and oilseeds. Eighty-four usable surveys were returned, representing handlers of maize and oilseeds. Among other things, an important objective of the survey was to estimate additional costs incurred when handling speciality crops compared with those for standard bulk commodities.

Seventy-six firms were identified as handling a specific type of speciality maize or soybeans, including high oil maize, STS soybeans (synchrony treated soybeans, a herbicide-tolerant, but not biotech variety), food-grade maize and food-grade soybeans. Segregation costs for high oil maize and STS soybeans are highlighted in Table 21.1 because the costs of segregation in this chapter are limited to non-biotech maize and soybeans, not including food-grade maize and food-grade soybeans that are typically segregated through IP.

The results of the Illinois study suggest that separation of speciality grains adds, on average, US$0.06 bushel^{-1} for high oil maize (HOM) and US$0.18 bushel^{-1} for STS soybeans (excluding purchasing premiums) above the customary costs of handling standard bulk commodities at each of those elevators or speciality grain firms (Table 21.1). Segregation costs include the additional costs of storage, handling, risk management (for example, if

Table 21.1. Additional costs incurred in handling high oil maize (HOM) and STS soybeans (source: Bender *et al.*, 1999).

Cost item	High oil maize (US$ bushel^{-1})	STS soybeans (US$ bushel^{-1})
Storage	0.01	0.02
Handling	0.02	0.06
Risk management	0.01	0.07
Transport	0	0
Analysis/testing	0.01	0.01
Marketing	0.01	0.02
Subtotal	**0.06**	**0.18**
Purchasing (including premium)	0.12	0.15
Total	**0.18**	**0.33**

[2] The 5% estimate is an informed but subjective judgement, not based on a scientific survey. Test kits available in the market are very sensitive and can detect the presence of biotech content at 1% or lower.

quality is not as high as specified in the contract), analysis and testing, and marketing (expenses associated with negotiating contract terms). Minimum oil content specified in the contract generally ranges from 6 to 8% (7%, on average) for high oil maize. In contrast, quality for STS soybeans is controlled by preserving the variety; growers are required to plant only the STS variety developed by DuPont and grain handlers are to prevent STS soybeans from being commingled with other varieties.

How do these additional costs of handling HOM or STS soybeans differ from empirical evidence available in the literature? A recent economic engineering study at the Kansas State University found that the costs of segregating wheat into different end-use quality mixes for country elevators that operate at 60–70% of receiving capacity average about US\$0.0191 to 0.0199 bushel^{-1} if the elevator has three pits and two bucket elevator legs (Herrman *et al.*, 1999). The average cost of segregation increases to US\$0.0237–0.0261 bushel^{-1} if the elevator has two pits and two bucket elevator legs, and to US\$0.0528–0.0530 bushel^{-1} if the elevator has two pits and one bucket elevator leg. In 1994, Hurburgh reported that the costs of testing and segregating soybeans was US\$0.02–0.03 bushel^{-1}, also based on an economic engineering study. Thus, the average additional cost of handling HOM (2 cents bushel^{-1}) reported by the Illinois study appears to be in line with empirical evidence in the literature based on the economic engineering analysis, especially if the elevator has multiple pits and two bucket elevator legs.

The Illinois study shows a higher additional cost of handling STS soybeans than the 2–3 cents bushel^{-1} cost of segregation reported by Hurburgh, in part because of the specific need for preserving the identity of STS soybeans by variety and possibly because of the small sample size (10). A more recent Illinois study (Good *et al.*, 2000) that confines the survey to elevators and speciality grain firms only in Illinois found that the additional cost of handling STS soybeans averages 3.48 cents bushel^{-1}, down from 6.0 cents bushel^{-1} obtained earlier by Bender, *et al.* (1999). This latest Illinois study had a sample size of 65.

Costs of Segregating Non-biotech Maize

The costs of segregating speciality grains (e.g. HOM) can be modified or adjusted to estimate the costs of segregation for non-biotech maize or soybeans because segregating non-biotech grains is essentially an extension of the segregation process for speciality grains. This section develops a scenario analysis under which each of the cost items in the Illinois study was examined at three points along the marketing chain – country elevator, subterminals and export elevator – to determine approximate segregation costs for non-biotech maize. Although the costs of segregation vary significantly among the surveyed elevators, results indicate that, across all elevators surveyed, costs for segregating non-biotech maize could be higher than for HOM because of higher costs of testing and handling.

Although the estimated costs are not small, they do not imply that disarray would occur in the grain marketing system if non-biotech maize were handled on a larger scale. If non-biotech maize remains a niche market, many elevators may choose to accept bulk grain and not attempt to segregate. This would be particularly true for those elevators handling maize and soybeans destined for domestic feed use.

Not all elevators that choose to distinguish between biotech and non-biotech would bear the costs identically. Some elevators currently handle niche market crops at relatively low cost, particularly if they are equipped with multiple pits and have bin space configured to facilitate segregation (Lin *et al.*, 2000). In addition, specialization across elevators (some handling biotech, others non-biotech) would also result in much lower added costs to the handling system. This specialization may lead to a cluster of non-biotech crop production. Further, adjustments in the grain marketing system would work to lower costs as economies of scale in handling are realized and new testing procedures are developed.

The cost estimates in this chapter, which should be taken as rough estimates given the limited data currently available, indicate that, on average across the 76 surveyed elevators, segregation could add about US\$0.22 bushel^{-1} (excluding premium to producers), or 12% of the average farm price for maize, to costs of handling non-biotech maize from country elevator to export elevator (Fig. 21.2). These estimates reflect costs at these elevators and may not represent costs incurred by any one elevator or other elevators in general. In addition, these cost estimates do not take into account any additional costs that could be associated with segregation at the farm level and shipment expenses beyond export elevators to foreign markets.

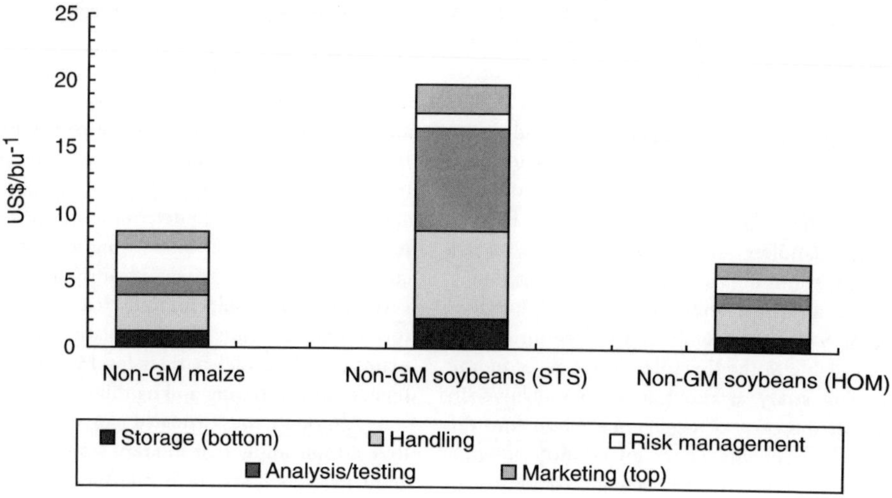

Fig. 21.2. Segregation costs for non-biotech maize and soybeans.

These cost estimates reflect a scenario analysis under the following assumptions: (i) risk management cost is not greater for non-biotech maize than for HOM (i.e. assuming a high tolerance level – generally around 5% – for biotech content); (ii) two-tier segregation is needed at the local elevator to guard against commingling (some elevators have already adopted this practice); and (iii) a multiple trait ELISA test kit will be introduced to detect biotech content for Roundup Ready and Liberty Link maize varieties.

In developing this scenario, two important adjustments to the Illinois cost estimates are made. First, the cost estimate for maize at the country elevator is adjusted to reflect a two-tier segregation requirement: to segregate biotech from non-biotech varieties, and to separate biotech varieties approved for shipment to the EU from EU-unapproved varieties (which reportedly account for less than 5% of US maize acreage). This two-tier segregation is necessary because most country elevators lack complete knowledge about the destination of maize shipments. For shipments to domestic markets, two-tier segregation might be necessary because some processors (such as ADM and A.E. Staley) accept only EU-approved maize varieties. Similarly, for shipments to the EU, no commingling with EU-unapproved varieties is permitted. To the extent that producers channel their maize to market outlets that accept EU-unapproved varieties (such as domestic feedlots), handling costs at local elevators could be lower.

Adjusting for two-tier segregation is estimated to increase handling costs for non-biotech maize at country elevators to \$0.03 bushel^{-1}, higher than the US\$0.02 bushel^{-1} reported in the Illinois study. Biotech segregation imposed no additional handling cost above the US\$0.02 bushel^{-1}, incurred at subterminals and export elevators for segregating speciality maize, because operators know the destination of grain shipments at those facilities. Thus, additional handling costs for segregating non-biotech maize from country elevator to subterminals, and then to export elevators add an average of US\$0.07 (0.03 + 0.02 + 0.02) bushel^{-1} above non-segregated maize as shown in Table 21.2, slightly higher than that for HOM (US\$0.06 bushel^{-1}). Also, segregation could cause the loss in potential savings that an exporter can now realize by blending at export points and shifting cargoes and customers because of vessel delays and other factors – a flexibility afforded by bulk commodity marketing.

The adjustment for testing costs reflects the higher cost of testing for biotech content, which is more complicated than testing for physical characteristics, such as oil content, for HOM. Grain handlers commonly use two testing methods: the DNA-based PCR and the protein-based ELISA. PCR takes 2–10 days at a cost of US\$200–450 per test, higher than most country elevators can afford because of the small volume per truck load, which typically is around 900 bushels. In contrast, an on-site ELISA microwell test takes 2 h and costs up to US\$10 per test. A faster and simpler ELISA

Table 21.2. Segregation costs for non-biotech maize and soybeans from country elevator to export elevator (source: adapted from Bender *et al.*, 1999).

Cost item	Non-biotech maize (HOM segregation) (US\$ bushel^{-1})	Non-biotech soybeans HOM segregation (US\$ bushel^{-1})	Non-biotech soybeans STS segregation (US\$ bushel^{-1})
Storage	0.03	0.03	0.06
Handling	0.07	0.06	0.18
Risk management	0.03	0.03	0.21
Transport	0.00	0.00	0.00
Analysis/testing	0.06	0.03	0.03
Marketing	0.03	0.03	0.06
Total	0.22	0.18	0.54

dipstick test to provide a 'yes–no' result takes 5–10 min and costs just US\$3.50 per test. At a 99% purity level, a typical ELISA test uses a sample of 50–60 kernels out of close to 900 bushels in a truck load. However, reliability of test results at this purity level becomes more of a concern; inconsistent results may occur from different tests of the same lot of grain. A smaller sample size (40–50 kernels) would be used for testing at a 95% purity level. The additional cost of testing for biotech content using ELISA test kits is estimated at US\$0.01 bushel^{-1} for one specific new trait (e.g. Bt maize) at country elevators. However, since current ELISA testing methods require a separate test for detection of each unique trait, several tests may be required to determine if a truck load of maize is free of biotech material. The analysis in this chapter assumes four separate ELISA tests for five biotech maize varieties at country elevators: three Bt varieties, plus Liberty Link and Roundup Ready.[3] While biotech content in the three Bt varieties can be detected technically in one test, multiple tests (usually two) are a common practice adopted by local elevators. This increases the cost of analysis and testing for non-biotech maize to US\$0.04 bushel^{-1} at country elevators, up from the US\$0.01 bushel^{-1} reported for HOM in the Illinois study.

At subterminals and export elevators, PCR testing is more common than ELISA because it is very sensitive and can be used to detect the presence of several gene modifications in one set of tests. However, PCR tests are generally conducted in commercial labs. It becomes more economical with the larger volume of grain being handled, remaining just US\$0.01 bushel^{-1} as estimated by the Illinois study. A typical sample size for testing is about 80 lb of grain in a river subterminal, which handles about 50,000–55,000 bushels of grain in a barge.

Thus, the costs of testing/analysis add, on average, about US\$0.06 bushel^{-1} for segregating non-biotech maize, from country elevators to export elevators. This doubles the US\$0.03 bushel^{-1} costs of testing and analysis for HOM throughout the marketing system as reported in the Illinois study.

Risk management costs for segregating grain into biotech and non-biotech conceivably could be greater than for handling HOM, because producers face significantly different risks. For example, a 1% lower oil content might reduce price premiums paid to HOM producers. However, 1% biotech content in a grain shipment could cause rejection, which has much more serious consequences for grain exporters. Because there is no way to quantify this extra cost, it is assumed that the risk management cost is the same as for HOM in the Illinois study, \$0.01 bushel^{-1} or \$0.03 from country elevator to export elevator.

Marketing costs for segregating non-biotech maize basically are the same as those for HOM (US\$0.03 bushel^{-1}) across the three elevator points. Similarly, storage costs require no adjustment, remaining at US\$0.03 bushel^{-1} across the typical grain marketing channel.

In considering segregation costs from production to marketing, this analysis excludes premiums to producers because the gain to producers offsets

[3] In autumn 2000, ELISA test kits available in the market focus on detecting the presence of biotech content in Bt varieties. However, a private firm plans to introduce new multiple-trait test kits in autumn of 2000, which, if realized, can extend the test to Roundup Ready and Liberty Link maize varieties.

the loss to country elevators. Thus, if segregation is viewed from the entire production and marketing system, not strictly from the grain handlers' perspective, premiums for non-biotech maize paid by country elevators are gains to producers. According to industry sources, the common range for purchasing premiums currently offered by a few elevators in 2000 was US$0.05–0.10 bushel^{-1} for non-biotech maize and US$0.10–0.15 bushel^{-1} for non-biotech soybeans.

Some US grain handlers are already segregating grain for certain export markets. For example, some grain companies are segregating non-biotech maize for food use (e.g. maize starch) in Japan, although without guaranteeing to meet a specific tolerance level for biotech material. Patterning maize segregation after handling procedures for HOM can usually meet the non-biotech requirements of Japanese buyers. To avoid commingling in shipments, grain handlers may also contract with producers to plant only non-biotech maize varieties and require adoption of specific production and harvesting practices.

How well do the estimates of the costs of segregation match with market reality? A credibility check of these cost estimates is to first deduct the average cost of segregation estimated from this study (US$0.22 bushel^{-1}) along with additional ocean freight expenses from price premiums foreign buyers are willing to pay for non-biotech maize, and then determine if the residual matches with price premiums offered to producers in the marketplace. According to a Japanese grain trading house, buyers in Japan in 2000 were willing to pay premiums for non-biotech maize, ranging from 40 to 50 cents bushel^{-1}. If Japanese buyers typically purchase a volume of 2000 t non-biotech maize for manufacturing niche products, instead of the entire cargo, then ocean freight expenses would be about US$0.13 bushel^{-1} higher than bulk commodity shipments. The residual after the deduction would average about US$0.10 bushel^{-1} if the mid-point of US$0.45 bushel^{-1} is taken as the average price premium for non-biotech maize. Because the US grain marketing system is very competitive, the fact that the US$0.10 bushel^{-1} residual matches fairly well with the price premiums being offered to producers (US$0.05–0.10 bushel^{-1}) for non-biotech maize in 2000 suggests that our cost estimates are in the ballpark.

Costs of Segregating Non-biotech Soybeans

This section extends the scenario analysis to estimating the costs of segregating non-biotech soybeans. Two separate scenarios are examined: (i) segregating non-biotech soybeans by patterning after the handling process for STS soybeans; and (ii) segregating non-biotech soybeans by following the handling process used for HOM.

The cost estimates indicated that, on average across the 76 surveyed elevators, segregation could add about US$0.54 bushel^{-1} (excluding premium to producers), or 12% of the average farm price for soybeans, to marketing costs of non-biotech soybeans from country elevators to export elevators under the STS segregation scenario (Fig. 21.2). In contrast, the cost estimates become smaller – an average of US$0.18 bushel^{-1}, or about 4% of the average soybean farm price – if segregation follows the same handling process used for HOM.

STS segregation

Two-tier segregation, which is necessary for segregating non-biotech maize, is not needed for segregating non-biotech soybeans because Roundup Ready soybeans is the only variety commercially grown currently in the US and that variety is EU-approved. As a result, the costs of handling remain at US$0.06 bushel^{-1} for each elevator point, or US$0.18 bushel^{-1} across the US grain handling system. However, the latest Illinois study (Good *et al.*, 2000) suggests that in Illinois, additional costs of handling add 3.48 cents bushel^{-1} for each elevator point, or US$0.10 bushel^{-1} from country elevators to export elevators, which is lower than the average cost nationwide reported in Bender *et al.* (1999).

Similarly, there is no complication in estimating the costs of testing biotech content for soybeans because Roundup Ready is the only variety of herbicide tolerant biotech soybeans. The costs of testing remain at US$0.01 bushel^{-1} at each of the elevator points, or US$0.03 from country elevators to export elevators. Additional costs of testing for biotech content in soybeans are lower in Illinois, 0.24 cents bushel^{-1} for each elevator point or US$0.01 bushel^{-1} across the three elevator points.

Additional costs of risk management, assumed to be identical to those for segregating STS soybeans, would average about US$0.07 bushel^{-1} for each elevator point, or US$0.21 bushel^{-1} throughout the grain handling system. However, risk management costs are lower in Illinois, US$0.04 bushel^{-1} for each elevator point or US$0.12 bushel^{-1} across the three elevator points (Good *et al.*, 2000).

Overall, segregation costs for non-biotech soybeans would add US$0.54 bushel^{-1} to US grain handlers if segregation is patterned after STS soybeans. For grain handlers in Illinois, however, the costs of segregation are lower – adding US$0.35 bushel^{-1} to marketing costs – in part because of higher efficiency in grain handling in Illinois than in other states.

HOM segregation

The costs of segregation for non-biotech soybeans can be substantially reduced if segregation is patterned after the handling process for HOM instead of the more rigorous STS segregation.

HOM segregation allows grain handlers to separate handling of non-biotech soybeans by following the process used for HOM segregation, where the focus is to meet minimum oil content specified in the contract, generally ranging from 6 to 8%. Some US grain handlers are already segregating non-biotech soybeans through this handling process, as was discussed earlier.

Table 21.2 shows that the costs of segregation for non-biotech soybeans add, on average, $0.18 bushel^{-1} for US grain handlers from country elevators to export elevators. The costs of segregation using this approach would be even lower for grain handlers in Illinois – about US$0.15 bushel^{-1} – from country elevators to export elevators.

Who Bears the Cost?

At the core of debates over the issues of segregation is the central question: 'Who bears the cost of segregation?' Some analysts take the position that in the long run, consumers would have to pay premiums for non-biotech crops. However, this is not supported by what we know about buyers from the EU; in general, European buyers are not willing to and currently do not pay premiums for non-biotech bulk grain shipments. As a result, other than a few niche, non-biotech products that receive premiums, European buyers purchase mostly non-segregated products. Overall, grain is being segregated primarily for Japan and other Asian markets.

In this section, we have identified four factors that can affect the distribution of the costs of segregation: (i) demand price elasticity of the commodity; (ii) competitive structure of the food industry; (iii) the proportion of ingredient in the value of the final product; and (iv) alternative sourcing of supply. The nature of their effects on the cost distribution is discussed below.

Demand price elasticity of the commodity plays a role in affecting the distribution of the costs of segregation. If the commodity is more price inelastic, either because of strong consumer preference for non-biotech crops, a lack of substitutes, or because of the nature of the commodity's market demand, then suppliers are in a better position to pass on the costs of segregation to consumers. An example is Japanese buyers' willingness to pay hefty premiums for food-grade soybeans, especially organic, food-grade soybeans. Japanese buyers have a strong preference for clear hilum, food-grade soybeans. Also, they regard soybeans from Brazil and Argentina, which have reddish tint, to be inferior in making tofu. If there is little or no substitute for the product, suppliers can more readily pass on the costs of segregation in the form of higher prices to consumers without incurring revenue losses.

Competitive structure of the industry plays an important role in affecting the distribution of the costs of segregation. If the grain handling industry and food chain are not highly competitive, then grain handlers, food manufacturers and retailers are in a better position to pass on the costs of segregation forward to consumers in the form of higher retail prices or backward to farmers in the form of lower farm prices.

The proportion of ingredient in the value of the final product also plays a role in determining the cost distribution. If the segregated commodity only accounts for a small proportion of the value of the final product, food manufacturers and retailers can more readily pass on the costs to consumers.

Finally, alternative sourcing of supply has proved to be an important factor in affecting the cost distribution. An example is the switching by Spain and Portugal to non-US sources for their maize since 1998, resulting from the moratorium on approving new biotech maize varieties in the EU regulatory approval process. Because of the availability of non-US sources of supply, buyers in these countries were less inclined to pay premiums for non-biotech maize. Thus, if US grain handlers had elected to segregate maize into biotech and non-biotech varieties, they would have incurred the costs of segregation and probably absorbed a good part of the costs.

Conclusions

There is a continuum of segregation for separating non-biotech from biotech maize and soybeans. Therefore, there is no one single cost estimate to represent the entire spectrum. A segregation process might be less rigorous and less costly, but it may also bring about lower price premiums. The tolerance level for biotech content plays a critical role in affecting the segregation process and the resultant costs of segregation at all points in the grain handling and distribution system.

The cost estimates in this chapter are meant to indicate general magnitudes and are likely to change as adjustments occur in the marketing system for specialized commodities and if the volume of segregated commodities expands and the grain handling industry realizes economies of scale. Also, the costs of segregation vary significantly among surveyed elevators, in part due to the small sample size in the earlier Illinois study. A comprehensive survey with a larger sample size of elevators and speciality grain firms would be useful in firming up the cost estimates.

In addition, the costs associated with segregation and the need to reduce risks of commingling might entice producers to bypass country elevators and sell their non-biotech crops directly to subterminals or processors. This might lead to more concentration of the grain handling industry.

This study focuses on estimating the costs of segregation for non-biotech maize and soybeans. Further research into the core issue 'who bears the cost?' would be very helpful in anticipating economic consequences of grain segregation into biotech and non-biotech varieties.

References

Agriculture and Agri-Food Canada (1998) *An Overview of Grain Segregation Issues*, Policy Development and Program Services, November 1998.

Bender, K., Hill, L., Wenzel, B. and Hornbaker, R. (1999) *Alternative Market Channels for Specialty Corn and Soybeans*, AE-4726, Dept. of Agri. and Con. Econ., University of Illinois.

Good, D., Bender, K. and Hill, L. (2000) *Marketing of Specialty Corn and Soybean Crops*, AE-4733, Dept. of Agri. and Con. Econ., University of Illinois.

Herrman, T., Boland, M. and Heishman, A. (1999) Economic feasibility of wheat segregation at country elevators. In: *The 2nd Annual National Wheat Industry Research Forum Proceedings*, Nashville, Tennessee, 4–5 February.

Hurburgh, C. (1994) Identification and segregation of high-value soybeans at a country elevator. *Journal of American Oil Chemistry Society* 71(10), 1073–1078.

Lin, W., Chambers, W. and Harwood, J. (2000) Biotechnology: U.S. grain handlers look ahead. *Agricultural Outlook*, AGO-270, April 2000, 29–34.

McKeague, D., Lerohl, M. and Hawkins, M. (1987) The Canadian grain grading system and operational efficiency within the Vancouver grain terminals. *Agribusiness* 3(1), 19–42.

US Department of Agriculture (2000) *Acreage*. CrPr 2–5, NASS, 30 June.

US Department of Agriculture (2001) *Acreage*. CrPr 2–5, NASS, 29 June.

22 Endogenous Demand and Optimal Product Regulation: the Case of Agricultural Biotechnology

Anthony Artuso
Department of Agricultural Food and Resource Economics, Cook College, Rutgers University, New Brunswick, New Jersey, USA

Introduction

No new bioengineered food or agricultural products have been approved for use in the European Union since October 1998 and five member countries have banned the sale of previously approved products (European Commission, 2000). The *de facto* moratorium in Europe has sent tremors through the market for agricultural biotechnology products. In 1999, several major food producers and agricultural commodity companies announced they would no longer purchase transgenic agricultural products. The rapid expansion in US acreage planted with transgenic crop varieties also slowed significantly in 2000. Stock prices of agricultural biotechnology companies have declined sharply since 1998 and several of the major life science companies have proceeded with plans to divest their agricultural biotechnology divisions.[1]

In the USA, the Food and Drug Administration (FDA) has steadfastly maintained the position that transgenic food products do not pose risks to human health that are qualitatively different from foods produced from traditional agricultural commodities. As a result, the FDA has not, until recently, required special regulatory treatment of bioengineered food products. The US Department of Agriculture (USDA) and Environmental Protection Agency (EPA) have required more extensive evaluation of the environmental impacts of bioengineered agricultural products. But the highly publicized reports of the effects of Bt maize on monarch butterflies and the recent controversy over the use in taco shells of a variety of transgenic maize not approved for human consumption indicates that USDA and EPA regulatory processes have not eliminated the possibility of unfortunate post-approval surprises.[2]

In May 2000, the FDA announced it would begin requiring companies to notify the agency of their intent to market food products developed using genetic engineering. Most companies had already been providing the FDA with pre-market notification of new products as well as information

[1] Pharmacia and Upjohn, which recently acquired Monsanto, has announced its intention to spin-off Monsanto's agricultural units. Novartis and Astra-Zeneca have recently completed the divestiture and merger of their agricultural operations.

[2] There is some evidence that the effects on butterflies of pollen from Bt maize was an issue that received evaluation in the regulatory review process. Nevertheless, the media treated the initial reports of potential harm to monarch butterflies as new information, and regulatory agencies were not able to provide an immediate response that satisfied public concerns.

(eds V. Santaniello, R.E. Evenson and D. Zilberman)

about their assessments of the safety of these products. The policy change simply makes this prior notification mandatory. The new policy will also require the FDA to issue a letter to the company regarding the regulatory status of the product. Both the company's submitted data and the FDA's analysis would be made available to the public through the FDA's web site.

The FDA's new policy was apparently prompted by growing public concern rather than any change in the agency's estimation of the risks of bioengineered food products. At the announcement of the proposed regulatory change, FDA Commissioner, Dr Jane Henney, stated that all bioengineered foods sold in the USA are as safe as their non-bioengineered counterparts (*Genetic Engineering News*, 2000). Nevertheless, the director of the FDA's Office of Premarket Approval in the Center for Food Safety and Applied Nutrition, acknowledged growing public concerns regarding bioengineered food products and indicated that the FDA had a responsibility to ensure that consumers continue to trust the safety of the US food system (Dutton, 2000).

When consumers face high costs and/or time delays in gathering information, product regulation can increase social welfare even if it adds significantly to production costs (Antle, 1996; Stiglitz, 2000). Economic analyses of food safety regulation have generally focused on compliance costs and the benefits of reducing food induced illnesses (Roberts *et al.,* 1996; Antle, 1999). Larson and Knudson (1991) examined the effects of alternative regulatory policy instruments on investment in agricultural biotechnology. However, their analysis did not examine the effects of changes in regulatory policy on demand for or net benefits of transgenic crop varieties. The recent developments in the market for transgenic agricultural products suggests the need for more detailed analysis of the interaction between regulatory policy, consumer risk perceptions and market expansion.

If consumer purchasing patterns are influenced by regulatory requirements, then regulatory policy formulation must take into account the demand altering effects of alternative policies. The welfare effects of regulations that influence consumer demand may be magnified over time if demand is also a function of prior sales. Vettas (2000) has shown that when demand is an increasing function of prior sales, private investment in a competitive market may be lower than optimal and, if expected demand is below a critical threshold, a potentially profitable market may not open at all due to the inability of initial entrants to benefit from their market expanding investment. In this chapter, I examine the implications for regulation of transgenic agricultural products when consumer acceptance and the rate of product innovation are affected by the stringency of regulatory requirements.

A Simple Model of Regulation Given Endogenous Demand

To permit comparative analysis of regulatory policies when demand is endogenous, I will adopt Marschack's (1978) production function framework and assume that changes in demand for transgenic food products reflect changes in consumer understanding about how best to maximize their welfare given fixed preferences for safe, healthy and environmentally benign food. This assumption leaves open the important issue of what information is available to consumers, while avoiding philosophical problems associated with welfare analysis when consumer preferences are changing (Pollack, 1970; Hammond, 1976; Grout, 1982).

Assume that demand is influenced by consumer perceptions of the safety, health and environmental effects of transgenic food and agricultural products. Past sales of transgenic agricultural products increase consumer confidence and future demand, while studies indicating the potential for adverse health and environmental effects decrease demand. Agricultural biotechnology firms are assumed to respond to these demand shifts by introducing new products and/or withdrawing existing ones.

More specifically, let P_t indicate the expected number of transgenic food and agricultural products on the market in year *t*. It is assumed that farmer and consumer experience with transgenic products increases demand, which results in the introduction of new products to the market at an annual rate of *g*. In addition, the number of products on the market declines by a factor, *d*, multiplied by the number of approved products which are the subject of studies showing potentially adverse health or environmental effects. The consumer response factor *d* can be interpreted as a reputational spillover. It is the number of products for which demand declines below the breakeven point for each product reported to have *potentially* detrimental health or environmental effects.

If p is the annual probability that a transgenic product will be the subject of reports of potential harm, then the expected number of products on the market at the end of year 1 is given by:

$$P_1 = P_0(1 + g - pd) \quad (22.1)$$

subject to $(1 + g > pd)$

and the expected number of products in year t is:

$$Pt = P_0(1 + g - pd)^t \quad (22.2)$$

Let B represent the expected annual benefits resulting from the sale of a transgenic agricultural product. This would include profits earned by the innovating firm, changes in net farm income resulting from adoption of these products and changes in consumer surplus prior to taking into account health or environmental damages.[3] Let D indicate the expected *real* damages caused by any product that has been reported to have potentially harmful effects.[4] If c is the annualized cost of regulatory compliance for transgenic products, then annual net benefits, (v_t), derived from transgenic products are:

$$v_t = P_t(B - pD - c) \quad (22.3)$$

for $B > pD + c$ else $v_t = 0$

which after substituting for P_t yields

$$v_t = P_0(B - pD - c)\,(1 + g - pd)^t \quad (22.4)$$

Given a discount rate r, the present value of the net benefit stream is:

$$V = \sum_{t=1}^{\infty} P_0(B - pD - c)\left(\frac{1 + g - pd}{1 + r}\right)^t \quad (22.5)$$

which reduces to:

[go to top of next column]

$$V = P_0\frac{(B - pD - c)}{(r - g + pd)} \quad (22.6)$$

s.t. $r + pd > g$

Equation 22.6 indicates that regulatory agencies seeking to maximize net benefits from transgenic food and agricultural products would need to take into account the effects of regulation on production costs and health and environmental damages as well as the more complex interaction between public risk perceptions and demand. Even if bioengineered products cause no harm to human health or the environment (i.e. $D \Rightarrow 0$), reports of potentially adverse impacts can result in welfare losses due to their demand-reducing effects.[5] The welfare effect of consumer reactions to perceived risks is captured by the third term in the denominator of Equation 22.6, which can be seen to have the same effect on the present value of net benefits as an increase in the discount rate.

If the stringency of regulation is represented by s, the regulator's problem can be defined as follows:

$$\max_s V = P_0\frac{(B - pD - c)}{(r - g + pd)} \quad (22.7)$$

s.t. $r + pd > g$ and $B > pD + c$ for some s

More stringent regulatory requirements would be anticipated to increase c, the cost of regulatory compliance for transgenic products, and reduce p, the probability that an approved product will be reported to have negative health and environmental effects. In addition, increases in c can be assumed to reduce g, the rate at which new products are introduced. Given these relationships, the first derivative of V with respect to s yields:

$$\frac{\partial V}{\partial s} = \frac{\left(\frac{-\partial p}{\partial s}D - \frac{\partial c}{\partial s}\right)P_0(r - g + pd) - (B - pD - c)P_0\left(-\frac{\partial g}{\partial s} + \frac{\partial p}{\partial s}d\right)}{(r - g + pd)^2} \quad (22.8)$$

[3] Moschini and Lapan (1997) present a conceptual model for estimating these benefits when innovations are protected by intellectual property rights. Falck-Zepeda *et al.* (2000a, b) and Moschini *et al.* (2000) provide empirical estimates of the magnitude and distribution of these benefits for transgenic cotton and soybean varieties.

[4] D could be further disaggregated into the probability that a product reported to have potentially harmful effects actually causes real damages and the expected magnitude of these damages.

[5] If documented damages from bioengineered products remain small or non-existent, consumers would become increasingly sceptical of reports of potential problems. But it could take a number of years before consumers significantly revise their prior expectations.

Setting the derivative equal to zero and rearranging, results in the following first order condition.

$$\frac{\partial c}{\partial s}-\left(\frac{B-pD-c}{r-g+pd}\right)\frac{\partial g}{\partial s}=D\left(\frac{-\partial p}{\partial s}\right)$$
$$+\frac{B-pD-c}{r-g+pd}d\left(\frac{-\partial p}{\partial s}\right) \qquad (22.9)$$

The first term on the LHS of Equation 22.9 is the change in expected regulatory costs due to more stringent regulation of bioengineered agricultural products. The second term on the LHS is the net cost (or benefit) to society of reductions in the rate at which new products are introduced due to more costly regulatory requirements. The first term on the RHS represents the reduction in health and environmental damages resulting from more stringent regulation. This much of Equation 22.9 captures the conventional rule in regulatory benefit–cost analysis; set regulatory requirements so that the marginal increase in regulatory costs plus the forgone benefits of new products is equal to the marginal reduction in health and environmental damages. The second term in parentheses on the RHS of Equation 22.9 adds an additional policy consideration, namely whether more stringent regulatory guidelines will affect demand by influencing consumer perceptions of product quality and safety. Some further specification of the model may be a useful means of exploring this additional consideration.

A Numerical Analysis

Initial estimates of the economic benefits associated with introduction of transgenic agricultural products have recently appeared in the literature (Falck-Zepeda *et al.*, 2000a; Moschini *et al.*, 2000). But there is still very little reliable information available on regulatory costs, the probabilities or magnitude of environmental or public health damages, or the effect that changes in regulatory policies might have on these variables. Nevertheless, by making some reasonable assumptions about key parameters in the model it is possible to generate some insights that are useful for policy analysis.

Assume that regulatory stringency (s) can be measured on a ten point scale, with zero indicating no regulation and higher values representing increasingly rigorous combinations of pre- and post-approval testing and monitoring. Annualized regulatory costs are assumed to be an increasing convex function of regulatory stringency as defined below.

$$c = bs^{\alpha} \text{ for } \alpha > 1 \qquad (22.10)$$

where b can be interpreted as the per product expenditures on public health and environmental risk assessment in the absence of regulation.

Let H indicate the probability that, in any given year, a product will be the subject of credible reports of potentially harmful effects in the absence of regulation. The probability of such reports is assumed to be a declining convex function of s as defined in Equation 22.11.

$$p=\frac{H}{(1+s)^{\beta}} \qquad (22.11)$$

Finally, the rate at which new products are introduced (g) is assumed to be a linear function of the cost of developing a product that can gain regulatory approval as shown below.

$$g=\frac{wB-c}{wB}m \text{ for } B \geq c, \text{ else } g=0 \qquad (22.12)$$

where w is the share of total product benefits captured by the innovating firm.

Table 22.1 summarizes a set of parameter values in which expected product benefits significantly exceed expected costs and marginal product development costs quickly exceed the marginal reduction in expected health and environmental damages as regulatory stringency is increased. Given these parameters, Fig. 22.1 presents the expected value of social benefits for different degrees of consumer reaction to reports of potentially harmful effects. If there is no need to take into account consumer risk perceptions (i.e. if $d = 0$), the optimal level of regulation would be quite low. However, if consumers react to reports of potentially harmful effects ($d > 0$), maintaining relatively low levels of regulatory oversight results in a significant reduction in social welfare. As d increases, the social cost of exceeding the optimal regulatory level also becomes progressively smaller.

What is also worth noting is that when $d = 0$, an increase in the expected real net benefits of transgenic agricultural products results in a decrease in the optimal level of regulation. As expected net benefits of each product increase this also increases the

Table 22.1. Parameter values for baseline analysis.

Number of products initially on the market	P_0 = 10
Expected annual benefits per product (US$millions)	B = 100
Expected annual damages from products reported to have potentially adverse effects (US$millions)	D = 200
Rate of product introduction in the absence of regulation	m = 0.075
Discount rate	r = 0.1
Exponent of increase in product development costs due to stronger regulation	α = 2
Per product risk assessment expenditures in the absence of regulation (US$millions)	b = 1.0
Probability of reports of potentially harmful effects in the absence of regulation	H = 0.25
Exponent of decrease in probability of negative reports due to stronger regulation	β = 2
Share of benefits realized by innovating firm	w = 0.5

forgone benefits of slowing new product introductions due to high regulatory costs. However, when

$d > 0$ and $\left(\frac{\partial p}{\partial s}\right)d > \frac{\partial g}{\partial s}$ an increase in expected

net product benefits results in an increase in the optimal level of regulation. The greater the expected social benefits of the products and the greater the consumer response to reports of potential harm, the more important it is to prevent consumer rejection due to incomplete or inaccurate information.

Figure 22.1 indicates that if expected net product benefits are positive, society would be better off if consumers would simply ignore unsubstantiated reports of potential harm. But if the distribution of benefits is taken into account, consumer reactions may not be as irrational as would appear at first glance. It is consumers that would bear the full costs of any public health effects of transgenic food products. Yet several studies have indicated that US consumers realized less than 10% of the total benefits attributable to initial adoption of the first generation

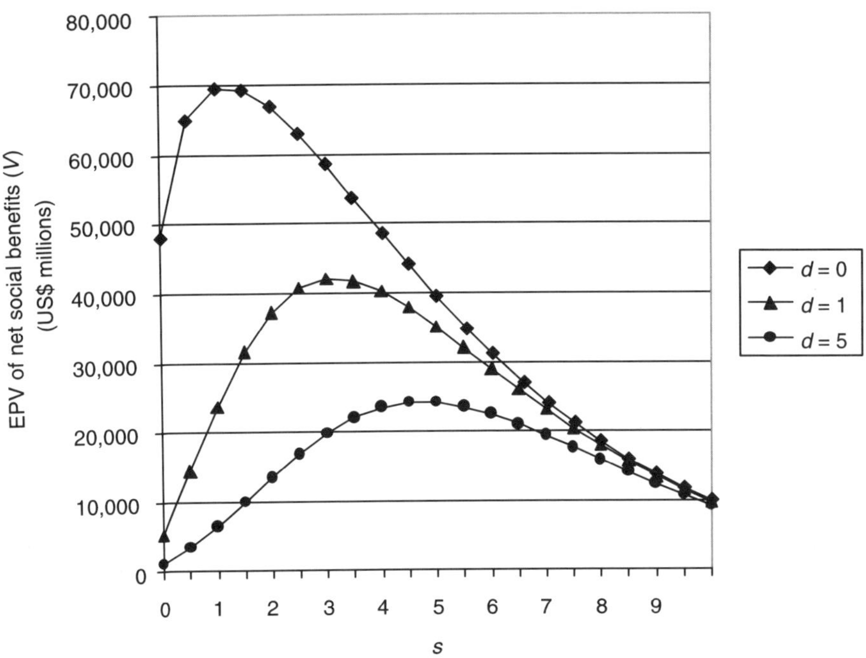

Fig. 22.1. Net social benefits as a function of regulatory stringency and consumer response to adverse information.

of transgenic crop varieties (Falck-Zepeda *et al.*, 2000a, b).[6] Faced with relatively small benefits and frequent reports of potential risks, lack of consumer support for these products is understandable.

Policy Implications

Taking into account the demand effects of consumer risk perceptions provides an economic rationale for increased product regulation beyond what might be justified by purely scientific risk assessments. This depends, of course, on the cost effectiveness of regulation as a means of maintaining consumer confidence. The model presented here indicates that if consumers are sensitive to adverse information, and regulatory policy changes will reduce the frequency of credible reports of potential harm with relatively little increase in product development costs, the social benefits of implementing these policy changes could be substantial.

The recently proposed changes in FDA regulatory policy towards transgenic food products can be interpreted as an effort to reduce consumer concerns with little or no increase in product development costs. It remains to be seen how effective the new policy will be. The commitment by the FDA to provide a written response to pre-market notifications, and to make any information contained in the notification and the FDA's response publicly available, should provide added incentives to ensure that health and safety claims can be defended from potential challenges. Nevertheless, the proposed regulatory changes do not require industry to undertake any new testing or evaluation processes and may therefore not be sufficient to ensure consumers of the safety of transgenic food products.

The analysis presented above also indicates that it may be in the interest of the US to encourage, rather than oppose, comparatively stronger regulatory controls in Europe. European consumers have demonstrated a greater sensitivity than their American counterparts to reports of potentially negative health and environmental effects of transgenic food products. If stronger regulations are effective in reducing the frequency of reports of potential problems, or the attention that European consumers give to these reports, agricultural biotechnology companies and US farmers producing transgenic commodities stand to benefit.

These policy implications are only strengthened if stochastic aspects of the model and potential feedback effects are taken into account. For a given level of regulatory stringency, the number of reports of potential harm in any given year is defined by a binomial distribution with mean of pP_{t-1} and standard deviation of $\sqrt{[pP_{t-1}(1-p)]}$. The occurrence of consecutive years in which a relatively high number of products are the subject of credible reports of potential harm could cause a sharp and long-term decline in the size of the market, particularly if the magnitude of consumer reactions is an increasing function of the frequency of such reports. If demand drops below a critical threshold, the market could even disappear entirely, requiring some demand-enhancing intervention to reopen. It could be argued that this has already occurred in Europe and government officials are now searching for ways to restore consumer confidence.

The model also provides a novel perspective on the political economy of product regulation. In markets where demand is influenced by prior sales and perceptions of health and environmental risks, public and industry perspectives on optimal regulation may converge. If B_p represents private revenues from transgenic food and agricultural products, then the expected present value of net benefits to industry are

$$V_p = P_0 \frac{(B_p - c)}{(r - g + pd)} \tag{22.13}$$

Figure 22.2 compares net private and social benefits as a function of increasing levels of regulation, using the parameter assumptions defined in Table 22.1. When consumers are (or are perceived to be) unresponsive to reports of adverse effects ($d = 0$), industry would prefer no regulation at all, while regulators acting in the public interest would prefer a higher level of regulation. This is the traditional view presented in the political economy literature. However, as illustrated in Fig. 22.2, if consumers are even moderately responsive to adverse information ($d = 1$), then both regulators and industry should prefer a significantly higher level of regulatory control, due to their shared interest in avoiding

[6] These first generation products were designed to reduce farm input costs and crop losses. Consumer benefits consisted largely of modest decreases in commodity prices.

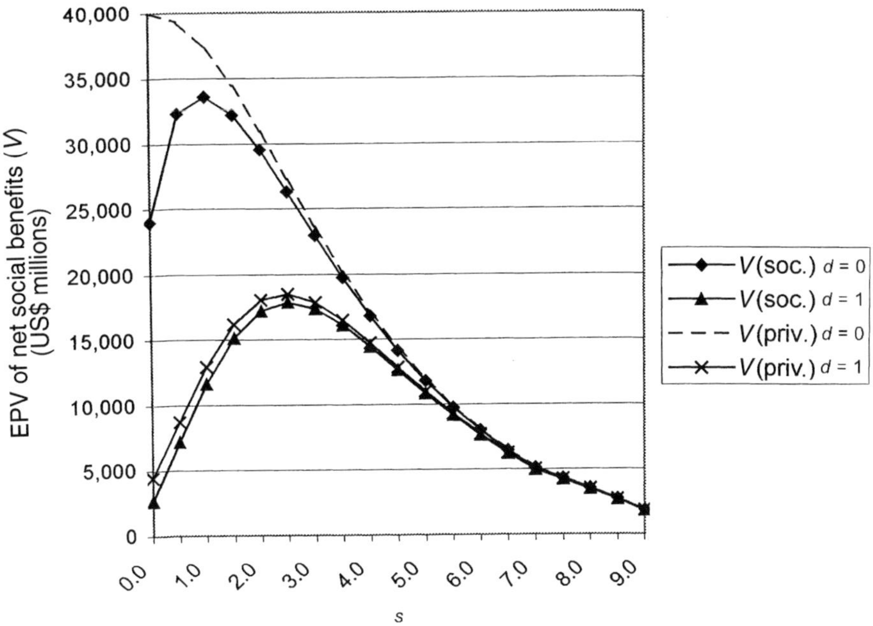

Fig. 22.2. Social vs. private benefits as a function of consumer response to reports of potential harm.

consumer rejection of profitable and socially beneficial products.[7]

Whether convergence of policy positions extends to environmental and consumer advocacy groups, depends largely on the perceived effectiveness of regulatory policy changes. If new, purportedly stronger, regulations are viewed as resulting in only a modest reduction in expected damages, consumer and environmental groups may prefer the status quo. The preceding discussion of welfare comparisons when consumer preferences are changing provides some insight into this apparently paradoxical response. Many groups concerned about the health and environmental effects of transgenic crops fear that consolidation in the agricultural input industry will make it increasingly difficult for farmers to switch back to conventional varieties once they have become reliant on transgenic seed. If regulatory policy changes are perceived as increasing consumer confidence, but are perceived to have little or no effect on the probability of real damages, these changes might be viewed as equivalent to promoting the use of harmful, addictive substances. Consequently, some environmental and consumer groups may continue to oppose any regulatory changes short of an outright ban on transgenic agricultural products.

Conclusions

The model developed in this chapter provides an economic rationale for increased product regulation in response to consumer perceptions of risk, even when these perceptions are perceived by experts to be unfounded. The model also indicates that recognition of the endogeneity of demand and consumer risk perceptions should promote convergence in the policy positions of regulatory agencies and industry. Recent developments in the market for transgenic agricultural products were used as a basis for illustrating the policy applications of the model. However, the analytical results can also be applied to other products and services for which it is difficult for consumers to determine safety or efficacy.

7 This result holds as long as expected benefits significantly exceed expected costs at the optimal level of regulatory stringency.

References

Antle, J. (1996) Efficient food safety regulation in the food manufacturing sector. *American Journal of Agricultural Economics* 78, 1242–1247.

Antle, J. (1999) Benefits and costs of food safety regulation. *Food Policy* 24, 605–623.

Dutton, G. (2000) FDA's proposals for GM foods generally accepted. *Genetic Engineering News* 20(11), 1, 47, 64, 75.

European Commission (2000) Commission takes initiative to restore confidence in GMO approval process. Press release, European Commission, Health and Consumer Protection Directorate General, Brussels. www.europa.eu.int/comm/dgs/health_consumer/library/press/press62_en.html

Falck-Zepeda, J.B., Traxler, G. and Nelson, R.G. (2000a) Surplus distribution from the introduction of a biotechnology innovation. *American Journal of Agricultural Economics* 82(2), 360–369.

Falck-Zepeda, J.B., Traxler, G. and Nelson, R.G. (2000b) Rent creation and distribution from biotechnology innovations: the case of Bt cotton and herbicide-tolerant soybeans in 1997. *Agribusiness* 16(1), 21–32.

Genetic Engineering News (2000) Overseeing biotech foods. *Genetic Engineering News* 20(10), 1.

Grout, P.A. (1982) Welfare economics of decision making with changing preferences. *The Review of Economic Studies* 49, 83–90.

Hammond, P.J. (1976) Changing tastes and coherent dynamic choice. *The Review of Economic Studies* 43, 159–173.

Larson, B.A. and Knudson, M.K. (1991) Public regulation of agricultural biotechnology field tests: implications of alternative approaches. *American Journal of Agricultural Economics* 73(4), 1074–1082.

Marschack, T.A. (1978) On the study of taste changing policies. *The American Economic Review* 68, 368–391.

Moschini, G. and Lapan, H. (1997) Intellectual property rights and the welfare effects of agricultural R&D. *American Journal of Agricultural Economics* 79(4), 1229–1242.

Moschini, G., Lapan, H. and Sobolevsky, A. (2000) Roundup Ready soybeans and welfare effects in the soybean complex. *Agribusiness* 16(1), 33–55.

Pollack, R. (1970) Habit formation and dynamic demand functions. *Journal of Political Economy* 78(1970), 745–763.

Roberts, T., Buzby, C.J. and Ollinger, M. (1996) Using benefit and cost information to evaluate a food safety regulation: HACCP for meat and poultry. *American Journal of Agricultural Economics* 78, 1297–1301.

Stiglitz, J.E. (2000) Imperfect information in the product market. In: Schmalensee, R. and Willigs, R.D. (eds) *Handbook of Industrial Organization*, Vol. 1. Elsevier Science, Amsterdam.

Vettas, N. (2000) Investment dynamics in markets with endogenous demand. *Journal of Industrial Economics* 48(2000), 189–203.

23 Tobin's *q* and the Value of Agriceutical Firms

Michael Boland
Department of Agricultural Economics, Kansas State University, Manhattan, Kansas, USA

The agriceutical (formerly called 'life science') industry has been founded on the premise that genetic engineering can be used to create plants and animals with characteristics that are healthier and environmentally friendly. Goldberg and Enriquez (2000) suggest that boundaries between food, health care, chemical and pharmaceutical companies are becoming blurred, and that the current food and agribusiness system will become an agriceutical industry. Thus far, the impact of agriceutical companies has been mixed. A large number of mergers, acquisitions and joint ventures have occurred among various firms in these industries. By June 2000, there were a number of firms that considered themselves agriceutical companies or had agriceutical segments among their various businesses.

A recent book by Smithers and Wright (2000) argues that stocks are vastly overpriced relative to the actual value of their assets. Using Tobin's q (ratio of a firm's market value to its assets), the authors report that average q is above 2 for publicly reported stocks. This suggests that investors recognize that many firms possess intangible assets such as brand value, intellectual property rights and goodwill, which are not reflected in their balance sheets. The research and development needed to discover and create agriceutical products is enormous. Public and private investments are needed to discover and extend the benefits of agriceutical products to countries with lower incomes per capita and/or at-risk populations. Although firms may possess intangible assets, implementing a successful strategy to use those assets profitably is required for private investment. Information on how agriceutical firms are valued currently can be useful to policy makers contemplating various public sector investments.

The objective of this study was to determine the value of q for pharmaceutical and chemical firms who operate in the consumer health, crop protection and animal health industries. Tobin's (1969) definition of q is used: the ratio of a firm's market value to its book value. A brief discussion of firm strategy is provided. Then, background information is presented for the firms included in the study. Finally, a discussion of q and its estimation is provided.

Agriceutical Firm Strategy

Agriceutical is a popular word to describe products created through the use of biology and biotechnology that are beneficial to consumers and the environment. For example, various components of feed and food grains (e.g. maize, soybeans, wheat) can be used to replace many types of petroleum by-products found in common household cleaners, which would enable manufacturers to replace the poison symbol on these labels with a biomass symbol. These by-products have little value (the price of petroleum is the main value in a barrel of crude oil) but also have little or no other use and would probably have environmental costs of disposal. In con-

 Market Development for Genetically Modified Foods (eds V. Santaniello, R.E. Evenson and D. Zilberman)

trast, the food and feed grain substitutes have costs associated with their development such as finding uses for all of the by-products.

Early uses of biotechnology have included modification of traits in maize, soybeans and other feed grains to reduce herbicide, insecticide or pesticide application costs. The successful adoption of pesticide- or herbicide-resistant cotton, soybean and maize seed threatened chemical companies, who viewed the companies selling seeds as direct competitors to their existing businesses. As a result, seed companies became more valuable in the 1990s, because they were regarded as essential assets for creating a competitive advantage.

Early entrants in this industry are having trouble implementing a broad corporate strategy founded on differentiation through agriceutical products (and services) or trying to become the low cost provider of these products. Thus far, the success of such strategies has been limited. Entrants are attracted by the potential returns, but low-cost substitutes exist for many proposed agriceutical products. And barriers to entry, such as research and development, exist in the development of those products.

Chemical and Pharmaceutical Firms

Mergers, acquisitions and joint ventures occurred among almost all chemical and pharmaceutical firms in the middle and late 1990s. At the beginning of the decade, many of these firms had various animal health, consumer health and crop protection assets. Agriceutical firms include Aventis (merger of Hoechst AG and Rhone-Poulenc SA); E.I. du Pont de Nemours and Company (acquisition of Pioneer Hi-Bred); The Dow Chemical Company (Union Carbide, Eli Lilly's plant science division and Mycogen); Pharmacia (merger of Monsanto and Pharmacia and Upjohn); and Syngenta (merger of AstraZeneca PLC and Novartis AG).

In addition, diversified chemical companies are involved with crop protection such as BASF AG, Bayer AG, ICI Americas and Mitsubishi Chemical Company. Finally, some pharmaceutical companies are involved with animal health products such as American Home Products, Eli Lilly, Pfizer, Roche Holding Ltd (until early 2000) and Schering AG.

Valuation of a Firm

One method used to measure the value of a firm is Tobin's q (commonly called q), which is the ratio of a firm's market value to the replacement cost of its assets and is calculated as:

$$q = \{P \times E + L\} \div K$$

Where P is the firm's stock price, E is its number of shares, L is the market value of liabilities and K is corporate assets. A q equal to 1 suggests that the market value of the firm is equal to the replacement value of its assets. If q is greater than 1, then the market values a firm's assets higher than the cost of those assets. For q to approach 1 a firm must increase its investment (e.g. investment costs of creating assets are less than their true worth) or the market value of those assets must decrease, which occurred during the 1929 to 1932 and 1969 to 1974 time periods.

Smithers and Wright suggest an alternate but equivalent form of q, which is used by the Federal Reserve, where q is calculated as:

$$q = (P \times E) \div (K - L)$$

The denominator simplifies to the firm's net worth or simply the sum of the book values of preferred stock and common equity. In either method of calculating q, the interpretation is the same. One advantage of this definition is that q can be broken into two parts: stock price and the value of net worth per share.

One criticism of q is that it does not value intangible assets such as patents and other intellectual property, the value of a firm's brands, and other similar goodwill assets. A second criticism of q is the maintained hypothesis that the Miller–Modigliani theorem always holds. Namely, in an efficient market, the market value of a firm should be unaffected by debt or equity financing. The q value is important to agriceutical firms and others involved in biotechnology, because intellectual property rights are important assets of these firms. Given the difficulty in implementing an agriceutical strategy, firms are likely to invest in technologies and products that offer the quickest return to the investors who have valued these assets accordingly. Alternatively, firms may use patents and licences to reap royalties for the use of these technologies. A firm's market valuation will determine the level of private investment necessary to conduct further research and development of other products.

Data to Calculate *q*

The methodology of Lindenberg and Ross (1981) and Smirlock *et al.* (1984) was used to estimate Tobin's *q* for each of the chemical and pharmaceutical firms listed earlier. Data for the year ending 31 December as reported by Standard & Poor's COMPUSTAT© database were used in this analysis.

To determine the numerator of *q* (*P* and *E*), the market value of the firm was determined for the last day of each year. COMPUSTAT© reports the amount of debt maturing in various years (2, 3, 4 or 5 years). Corporate bonds were assumed to have the industrial average yield as reported in Moody's Composite average. Debt was valued each year by calculating the price of the remaining debt at the current yield as reported by Moody's. Other liabilities were recorded at their book values. Deferred taxes were not used. Common equity was recorded using the value provided by COMPUSTAT© at 31 December of each year, which was the product of the number of shares outstanding and the stock price. Preferred stock was recorded at its book value.

To determine the denominator of *q* (*K* and *L*), the replacement value of assets was determined on the last day of each quarter ending in April, August and December. If quarterly data were not available, then calendar year data were used. Assets were divided into three categories: plant and equipment, inventory and other. Plant and equipment were valued by determining an acquisition schedule. The value of plant and equipment during the first year of data was assumed to be its book value. A 10-year depreciable life was assumed, and the book value was reduced by 10% (standard depreciation) each year and then adjusted to its new market value by multiplying by the GDP price deflator for that year. Changes in assets (sales or purchases) were added at book value. Inventory was adjusted according to whether the firm used FIFO or LIFE (COMPUSTAT reports this figure for each firm). Book values were used for other inventory and other assets.

Results

Table 23.1 reports the average value of *q* and its range over the time period for various seed, chemical and pharmaceutical companies. Note that pharmaceutical firms have the highest *q* over the decade, which is not surprising. Freberg's (2000) analysis of 112 food and agribusiness firms in 16 different Standard Industrial Classification codes found that the pharmaceutical industry had the highest returns on equity and on investment and had less variability in returns relative to the average over the 1980 to 1998 time period. The intangible assets of seed companies also were valued at more than two times their book value by investors. Similarly, investors valued the intangible assets of the agriceutical firms higher than their book value.

The results suggest that these firms have undervalued various intangible assets (assuming the market valuation is correct), which is not surprising because they often have great difficulty valuing such assets. Investors have even greater difficulty, because

Table 23.1. Estimates of *q* for selected pharmaceutical and chemical firms, various years.

Firm	Average *q*	Range
Delta and Pine Land[a]	3.32	1.55 to 7.11
Pioneer Hi-Bred International[a]	1.99	1.15 to 3.19
AstraZeneca[b]	2.70	0.68 to 5.27
Aventis[c]	1.06	0.78 to 1.43
The Dow Chemical Company[a]	1.43	1.15 to 1.88
E.I. du Pont de Nemours and Company[a]	1.89	1.21 to 2.90
Novartis AG[a]	2.98	1.91 to 2.78
Pharmacia[a]	1.78	1.21 to 3.34
American Home Products[a]	2.97	1.61 to 4.84
Eli Lilly[a]	3.88	1.61 to 6.68
Pfizer[a]	3.31	1.91 to 8.37

[a]1990 to 1999.
[b]1996 to 1999.
[c]1997 to 1999.

the future is so uncertain with the current controversy over genetically enhanced food and feed grains. No one knows exactly how the market for agriceutical products will evolve.

Finally, some of these intangible assets have yet to achieve their full economic benefits. Economies of scale are needed in research and development in order to realize their full value. However, pharmaceutical and chemical firms have used an enormous amount of resources to finance the mergers and acquisitions to achieve these economies of scale.

Summary

Chemical and pharmaceutical firms recently have invested tremendous resources to achieve the economies of scale needed to carry out research on agriceutical products. Although the long-run future for these companies is probably positive, the public may perceive in the short run that these firms will be unable to achieve all that has been promised. Private and public investments will be needed to enable the full benefits of biotechnology to be shared by all consumers.

References

Freberg, C. (2000) Competition, strategy, and profitability in food and agribusiness industries. MS thesis, Department of Agricultural Economics, Kansas State University, Manhattan, Kansas.

Goldberg, R.A. and Enriquez, J. (2000) Transforming life, transforming business: the life-science revolution. *Harvard Business Review* March–April, 96–104.

Lindenberg, E.B. and Ross, S.A. (1981) Tobin's *q* and industrial organization. *Journal of Business* 54, 1–32.

Smirlock, M., Gilligan, T.W. and Marshall, W.J. (1984) Tobin's *q* and the structure-performance relationship. *American Economic Review* 74, 1051–1060.

Smithers, A. and Wright, S. (2000) *Valuing Wall Street: Protecting Wealth in Turbulent Markets.* McGraw-Hill, New York.

Standard & Poor's COMPUSTAT© (2000) McGraw-Hill Companies, Englewood, Colorado.

Tobin, J.A. (1969) General equilibrium approach to monetary theory. *Journal of Money, Credit, and Banking* 1(1), 15–29.

24 The Structure of the European Agro-food Biotechnology Industry: Are Strategic Alliances Here to Stay?

W. Bruce Traill and Christopher E. Duffield
Department of Agricultural and Food Economics, The University of Reading, Reading, UK

Introduction

In the pharmaceuticals industry, the biotechnology revolution has produced a core of large companies, often looking to consolidate, with a large and vibrant fringe of dedicated biotechnology firms (DBFs). These DBFs form the middle ground between the large firms and universities and other research groups. The industry is characterized by alliances between these three groups, which is believed to enhance the level of innovation within the industry. The agro-food biotechnology sector has likewise seen consolidation among the downstream 'life science' companies and the development of alliances between these and agro-food DBFs and universities and other public sector research organizations. It has furthermore seen substantial consolidation in the seed industry and the (vertical) integration of independent seed companies, at least American ones, into the life science companies. When conceived, the life science companies were argued to be taking advantage of synergies in biotechnology research between pharmaceutical and agro-food applications, but recently Monsanto has combined with Upjohn and spun off its agro-food division, as have AstraZeneca and Novartis in spinning off their agribusiness divisions to form Syngenta. This suggests that at the downstream level the expected synergies between agro-food and pharmaceuticals research have not emerged, though the continued presence within agro-food life-science companies of seeds, crop protection chemicals and, in some cases, animal health products, suggests some synergies in the delivery of inputs to farmers: 'plant biotechnology did deliver synergy between Seeds and Crop Protection, but as our experience grew, so did the realisation that the synergies between Agribusiness and our other activities would remain marginal' (Novartis Operational Review, 1999: 2).

The focus of plant biotechnology has so far delivered mainly input traits, though the next generation of products is expected to emphasize output traits, targeted at end product attributes desired by food manufacturers (e.g. processing quality) and consumers (e.g. taste, safety and health). With respect to health, such products will be in competition with a fast-growing high-tech sector of the food industry, functional foods. This area is beginning to show the development of alliances between traditional food and pharmaceutical companies such as those between Quaker and Novartis, and Raisol and Johnson and Johnson.

There has been tentative media speculation about firms wishing to integrate vertically through the full food chain, producing high tech seeds, selling them through identity preserved vertical supply channels involving primary processing (e.g. the

(eds V. Santaniello, R.E. Evenson and D. Zilberman)

alliance between Monsanto and Cargill giving the former access to the latter's grain elevators and primary processing supply channels) and, potentially, finished food manufacturers. The *Wall Street Journal* (27 May 1998) refers to Monsanto and DuPont 'racing to build "dirt-to-dinner" biotechnology pipelines' (p. B1) and 'DuPont hopes to be able to take orders for a new type of crop from food companies such as Nestlé SA or ConAgra Inc., create it in the laboratory, contract with farmers to grow millions of acres and process it into a food ingredient' (p. B1). Food retailers, often thought nowadays to be the food chain channel captains (particularly in Europe), have not yet entered the fray, except to ban genetically modified (GM) products from their shelves.

The present structure of the agro-food biotechnology industry may be portrayed as a large number of DBFs primarily owned and located in America connected via a series of alliances to life science firms which are either American owned (Monsanto, DuPont, Dow) or European (Novartis/AstraZeneca, Aventis). Whatever their origins, these firms operate on a global scale. To date the forward linkages to food manufacturing are minimal. The food manufacturing giants with a global presence and a high degree of research intensity are predominantly European (Unilever and Nestlé), though research and production activities are globally located.

A number of questions emerge: what will be the future vertical structure of the agro-food industries under the influence of agricultural biotechnologies and a host of other pressures for structural change? Who will own the industry and where will it be located? From a European perspective, do the combined disadvantages of a low number of DBFs, strict regulation and consumer antagonism marginalize the European agro-food system and leave it producing expensive, unhealthy food targeted solely at the (isolated) home market?

Regretably this chapter does not provide answers to all of these questions. Indeed, it is more of a question raising than answering chapter and in this sense represents work in progress (or just beginning). The main focus of the chapter is on the 'make or buy' nature of the upstream relationships between life science and seed companies and DBFs. At this stage the research does no more than expose some of the theoretical predictions of the economics and business strategy literatures to statements appearing in the annual reports of the life science companies. Based on these, the conclusions section speculates on some of the broader issues raised above.

The theories of market structure considered here are: transaction cost economics, which indicates a move towards vertical integration (Pisano, 1991); a real option model (Lavoie and Sheldon, 2000a, b), again suggesting only a transitory role for alliances; complementary assets, which suggests horizontal and vertical consolidation (Graff *et al.*, 1999); and dynamic capabilities, which suggests that the continuation of alliances is to be expected in both the short and the long runs (Teece *et al.*, 1997).

Transaction Cost Economics

Transaction cost economics (TCE) is most commonly used to explain/predict the boundaries of a firm. It argues that the market may not be the ideal mechanism for some economic transactions due to the transaction costs of using the price mechanism. The firm is viewed as a suppression of the price mechanism to reduce the costs of the many transactions required in order to provide even the simplest of goods. TCE is concerned with static efficiency (Teece and Pisano, 1994) and examines whether the firm should 'make' (internally) or 'buy' (externally) each of the constituent products or stages of production. In the TCE framework, internal and external resources are substitutes and these are utilized according to their relative cost efficiency. Gulati *et al.* (2000) state that 'the transaction cost perspective stresses the efficiency benefits from reducing the governance cost of a transaction' (p. 204).

Thus TCE treats the development by firms of internal competences as a substitute for outsourcing and alliances are temporary and unstable, stepping stones to integration (Pisano, 1991; Williamson, 1995). The biotechnology industry exhibits many characteristics that favour make rather than buy with respect to R&D:

- Much biotechnology research is characterized by asset specificity, exposing the DBF to *ex post* small number bargaining.
- The case of high technology industries is complicated by bounded rationality; the firm purchasing R&D or technology may be unable to acquire the information needed in order to decide a suitable price. It is difficult to specify the characteristics of the good or it may be that this requires full disclosure, which then removes

the need for purchase. Geroski described this as the 'difficulties that buyers have in trying to value an innovation, and that sellers have in helping them without giving away crucial secrets' (1992: 139). This raises the possibility of opportunistic action on both sides, and again integration is considered to be the solution. In biotechnology, however, there may be less scope for opportunism than sometimes envisaged in TCE. The limited number of players in the field and the strong links in the scientific community may act as a deterrent (Dietrich, 1994).

- From the large company perspective, it is difficult to define contracts effectively because it is difficult to define precisely the performance and output of DBFs.
- High technology industries are characterized by tacit knowledge, knowledge that is acquired but cannot be easily transferred as it resides either within firms' human capital or can be part of firms' traditions or routines. The existence of tacit knowledge makes it even harder for the firm to accurately determine what it is purchasing and to contract in order to accurately specify its intentions. This added problem in contracting will lead to increased cost of governance for the transaction. The difficulty of transferring tacit knowledge across firm boundaries, especially in the absence of complete trust also increases the relative efficiency of vertical integration.
- Appropriability of research output is also a concern given the imperfect protection afforded by patents (Pisano, 1991).

These various risks and uncertainties are all exacerbated in an international setting. They suggest over time the integration of DBFs into the R&D activities of the life science companies. Note that these TCE predictions derive from a static efficiency maximization/cost minimization view of the world.

Real Options

Lavoie and Sheldon (2000a) use a real options model to explain the differences in relative R&D levels of US and European firms. They suggest that it is differences in the external environment, that is capital availability and the regulatory regime, which alter the firms' actions and therefore their comparative advantage. The R&D process is lengthy and costly, with the time period and cost unknowable *ex ante*. The costs incurred are paid up front and are considered at least partially irreversible. The partially random nature of returns to R&D investment favours American companies which have access to larger capital supplies than European firms. The US firms, given the additional capital, can invest more over a shorter time period, reducing the likelihood of the project suffering from unforeseen delays. The regulatory environment is also seen as more favourable in the USA, lowering the risk of investment in R&D. The capital market is also thought to favour early entry by DBFs; the large multinationals monitor DBF performance with a view to determining whether to exercise an option to invest later, either internally or by acquisition (Lavoie and Sheldon, 2000b).

Lavoie and Sheldon suggest a European catch-up following the lagged entry of life science multinationals and growth in the European DBF sector. Based only on what is written in the preliminary abstract Lavoie and Sheldon (2000b), we are unsure whether this prediction arises from an easing of the capital constraint on investment as the profitability of the industry becomes more evident over time; an easing of the regulatory environment in Europe; or an increasing globalization of the industry due to the growing presence of the life science multinational enterprises (MNEs), resulting in the growing irrelevance of the home country's regulatory environment or the location of DBFs. In any case, the model appears to predict the consolidation over time of R&D within MNEs, an outcome in line with TCE.

Complementary Assets

Graff *et al.* (1999), using a complementary assets model, also predict consolidation of agro-food biotechnology research. They list three important categories of assets in the development of new plant varieties by genetic transformation:

1. Process technologies for plant transformation.
2. Genetically coded traits and enhancements.
3. Elite germplasm.

Each of these assets is subject to economies of scale and, pair-wise, each is complementary to the others (i.e. there are synergies between them). If this is the case then the industry will slowly consolidate until there are only a few large integrated life science companies. These firms will span the whole spectrum of industries for which biotechnology is an enabling technology: pharmaceuticals, chemicals, agriculture and food. DuPont is a prime example of this type of large, integrated firm using biotechnology to produce goods from food to man-made fibres. The multinationals will swallow the smaller DBFs and then enter a process of mergers in order to further consolidate the industry. Artuso (1999) indicates that this is already the case in the pharmaceutical industry.

The model is similar in nature to TCE in the sense that it is static efficiency that determines the optimal size and scope of the firm.

Dynamic Capabilities

The literature on dynamic capabilities provides a rationale for the existence and persistence of the networks observed so far in biotechnology. Like the complementary assets model, this model has its origins in the 'resource-based theory of the firm' (Barney, 1996). Firms are viewed as bundles of resources/competences/capabilities, which may be valuable if they are rare and hard to imitate, the latter condition applying particularly to intangible assets such as tacit knowledge and firm culture, which are often path dependent. 'The accumulation of know-how is viewed as an incremental and path-dependent process with the firm's overall knowledge base and hence capabilities being the evolutionary outcome of the nature and pattern of its experiences' (Anon., 2000: 3).

Within biotechnology, pharmaceutical and life science companies have competences in testing, regulation, manufacture and marketing, all subject to economies of scale and beyond the reach of DBFs who rarely bring products to market (Anon., 2000). DBFs, often created by entrepreneurial academics using risk capital have expertise in specific areas of biotechnology.

> In industries characterised by rapid technological change such as biotechnology, it is unlikely that any firm's internal knowledge base is adequate … collaboration with external actors who possess complementary skills becomes an attractive proposition, perhaps an imperative one.
>
> (Anon., 2000: 9)

Large firms continue to invest in their knowledge base in order to assimilate external knowledge and make better informed investments, but they form alliances with DBFs with specific competences. Given the dynamic nature and uncertainty over future competence needs, alliances are less risky than outright assimilation of DBFs by MNEs so long as the technological base of the industry remains dynamic.

Forward thinking firms wanting to compete look for partners with the competences/knowledge they lack in order to open up their range of options. Coombs and Tomlinson state, 'the most innovative firms are characterised by a sharp awareness of the gaps that may exist in their capabilities' (1998: 24) and so these are the most likely to seek out partners with complementary competences. Gutterman (1997) states 'an innovative firm need not have the internal capacity to independently perform all of the steps' (p. 100). The firm is no longer viewed as a stand-alone entity but as part of a network of firms involved in symbiotic competition, with strategic alliances being used to 'extend their own pool of resources and capabilities' (Pena *et al.*, 1999: 3).

In biotechnology and other high technology industries these networks of companies are thought to be especially important for innovation. This is summed up by Arora and Gambardella (1990): 'The locus of innovation should be thought of as a "network" of inter-organizational relations' (p. 374). The networks are viewed as increasing the output of R&D (Bartholomew, 1997; Padmore and Gibson, 1998) and of competitiveness overall (Bureth *et al.*, 1999).

Cohen and Levinthal's idea of absorptive capacity is part of this and also provides an insight into the pattern of innovation in the biotechnology industry. They state 'most innovations result from borrowing rather than invention' (Cohen and Levinthal, 1990: 128) and that this use of external sources of knowledge and ideas is vital. Padmore *et al.* (1998: 622) argue that 'Internal R&D capacity has increased, both to meet the increased demand for step innovation, and to increase the firms capacity to absorb external innovation'. This an example of how important many studies view this (Llerena and Matt, 1999). Internal R&D has even been described as an entry ticket or currency for entering into network agreements (Powell *et al.*, 1996).

The DBFs and the universities and other not-for-profit research establishments fit into this view of the industry structure by filling in the gaps of the larger firms' competences. The biotechnology industry is very closely linked to the basic research undertaken by these research centres and the pace of that research requires the larger firms to keep a careful watch on developments in these centres. Indeed 'technological innovation ... relies heavily on the scientific discoveries made by non-profit institutions' (Malo and Geuna, 1999: 15). The DBFs, often set up by academics to exploit their IPRs (Oehmke *et al.*, 2000), are a vital link between these research centres and the large firms (Sharp and Senker, 1999).

Anon. (2000) uses the North Carolina Biotechnology Center 'Actions Database' on international alliances to suggest that, for biotechnology as a whole, two-thirds of international transactions were organized through collaborative modes and that this share remained constant over 12 years (to 1992), supporting the dynamic capabilities view that the threat of technological obsolescence results in firms' reluctance to commit to vertical integration. The data are for all biotechnology, meaning that they are dominated by the pharmaceuticals sector and it is unclear whether the picture is the same for the agro-food sector.

Research Questions

The four theories presented above predict differing structures for the agro-food biotechnology industry.

1. Transaction cost economics: predicts that the industry will vertically integrate in order to overcome the inherent costs and problems of effective transaction governance.
2. Real options theory: predicts that the industry will show an initial burst of DBF set-ups followed by the entry of multinationals, some of which will enter by a process of acquisition. This model does not predict the persistence of alliances.
3. Graff *et al.*'s theory of complementary intellectual assets: predicts the consolidation and vertical integration of the industry over time as firms attempt to achieve economies of scale and to make use of complementary assets.
4. Dynamic capabilities: predicts that alliances will be a permanent feature of the industry, at least so long as it is in a period of rapid technological change, as each firm sits within a web of changing linkages, allowing each to gain from the others' competences.

Review of Life Science Firms' Research Strategies

At this early stage in the research we are unable to test the above model predictions scientifically. Rather we obtain an impression of what the life science firms themselves believe as suggested within their annual reports and also some press reports.

The statements support the accepted wisdom that horizontal consolidation is taking place within both the seed and life science companies and that the two are becoming vertically integrated.

> The global market for vegetable and flower seeds has become much more concentrated over the last 10 years.
>
> Several companies have joined larger groups, mainly seed groups in the case of vegetables, unlike what has been happening in field seeds, where it was mainly agro-chemists and agro-industrialists who were doing the take-over.
>
> (Vilmorin, Clause and Cie Annual Report, 1998/1999: 30)

The Novartis Operational Review 1999 agrees saying 'crop protection companies have acquired many formerly independent seed companies in order to gain access to germplasm, technological expertise and seed distribution networks that are necessary to produce genetically enhanced crops' (p. 5). This statement could be viewed as providing backing for either the complementary assets or transaction costs viewpoints. Notably, Graff *et al.*'s (1999) statements 'the more varieties into which a given gene can be incorporated, the higher the value of owning a patent on the gene' and 'the more genes that can be added to a plant as extra enhancements (stacking), the higher the value of the base variety' are persuasive in this respect, though given the regulatory and consumer acceptance environments in Europe, the consolidation of the seed industry in that continent must be based on a very long-run perspective.

Further evidence of vertical integration can be found, again in the Vilmorin, Clause and Cie Annual Report 1998/1999 (p. 42), suggesting that dealing with problems in managing the value chain leads to vertical integration:

> The increasing complexity of managing these products, which are the result of more and more sophisticated research, means that total integration of the whole supply chain is necessary – research, production, factory processing and distribution – which ensure the ultimate goal: quality.

This statement seems to back up the central arguments of the transaction cost economics school of thought.

The majority of comments, however, refer to the importance of alliances for R&D, either as a means of entry (as predicted by the real options theory), or as a permanent feature of the industry. The Financial Times Mastering Strategy Series notes that

> Monsanto, the life sciences powerhouse, is an excellent example of an organisation that has used an alliance network to spearhead the transformation from an old-line chemicals concern to the cutting-edge of biotechnology, and then to cope with the recent rapid technological convergence in the life sciences (p. 4).

The use of alliances is echoed by Novartis in their Operational Review 1999 (p. 5)

> An industry-wide web of collaborations has been established to improve the productivity of R&D investments as the leading companies seek access to the full range of new platform technologies from niche companies at the forefront of their particular technology areas

and in the Bayer Annual Report 1999, 'We are building a position at the leading edge of technology through a network of research alliances' (p. 13) and

> key technologies like genomics, bioinformatics, combinatorial chemistry and high-throughput screening are indispensable in crop protection research too. We are utilising these technologies intensively by way of cooperation agreements with leading research companies (p. 25).

The importance of alliances are also stressed by: Vilmorin, Clause and Cie, Annual Report 1998/9 'the firm must ... multiply partnerships to remain at the forefront of high-speed scientific progress' (p. 3); 'BASF has established two research joint ventures and begun cooperating with partners in plant biotechnology' (BASF, 1999: 41); 'to achieve new product breakthroughs, we will continue to invest in new technologies and leverage our strengths by cooperating with innovative, cutting edge biotech companies' (Aventis, Brief Annual Report, 1999: 9); 'R&D expenditure increased to $297m; the increase was largely associated new collaborations in biotechnology research' (AstraZeneca Annual Report, 1999: 41); 'In R&D, close collaboration with external partners is essential' (DSM Annual Report, 1999: 12).

While these statements do not exclusively relate to agro-food biotech, nor do they prove that collaboration is a permanent rather than transitory state, they are suggestive of this. The view is partially supported by Ernst & Young (2000) showing that in Europe for the 4 years from 1996 to 1999, the biotech industry (dominated, it is true, by pharmaceuticals) saw strategic alliances and joint ventures double in number from 123 to 261 whereas mergers and acquisitions increased only from 40 to 46. Incidentally, the majority of the alliances made by European firms were with American DBFs and pharmaceutical companies. Ernst & Young (1999) also make the point that, in earmarking up to 30% of its R&D budget to external collaborations,

> companies must ensure that sufficient discovery capability remains in-house to enable them to make value judgements on what is on offer while ensuring that they do not become overly reliant on external organisations at critical stages of the R&D process.

Also that 'consolidation in biotech goes on all the time – it's just not enough to offset the formation of new companies' (p. 25).

While collaboration is seen as necessary to the large life science companies, it is the lifeblood of the DBF: Diversa corporation saw 'the increase in revenues [688%] resulted from several agreements signed in 1999' (Agra Food Biotech, 2000: 20). These 'collaborations include Novartis Agribusiness Biotechnology Research, Inc. for new seed products to enhance crop production and provide improved performance and quality and output traits, and Dow Chemical company to develop novel enzymes' (Agra Food Biotech, 2000). Of Genzyme Transgenics corporation's first quarter revenues of US$18.9 million, US$3.6 million came from transgenic collaborations (Agra Food Biotech, 2000).

It is not only the agro-chemical and life science companies which are creating alliances in this field; Unilever's Annual Review 1999 says that 'Collaboration with external agencies is an integral part of our research' (p. 19). Nestlé's booklet 'Research and development at the dawn of the 21st century' states

> The whole of the Nestlé R&D system benefits from co-operation with leading and reputed research institutions and universities on the international scene and on all continents who contribute their skills in highly specific areas of science and technology. Even as big a company as Nestlé cannot perform all the tasks in-house (p. 8).

Conclusions

The limited evidence presented in this chapter tends to support the dynamic capabilities view of the world, suggesting that so long as the agro-food biotechnology industry remains dependent on high levels of R&D expenditure and continues to draw on a diverse and unpredictably changing range of highly specialized scientific disciplines, alliances between DBFs, universities and life science companies will continue. The major seed companies, though, have already been integrated into the life science firms, in Europe as well as the USA; synergies (complementary capabilities) being a likely explanation, though it would be unwise to rule out completely market power as a motive. The question of location (and the related question of international competitiveness) has not been directly addressed. On the one hand, it is clear that life science firms are creating alliances and joint ventures with what they perceive to be the firms with the best skills wherever they are located globally. On the other hand, personal discussions with those in life science firms responsible for managing joint ventures suggest that they see advantages in proximity and, all else being equal, would prefer local partners. If so, this would suggest that the creation of an environment attractive to vibrant DBFs and public sector research would create conditions favouring the nearby location of life science companies' R&D facilities (and likewise, the existence of life science R&D facilities could be one factor encouraging the establishment of local DBFs). Another open question is the extent to which adaptation to local environmental conditions requires some sort of local genetic research capability (in the same way that food manufacturers adapt products to local consumer preferences). If so, this more applied and standard research is likely to take place internally to the large seed manufacturers (within the life science companies) as it will not require the specialist skills of the DBFs. Finally, there is no evidence at this stage concerning the possible further vertical integration of the food chain to include the high-tech end of food manufacturing. Being more sensitive to final consumer demand, we would anticipate that such firms will be reluctant (at least in Europe) to form partnerships, but over the longer term it is a possible development we will watch with interest.

References

Agra Food Biotech (2000) No. 29. Agra Europe (London) Ltd, London.

Anon. (2000) The international organisation of the biotechnology industry. Paper submitted to *International Business Review.*

Arora, A. and Gambardella, A. (1990) Complementarity and external linkages: the strategies of the large firms in biotechnology. *Journal of Industrial Economics* 38(4), 361–379.

Artuso, A. (1999) Life science industries: comparative analysis of structure, performance and regulatory policy in the agricultural, biotechnology and pharmaceutical sectors. Paper prepared for the ICABR Conference on *The Shape of the Coming Agricultural Biotechnology Transformation: Strategic Investment and Policy Approaches from an Economic Perspective.* University of Rome Tor Vergata, Italy, 17–19 June 1999.

AstraZeneca Annual Report (1999) AstraZeneca.

Aventis Brief Annual Report (1999) Aventis.

Barney, J.B. (1996) *Gaining and Sustaining Competitive Advantage.* Addison-Wesley, Reading, Massachusetts.

Bartholomew, S. (1997) National systems of biotechnology innovation: complex interdependence in the global system. *Journal of International Business Studies* 2nd Qtr, 241–266.

BASF (1999) *Facts and Figures.* BASF.

Bayer Annual Report (1999) Bayer.

Bureth, A., Wolff, S. and Zantei, A. (1999) Co-operative learning and the evolution of inter-firm agreements in the European electronics industry. In: Gambardella, A. and Malbera, F. (eds) *The Organization of Economic Innovation in Europe.* Cambridge University Press, Cambridge.

Cohen, W.M. and Levinthal, D.A. (1990) Absorptive capacity: a new perspective on learning and innovation. *Administrative Science Quarterly* 35, 128–152.

Coombs, R. and Tomlinson, M. (1998) Patterns in UK company innovation styles: new evidence from CBI innovation trends survey. *CRIC Discussion Paper,* 7. CRIC, The University of Manchester, Manchester.

Dietrich, M. (1994) *Transaction Cost Economics and Beyond: Towards a New Economics of the Firm.* Routledge, London.

DSM Annual Report (1999) DSM.

Ernst & Young (1999) *13th Biotechnology Industry Annual Report.* Ernst & Young International, London.

Ernst & Young (2000) *European Life Sciences Seventh Annual Report.* Ernst & Young International, London.

Financial Times Mastering Strategy Series (undated) *The Financial Times.*

Geroski, P.A. (1992) Vertical relations between firms and industrial policy. *The Economic Journal* 102, 138–147.

Graff, G.D., Rausser, G.C. and Small, A.A. (1999) Agricultural biotechnology's complementary intellectual assets. Paper prepared for the ICABR Conference on *Shape of the Coming Agricultural Biotechnology Transformation: Strategic Investment and Policy Approaches from an Economic Perspective.* University of Rome Tor Vergata, Italy, 17–19 June 1999.

Gulati, R., Nohria, N. and Zaheer, A. (2000) Strategic networks. *Strategic Management Journal* March, 203–215.

Gutterman, A. (1997) *Innovation and Competition Policy.* Kluwer Law International, London.

Lavoie, B.F. and Sheldon, I.M. (2000a) The source of comparative advantage in the biotechnology industry: a real options approach. *Agribusiness: an International Journal* 16(1), 56–67.

Lavoie, B.F and Sheldon, I.M. (2000b) Market structure in biotechnology: implications for long run comparative advantage. Paper presented to 4th ICABR Conference, on *The Economics of Biotechnology,* Ravello, Italy, 24–28 August 2000.

Llerena, P. and Matt, M. (1999) Inter-organizational collaboration: the theories and their policy implications. In: Gambardella, A. and Malbera, F. (eds) *The Organization of Economic Innovation in Europe.* Cambridge University Press, Cambridge.

Malo, S. and Geuna, A. (1999) Science–technology linkages in an emerging research platform: the case of combinatorial chemistry and biology. *SPRU Electronic Working Paper Series Paper,* 37. The University of Sussex, Brighton.

Nestlé (undated) *Research and development at the dawn of the 21st century.* Nestlé.

Novartis Operational Review (1999) Novartis.

Oehmke, J.F., Weatherspoon, D.D., Wolf, C.A., Naseem, A., Maredia, M. and Hightower, A. (2000) Is agricultural research still a public good? *Agribusiness: an International Journal* 16(1), 68–81.

Padmore, T. and Gibson, H. (1998) Modelling systems of innovation 2: a framework for industrial cluster analysis in regions. *Research Policy* 26(2nd Qtr), 625–641.

Padmore, T., Schuetze, H. and Gibson, H. (1998) Modelling systems of innovation: an enterprise-centered view. *Research Policy* 26(2nd Qtr), 605–624.

Pena, I., Boehje, M. and Akridge, J. (1999) Collaborative agreements in the ag-biotechnology industry: the importance of transaction costs and investment. Strategy paper prepared for the ICABR Conference on *Shape of the Coming Agricultural Biotechnology Transformation: Strategic Investment and Policy Approaches from an Economic Perspective.* University of Rome Tor Vergata, Italy, 17–19 June 1999.

Pisano, G. (1991) The governance of innovation: vertical integration and collaborative arrangements in the biotechnology industry. *Research Policy* 20, 237–249.

Powell, W.W., Koput, K.W. and Smith-Doerr, L. (1996) Inter-organizational collaboration and the locus of innovation: networks of learning in biotechnology. *Administrative Science Quarterly* 41, 116–145.

Sharp, M. and Senker, J. (1999) European biotechnology: learning and catching up. In: Gambardella, A. and Malbera, F. (eds) *The Organization of Economic Innovation in Europe.* Cambridge University Press, Cambridge.

Teece, D.J. and Pisano, G. (1994) The dynamic capabilities of firms: an introduction. *Journal of Economic Behaviour and Organization* 3, 451–489.

Teece, D.J., Pisano, G. and Shuen, A. (1997) Firm capabilities, resources and strategy. *CCC Working Paper* 908. Center for Research in Management, University of California at Berkley.

Unilever's Annual Review (1999) Unilever.

Vilmorin, Clause and Cie Annual Report (1998/1999) Vilmorin, Clause and Cie.

Wall Street Journal (1998) 27 May.

Williamson, O.E. (1995) Hierarchies, markets and power in the economy: an economic perspective. *Industrial and Corporate Change* 4(1), 21–49.

25 Market Structure in Biotechnology: Implications for Long-run Comparative Advantage

Brian F. Lavoie and Ian M. Sheldon
Department of Agricultural, Environmental and Development Economics, The Ohio State University, Columbus, Ohio, USA

Introduction

In a recent paper, Lavoie and Sheldon (2000) advance the hypothesis that observed US comparative advantage in biotechnology can be explained by sources of heterogeneity within the biotechnology R&D investment process. Using a real options approach, they illustrate how international differences in the per-period rate of investment and level of domestic regulatory uncertainty could be sufficient to explain the emergence of US biotechnology firms as world leaders in the industry.

Specialization in high technology implies that domestic firms should enjoy excess returns. However, the accumulation of rents stemming from innovative activity could be diminished by the growing presence of foreign-based multinationals in the domestic industry. In particular, foreign multinationals can enter into alliances with or even acquire established domestic start-ups and, therefore, obtain access to proprietary knowledge stocks generated from ongoing domestic R&D. The current and future rents embodied in these knowledge stocks can, therefore, be partially or fully appropriated by overseas entities.

Recent dynamics in the biotechnology industry suggest that this process may be underway. Sharp (1996: 2–3) observes,

> European multinationals are penetrating and exploiting American capabilities in biotechnology ... The large European-based multinationals in chemicals and pharmaceuticals, in pursuit of the necessary knowledge and skills in biotechnology, have through arrangements of one sort or another widely penetrated the American knowledge base.

Recent examples of this process include: a US$45 million deal between Hoechst Schering AgrEvo and Gene Logic to discover genes useful for crop protection and improvement products; an agreement between Hoffman-LaRoche and Agouron to develop anti-cancer drugs; and the establishment of a 5-year R&D collaboration between BASF AG and Mitotix.

In this chapter, the possible reasons for and implications of foreign-based multinational activity in the US biotechnology industry are considered. In particular, a two-tiered industry structure for biotechnology is posited, consisting of start-ups, which pioneered the industry, and multinationals, which are relatively late entrants. Sources of heterogeneity within the biotechnology R&D process led to US start-ups emerging as world leaders in biotechnology. Although many of these start-ups are yet to be profitable, they have accumulated valuable assets in the form of proprietary knowledge stocks originating from ongoing R&D projects. By either forming alliances with or acquiring American start-ups, European multinationals can gain access to knowledge stocks arising from US R&D and, hence, establish a claim on the future excess returns these stocks may produce.

In the next section, a characterization of industry dynamics in a biotechnology sector populated by start-ups and multinationals is developed. This allows an examination of how late entry by European multinationals could result in a re-allocation of ownership rights to potential excess returns from innovation embodied in proprietary US knowledge stocks. In the third section, the entry decision of start-ups and multinational firms is examined in more detail, then some of the implications of this characterization are illustrated using computer simulation. Specifically, the model includes the possibility that multinational firms may acquire R&D undertaken by domestic start-ups. Finally, policy and trade implications stemming from European penetration of the US biotechnology industry are discussed.

A Characterization of Industry Dynamics in Biotechnology

The basic premise of Lavoie and Sheldon is that traditional explanations for the pattern of specialization in high technology industries are poor indicators, *a priori*, of the eventual emergence of US firms as world leaders in biotechnology. This point can be illustrated by considering Grossman and Helpman's (1991) model, where comparative advantage in high technology industries arises from one of two sources: relative factor endowments or initial knowledge stocks. These elements have an impact on innovation by creating asymmetries in the cost of R&D. If firms in all countries begin with identical knowledge stocks, and knowledge spillovers from R&D are international in scope, a traditional Heckscher-Ohlin result is predicted: relative factor endowments determine the pattern of specialization. The country that is relatively abundant in the factor used intensively in R&D, and in high technology production, emerges as the leader in high technology industries. Alternatively, if relative factor endowments are similar across countries, the pattern of trade can still be fixed if one country begins with a larger knowledge stock than its rivals. This country enjoys a 'head start' in innovation in the form of lower R&D costs, and if knowledge spillovers from R&D are restricted to be national in scope, its lead is perpetuated over time.

Neither source of heterogeneity seems especially appealing as an explanation for US comparative advantage in biotechnology. The convergence of industrialized countries since 1945 in terms of traditional sources of comparative advantage has been well documented. Furthermore, there is no indication that the USA enjoyed any form of advantage in initial knowledge stocks as the biotechnology industry began to develop in the late-1970s. Initial knowledge stocks probably took the form of basic scientific research, much of which occurred outside the USA, and was readily available internationally in scientific journals.

As an alternative, Lavoie and Sheldon develop a real options framework to explain US comparative advantage in biotechnology. They formulate a firm's decision to invest in biotechnology R&D as analogous to holding a financial option; that is the right, but not the obligation, to invest in an R&D programme. Comparative advantage is then a question of option management: what incentives caused US biotechnology firms to exercise their options to invest earlier than European firms? Lavoie and Sheldon find that the presence of international differences in the form of a higher US per-period rate of investment and a less uncertain US regulatory environment yield the result that US biotechnology firms, on average, exercise their options sooner. Consequently, they initiate more R&D projects, begin investment sooner, innovate more rapidly, persevere longer in the face of mounting R&D costs and successfully complete more projects than European firms.

Typically, it is assumed that a country specializing in a high technology sector will enjoy excess rents associated with innovation. Indeed, public policy has often stressed the need to promote high technology industries for this very reason (Krugman, 1984). However, recent history of the biotechnology industry suggests that this assumption may be too simplistic. The industry is currently undergoing a period of consolidation, in which multinational corporations have acquired or formed alliances with many of the smaller start-up companies who pioneered the industry. It is significant to note that many of these alliances and acquisitions have been transatlantic in nature, in particular, European multinationals operating in the US biotechnology industry.

Given these conditions, it can be hypothesized

that penetration of the US biotechnology industry by foreign multinationals serves to dilute the concentration of current or future rents associated with biotechnological innovation in the USA. In other words, returns to innovation expected to accrue to domestic firms specializing in high technology industries like biotechnology may be dissipated by the increased presence of foreign-based multinationals in the domestic market. We elaborate on this theme with the following characterization of the dynamics of the biotechnology industry.

There are two distinct classes of firms present in the biotechnology industry: start-ups and multinationals. Start-ups are relatively small, un-diversified firms built 'from the ground up' for the sole purpose of exploiting opportunities in the commercialization of biotechnology. Typically, their capital is obtained from external sources. Start-ups are often the result of a union between bench scientists and venture capitalists; the latter provide seed capital to form the company and begin operations. After a time, it is not uncommon for a start-up to be taken public through an initial public offering, which provides a further infusion of capital for its research efforts.

In contrast, multinationals are relatively large, diversified firms with operations in more than one country, and access to internally generated capital. In general, start-ups were the earliest entrants to the biotechnology industry, followed by entry of multinationals, the latter currently initiating consolidation in the industry through acquisition of and alliances with established start-ups and other multinationals.

Early entry of the start-ups into biotechnology may be partially a consequence of competition for external capital combined with a greater flexibility to undertake R&D projects that pioneered the biotechnology industry. As Lavoie and Sheldon note, however, conditions in the USA have fostered more rapid growth in start-ups than in Europe, such that US biotechnology firms innovated more rapidly and on a larger scale than firms in other countries.

Multinationals, on the other hand, have, in general, been late entrants to the biotechnology industry. Sharp (1996: 6–7) notes,

> [a] combination of uncertainty, skepticism, and inexperience led to what may be called a 'a minimalist strategy' on the part of most large firms. While avoiding large investments most of the companies built up teams of researchers large enough to keep abreast of the science and to monitor developments and competitors... One consequence of this strategy of 'watching and waiting' was that it conceded leadership in development of the new technology to the small companies which were so closely linked to the academic base.

The biotechnology industry can thus be characterized by the following dynamics. Start-ups entered first, with US-based start-ups emerging as world leaders in the industry. Start-ups were then followed by entry of multinationals. Given this characterization, two questions may be posed: what precipitated the respective entry decisions of start-ups and multinationals, and what are the implications of this two-tiered industry structure?

Firms' Entry Decisions

In order to understand entry decisions of start-ups versus multinational firms, it is useful to restate the approach developed by Lavoie and Sheldon. They construct a model of biotechnology R&D investment based on Pindyck's (1993) real options model of uncertain investment cost. A biotechnology firm acquires an opportunity to invest in a new R&D project. When completed, the project yields a product or process innovation worth V with certainty. However, the cost to complete the project is uncertain. The firm holds an option to invest in this project, which it has the right, but not the obligation, to exercise. The expected cost to completion, K, evolves according to:

$$dK = -Idt + \beta(IK)^{1/2}dW + \gamma KdZ \qquad (25.1)$$

I is the per-period rate of investment, β and γ are scalars representing the level of technical uncertainty and regulatory uncertainty, respectively, and dW and dZ are increments of standard Wiener processes, with mean zero and variance dt. Equation 25.1 represents the law of motion for expected cost to completion, driven by investment activity of the firm (the first term), evolution of technical uncertainty (the second term) and evolution of regulatory uncertainty (the third term).

Lavoie and Sheldon extend Pindyck's model to include the possibility of a termination event, which corresponds to abrupt cessation of R&D. The value of the investment opportunity, $F(K,q)$, is affected by the possibility of a random Poisson termination event, q, which takes the form:

$$\xi dq \tag{25.2}$$

where, $\xi = -F$, and $dq = 1$ with probability λdt, and 0 with probability $(1 - \lambda dt)$. λ is the constant mean arrival rate of a termination event. A termination event may be attributed to the results of basic research conducted in an external scientific community, which reveal that the scientific principles on which the research is based are in error. According to Equation 25.2, occurrence of the event implies that the value of the project instantaneously falls to zero, and the project is, therefore, immediately abandoned.

To determine its optimal investment strategy, the firm solves the following infinite horizon optimal stopping problem using dynamic programming:

$$F(K,q) = \max E_0 \left[V_e^{-\mu T} - \int_0^T I(t) e^{-\mu t} dt \right] \tag{25.3}$$

where time to build, T, is stochastic. Asset valuation in a risk-neutral economy is subject to the following relation:

$$rF = -I + E[dF/dt] \tag{25.4}$$

which states that the risk-free return from holding the asset must equal the expected net cash flow plus the expected capital gain. Applying Ito's Lemma yields:

$$E[dF/dt] = -IF_K + 1/2\beta^2 IKF_{KK} + 1/2\gamma^2 K^2 F_{KK} - \lambda F \tag{25.5}$$

Therefore:

$$(r + \lambda)F = -I - IF_K + 1/2\gamma^2 K^2 F_{KK} + 1/2\beta^2 IKF_{KK} \tag{25.6}$$

which is subject to the boundary conditions: $F(0) = V$; $\lim (K \to \infty) F(K) = 0$; $1/2\beta^2 K^* F_{KK}(K^*) - F_K(K^*) - 1 = 0$; value matching condition: $F(K)$ continuous at K^*. Equation 25.6 is then solved numerically for K^*, which is the critical cost to completion.

The model presented above can be summarized as follows. The biotechnology firm acquires an option to invest in an R&D project of certain value V. Investment is constrained to proceed at the maximum per-period rate I. Expected cost to completion K evolves stochastically according to uncertainty in the investment environment. Technical uncertainty and regulatory uncertainty are represented respectively by the parameters β and γ, while uncertainty in the scientific environment is represented by a Poisson process with mean arrival rate λ. The risk-free rate of interest is given by the parameter r. Given values for V, I, r, λ, β and γ, the model can be solved numerically for the firm's critical cost to completion K^*: the maximum level of cost to completion for which it is economically feasible to either initiate investment or continue an ongoing R&D project. If initial expected cost to completion K exceeds K^*, the firm delays investment. If investment has already been initiated when the evolution of K exceeds K^*, the firm abandons the project midstream.

Heterogeneity is introduced into the model by noting that, on average, US biotechnology firms invest at a higher per-period rate than their European rivals. Furthermore, the level of regulatory uncertainty pertaining to biotechnology appears to be lower in the USA. These factors both work to create a higher critical cost to completion K^* for US firms. This in effect results in a looser decision criterion for US firms as they evaluate R&D opportunities in biotechnology, both when the investment option is initially acquired and once the option is exercised, as ongoing R&D proceeds.

This analysis suggests that the entry decision of start-ups has been influenced by technical and regulatory uncertainty. Technical uncertainty is endemic to the R&D process itself; in other words, it exists as a component of the firm's ongoing R&D activity. Thus, the firm can (partially) resolve technical uncertainty only by undertaking investment; as the stages of R&D are incrementally completed, the firm gains more and more insight into the technical difficulties of actually completing the research. Pindyck notes that technical uncertainty may be especially important for R&D projects such as those undertaken in the biotechnology industry.

Regulatory uncertainty, on the other hand, stems from factors that are independent of the firm's investment activity. Changes in the regulatory regime governing the commercialization of biotechnology will occur regardless of whether the firm is investing or not; therefore, the firm can partially resolve this uncertainty without initiating investment. This creates the opposite incentive from that corresponding to technical uncertainty: in particular, it encourages the firm to delay investment in order to observe changing conditions in the regulatory environment.

The evolution of expected cost to completion, expressed in Equation 25.1, indicates that the technical uncertainty term β is not independent of I, the rate of investment, while the regulatory uncertainty term γ is independent of I. A rise in β, *ceteris paribus,* increases a firm's incentive to initiate investment. This would be manifested in a higher critical cost to completion, K^*. Alternatively, a higher γ, *ceteris paribus,* encourages a firm to delay investment, in other words, lowers K^*.

Can the uncertainty pertaining to biotechnology R&D also be used to explain the disparity in investment behaviour observed between start-ups and multinationals? It is possible that inter-firm differences in *technical uncertainty* may drive the difference in investment strategies adopted by start-ups and multinationals. In particular, we posit a scenario whereby the accumulation of proprietary knowledge stocks on the part of early entrants into the industry, start-ups, can be at least partially utilized by late entrants, the multinationals, to contribute towards the resolution of technical uncertainty.

It was noted earlier that start-ups began with the same baseline 'knowledge stock': information published in the scientific literature. However, as R&D programmes commence, and projects are gradually completed, research results and experiences accumulate. These results and experiences are closely guarded proprietary assets of firms that produce them. Thus, conduct of R&D is synonymous with the creation of the type of proprietary knowledge stocks to which Grossman and Helpman refer in their model of endogenous innovation. The value of a firm, especially in a relatively young industry such as biotechnology, is often heavily based on the creation of these proprietary knowledge stocks, in that they serve as an indicator of the firm's future ability to successfully commercialize valuable biotechnology products. These knowledge stocks also contribute towards resolving, at least partially, technical uncertainty faced by firms undertaking biotechnology R&D.

What precipitated the multinationals' decision to delay entry? In comparing different types of firms, rather than identical firms across countries, *technical,* rather than regulatory uncertainty may have been the source of heterogeneity responsible for asymmetric investment strategies exhibited by the two types of firm. Inter-firm differences in technical uncertainty can be manifested in two ways. First, the level of technical uncertainty, β, may have been lower for multinationals than for start-ups, *ceteris paribus.* This explanation might make sense if the level of scientific talent were greater in multinational corporation research laboratories relative to that available to start-ups. However, the evidence contradicts this hypothesis: start-ups have been typically formed around the talents of university scientists working on the cutting edge of biotechnology research. Indeed, large corporations have had trouble attracting these scientists to work in their own laboratories.

A more plausible explanation is that technical uncertainty has been relatively similar for both types of firms, but the way in which it has been manifested is different. In the formulation of the start-ups' investment decision, it was noted that technical uncertainty could only be resolved by actually undertaking investment. This creates an incentive to commence investment right away. Regulatory uncertainty, however, evolves independently of the firm's investment activity, and creates an incentive to delay. On balance, technical uncertainty has been the stronger factor and start-ups have tended to invest early, under extremely uncertain conditions and with little or no cash flow.

In the case of multinationals, it may be that technical uncertainty has been at least partially resolvable without actually undertaking investment, that is when $I = 0$. Some technical uncertainty may still exist that can only be resolved by actually commencing investment, but the remainder can be observed through monitoring the performance of start-ups; the 'watch and wait' strategy noted by Sharp. This implies that in comparing multinationals and start-ups, β has been lower and γ has been higher for the former than for the latter, where interpretation of γ is expanded to include not only regulatory uncertainty, but also any form of uncertainty independent of the firm's investment level. This in turn would create more incentive, represented by a lower K^*, for the multinationals to delay entry in order to obtain more information contributing towards partial resolution of the uncertainty stemming from sources independent of the firm's R&D activity. As a consequence, multinationals would be late entrants to the industry in comparison with start-ups.

Given this scenario, one question remains: why should technical uncertainty assume a different structure for multinationals than for start-ups? The answer pertains to the knowledge stocks developed by start-ups during early stages of the industry. In particular, the multinationals possess an option that start-ups probably do not: the ability to tap into the knowledge base established by start-ups engaged in ongoing R&D.

To see how this process may unfold, a heuristic formulation of the multinationals' entry decision can be developed. In the early stages of the industry, multinationals, in addition to start-ups, acquired options to invest in biotechnology R&D. Both types of firm faced an R&D investment structure similar to that formulated by Lavoie and Sheldon. However, there was one significant difference between the types of firms. Multinationals, being relatively well funded compared with start-ups, could afford to 'watch and wait', observing conditions in the investment environment, including the performance of start-ups. Most start-ups were not profitable, so it is unlikely that multinationals used profitability as the performance metric; rather, they monitored developments in start-ups' ongoing R&D, or equivalently, the accumulation of proprietary knowledge stocks. Multinationals were able to access these stocks over time in an incremental fashion, perhaps by hiring experienced talent away from start-ups, or by purchasing the partially completed R&D of insolvent start-ups. This option is not likely to have been available to most start-ups, whose funds are almost entirely consumed in maintenance of ongoing R&D programmes. These inroads into the knowledge stocks of start-ups assists in resolving some technical uncertainty surrounding commercialization of biotechnology, without engaging in a full-scale commitment to a biotechnology R&D programme.

If expected cost to completion eventually falls below the critical value pertaining to multinationals, which is lower than that corresponding to start-ups, the multinational will exercise its investment option and invest directly in biotechnology R&D, through in-house research programmes, and often through strategic alliances with or acquisition of established start-ups. In the case of alliance or acquisition, the multinational obtains full access to the proprietary knowledge stock of the start-up.

The industry dynamics of biotechnology can now be considered in the context of this description of the start-ups' and multinational's investment decisions. In the industry's earliest stages, technical uncertainty was the dominant form of uncertainty surrounding biotechnology R&D. As such, it encouraged start-ups to exercise their option to invest early, and engage in large-scale investments in the face of no significant cash flow. In this environment, the sources of comparative advantage favoured the USA, as US firms invested at a higher per-period rate and were governed by a less uncertain regulatory environment. In engaging in this early investment, start-ups began to accumulate valuable proprietary knowledge stocks, stemming from ongoing research. These knowledge stocks served to reduce technical uncertainty as the commercialization of biotechnology proceeded.

Multinationals, on the other hand, delayed exercising their option to invest, since the R&D investment structure facing them was slightly different from that of the start-ups. In particular, technical uncertainty was at least partially resolvable without actually undertaking direct investment, through the expediency of access to the knowledge stocks established by start-ups engaged in ongoing R&D. This implies that the critical cost to completion has been lower for multinationals relative to start-ups and, thus, the incentive to exercise the option to invest has been less.

Multinationals entering the industry often do so by either forming alliances with or acquiring established start-ups. In doing so, they obtain direct access to the knowledge stocks that these firms have developed over time. It seems likely that the selection of which start-ups to form strategic alliances with or to acquire will be based on the pattern of specialization established in the industry through the R&D activity of the start-ups. In particular, it is hypothesized that multinationals will be more likely to invest in start-ups located in the country holding the comparative advantage in biotechnology R&D and production. Two explanations can be used to justify this behaviour. First, since the USA has established a comparative advantage, investment conditions must be more favourable in that country than elsewhere. Second, the fact that US firms are world leaders in the industry suggest that the knowledge stocks they have accumulated exceed, on average, those of their European counterparts. Therefore, it may be expected that multinational penetration of the American biotechnology industry will be greater than that observed in the European industry.

Implications for Rent Distribution

In this section, an illustration is offered of how acquisition and alliance activity of foreign-based multinationals can reduce the concentration of excess returns accruing to start-up firms in the country specializing in biotechnology. The implications of the hypothesized industry dynamics discussed above are examined using a refinement of the simulation techniques employed in Lavoie and Sheldon (2000).

The implication that European multinationals will have a greater propensity to acquire US start-ups than European ones would result in a cross-country pattern of ownership of the assets of US start-ups in the form of European claims on the proprietary knowledge stocks of US-based start-ups. This has the ancillary effect of re-allocating the current and future rents embodied in these knowledge stocks from their originators, US start-ups, to European multinationals. Thus, an asymmetry emerges in the long-run structure of the industry, in that biotechnology R&D and production is concentrated in the USA, based on the comparative advantage established by US start-ups, but the long-run allocation of the excess returns arising from this specialization is more evenly distributed across US and European enterprises.

First, an industry populated solely by American and European start-ups is considered. Estimates place the number of start-ups in the USA in 1996 at 1287, compared with 716 European firms: these estimates are utilized in the simulation (Ernst & Young, 1997a, b). At time $t = 0$, each biotechnology firm acquires an option to invest in an uncertain R&D project. The model is parametrized using the combination of 1996 industry aggregates and ad hoc values employed by Lavoie and Sheldon. In particular, the value of R&D, the risk-free rate of interest, λ and β are assumed to be the same for both types of firms, and are parametrized as US$262 million, 0.055, 0.067 and 0.5, respectively. Heterogeneity takes the form of the maximum per-period rate of investment, set to US$16 million year^{-1} for the US firm, and US$6 million year^{-1} for the European firm, and the level of regulatory uncertainty, γ, set to 0.1 for the US firm and 0.2 for the European firm.

An iteration of the simulation begins with random draws to obtain an initial expected cost to completion K, and a waiting time for the first occurrence of a Poisson termination event. At time $t = 0$, the firm determines if the initial K exceeds K^*: if so, the firm delays investment to observe the stochastic evolution of K, driven by the random component associated with regulatory uncertainty. If the current value of K eventually falls below K^*, the firm exercises its option to invest at that time. Otherwise, the firm observes K until the termination event occurs, rendering the investment option worthless.

If the option is exercised, investment proceeds as follows. For each time period, the expected cost to completion is incremented by subtracting the firm's current investment, and adding on the (positive or negative) random components embodied in the technical and regulatory uncertainty. The firm compares the current K to K^*; if K exceeds K^*, the project is abandoned midstream; else, incremental investment continues until expected cost to completion equals zero, at which point the R&D project is considered successfully completed. If at any time the current period coincides with the random time corresponding to the occurrence of the termination event, the project is terminated immediately.

Heterogeneity is introduced into the simulation by designating different values for I and γ for US and European firms. Simulation suggests the representative US firm on average initiates more R&D projects, commences investment sooner, innovates more rapidly, perseveres longer in the face of mounting R&D costs and, ultimately, successfully completes more projects than the representative European firm. This suggests that US biotechnology firms will eventually emerge as the world leaders in the industry and, by extension, acquire the assets and excess returns associated with innovation in high technology industries.

Using the parametrization detailed above, 1287 iterations were run representing the number of US biotechnology start-ups in 1996; in addition, 716 iterations were conducted to represent the number of European start-ups that same year.

Table 25.1. Share of successfully completed R&D projects (% owned).

US start-ups	91
European start-ups	9

Given these 2003 total iterations, Table 25.1 reports the share of successfully completed R&D projects owned by US and European start-ups. In an industry populated solely by start-ups, the sources of heterogeneity, embodied in the per-period rate of investment and the level of regulatory uncertainty, result in the majority of successful R&D projects belonging to US firms. By extension, and absent the presence of multinationals, it is also the case that excess returns from innovation embodied in these projects are concentrated in the USA.

Multinationals are introduced into the analysis in the form of a random process representing multinational penetration of the biotechnology industry. This process is intended to be suggestive of the option management problem underlying the multinationals' entry decision. A function $f(t)$ is defined to characterize the probability at time t that a start-up's R&D will be acquired by a multinational, where $f(t) = 1/[(\gamma_F/\gamma_D)\rho t+1]$, defined on the interval [0, 1]. In this equation, γ_F and γ_D are the levels of regulatory uncertainty in the foreign and domestic biotechnology industries, and ρ is a constant scalar set to 0.001. The function $f(t)$ equals one at time $t = 0$, and converges to zero in the limit as t goes to infinity. This function is used to characterize the probability that a multinational acquires the R&D of a start-up. At the commencement of each iteration, a random draw u is made from the uniform distribution, on the range [0,1]. For each time period t, u is compared with the contemporaneous level of $f(t)$. If $u > f(t)$, then the R&D is acquired by a multinational; else, the start-up continues to retain ownership. Note that the probability of a multinational acquiring the R&D increases with t, corresponding to the idea that the value of ongoing R&D increases over time, as proprietary knowledge stocks develop and mature.

The ratio (γ_F/γ_D) is positively correlated with $f(t)$. In other words, an increase in the level of regulatory uncertainty in the foreign industry, or a decrease in the level of uncertainty in the domestic industry, increases the probability that a domestic start-up's R&D will be acquired by a multinational. This is because, *ceteris paribus*, a multinational would prefer to minimize the risk associated with its investment by operating in a relatively certain regulatory environment.

If the random process determines that a start-up's R&D is acquired by a multinational, a second random draw, uu, from the uniform distribution determines the geographical origins of the purchaser. For simplicity, it is assumed that US and European multinationals are equally likely to purchase ongoing start-up R&D. Therefore, if $uu > 0.5$, the multinational is considered US-based; else, the multinational is considered European-based.

Table 25.2. Share of successfully completed R&D projects (% owned).

	USA	European
Start-ups	51	9
Multinationals	21	19

It is assumed that multinationals do not perform R&D internally, and that the potential for multinational acquisition does not enter the start-up investment decision process. Given this characterization of the activity of multinationals, the simulations for US and European biotechnology firms are re-run, with the added refinement of the possibility of multinational acquisition.

The results in Table 25.2 show that introduction of multinationals into the industry has the effect of transferring control of a portion of the successful R&D projects to multinationals: approximately 60% of the projects are owned by the start-ups who originated them, while the remaining 40% have been acquired by multinationals. The penetration of multinationals into the market, however, is not symmetric across countries. In Table 25.3, the level of multinational penetration in the US and European industries is reported.

Table 25.3. Level of multinational penetration (%).

US industry	42
European industry	26

Penetration of the US biotechnology industry by multinationals is greater than that of the European industry, brought about by the higher level of regulatory uncertainty associated with the latter; in other words, multinationals are more likely to acquire the R&D of US-based start-ups than their European counterparts. This result is made sharper by examining cross-country ownership of biotechnology assets (Table 25.4).

Table 25.4. Cross-country multinational penetration (%).

R&D originated by US start-up, owned by European multinational	19
R&D originated by European start-up, owned by US multinational	10

These results suggest that concentration in the USA of assets and returns generated from successful R&D projects has been diluted compared with the case of a start-ups-only industry, as control of R&D shifts from US start-ups to European-based multinationals. Thus, despite US comparative advantage in biotechnology, European entities control a substantial portion of the market. Specifically, US firms, start-ups and multinationals, control 72% of the R&D, while European firms control the remaining 28%. This can be compared with the benchmark start-ups-only case, where US firms controlled over 90% of R&D. This re-allocation of returns is, ironically, the result of one of the very sources of heterogeneity that established US comparative advantage in the first place.

To sharpen understanding of the effect of regulatory uncertainty on multinationals' decisions to acquire ongoing start-up research, a comparative static analysis is undertaken in the form of increasing the level of European regulatory uncertainty to 0.5.

Table 25.5. Share of successfully completed R&D projects (% owned).

	US	European
Start-ups	34	15
Multinationals	24	27

As shown in Table 25.5, re-running the simulations with this change in place has the effect of further reducing the share of the industry controlled by US start-ups, from 51 to 34%, and simultaneously increasing the share held by multinationals. Multinational penetration of the US and European industries is reported in Table 25.6.

Table 25.6. Level of multinational penetration (%)

US industry	60
European industry	7

The increase in the level of regulatory uncertainty in the European industry has the effect of increasing the presence of multinationals in the US market from 42 to 60% while, at the same time, reducing penetration of the European market from 26 to 7%. This point is corroborated through an examination of the cross-country ownership of biotechnology R&D (Table 25.7).

Table 25.7. Cross-country multinational penetration (% owned).

R&D originated by US start-up, owned by European multinational	32
R&D originated by European start-up, owned by US multinational	5

The presence of European-based multinationals grows from 19 to 32%, indicating a significant transfer of assets and returns from US entities to European ones. Specifically, US firms now control only 58% of the global industry, while the European share has climbed to 42%. As regulatory uncertainty increases in Europe relative to the USA, start-ups in the USA become correspondingly more attractive relative to European start-ups as acquisition targets for multinationals. The implication is that as the European Union continues to tighten its regulatory regime governing biotechnology, this translates into a higher value of γ for the European industry. As the results above suggest, a higher European γ has the dual effect of increasing both multinational penetration of the US industry, and also the number of alliances and acquisitions effected between European multinationals and US-based start-ups.

Implications for Trade Equilibrium and Policy

The presence of European multinationals in the US biotechnology industry raises an interesting policy issue. As the regulatory environment in the European Union toughens, the alliance and acquisition activity of European multinationals in the US biotechnology industry should increase. This implies that even as the European Union restricts biotechnology activity within its own sphere of influence, it still enjoys the fruits of biotechnology in the form of European multinationals' claims on the excess returns from innovation embodied in the proprietary knowledge stocks originating from American start-ups. This suggests the following question: should the US impose policies to protect ownership of its biotechnology assets in response to Europe's tightening of its regulatory restrictions and, if so, what form should these policies take?

Another issue is the long-term trade equilibrium that will emerge in biotechnology. It might be expected that, given the favourable R&D conditions, firms located in the USA will continue to specialize in biotechnology R&D and production. These firms will be either independent start-ups, start-ups in R&D alliances with multinationals, or start-ups that are wholly or partially owned by multinationals. This pattern of specialization suggests that in the long run, the USA will be a net exporter of biotechnology products. However, the rents commonly associated with high technology production will not be entirely captured by US enterprises: rather, a portion will be appropriated by foreign-based multinationals, in the form of intra-firm transfers from US subsidiaries or profits shared with US partners.

It is interesting to note that in the food and agricultural sector, two trading patterns could arise. In the first, the USA exports 'intermediate' biotechnology products in the form of genetically modified (GM) seeds, which embody the proprietary R&D, and subsequently imports the 'finished' product in the form of food containing GM ingredients derived from the seeds. In this case, the rents from innovation accruing to US entrepreneurs are dissipated on two levels: first, through capture of rents brought about by European multinationals' penetration of the US biotechnology industry and, second, through capture of rents generated in the final consumer market in the form of premiums commanded by differentiated goods, specifically, genetically engineered 'designer' foods.

In the second case, the USA exports either agricultural commodities and/or finished food products that embody the proprietary R&D. In this case, the rents from innovation accruing to US entrepreneurs are dissipated on two levels: first, through capture of rents brought about by European multinationals' penetration of the US biotechnology industry and, second, through the extent to which European multinational firms have also acquired the ability to deliver biotechnology via either acquisition or alliance with US seed and food processing firms.

References

Ernst & Young (1997a) *Biotech 97: The Eleventh Industry Annual Report.* Ernst & Young LLP, Palo Alto, California.

Ernst & Young (1997b) *European Biotech 97: a New Economy: the Fourth Annual Ernst & Young Report on the European Biotechnology Industry.* Ernst & Young LLP, Stuttgart, Germany.

Grossman, G.M. and Helpman, E. (1991) *Innovation and Growth in the Global Economy.* MIT Press, Cambridge, Massachusetts.

Krugman, P.R. (1984) Import protection as export promotion: international competition in presence of oligopoly and economies of scale. In: Kierzkowski, H. (ed.) *Monopolistic Competition and International Trade.* Oxford University Press, Oxford.

Lavoie, B.F. and Sheldon, I.M. (2000) The source of comparative advantage in the biotechnology industry: a real options approach. *Agribusiness – an International Journal* 16, 56–67.

Pindyck, R. (1993) Investments of uncertain cost. *Journal of Financial Economics* 34, 53–76.

Sharp, M. (1996) The science of nations: European multinationals and American biotechnology, *STEEP Discussion Paper* No. 28. Science Policy Research Unit, University of Sussex, Brighton. www.sussex.ac.uk/spru/docs/steepdps/28/steep28.txt

26 Biotechnology in the Supply Chain: Managing a Product Differentiating Technology

Robert D. Weaver and Taeho Kim
Pennsylvania State University, University Park, Pennsylvania, USA

Background

Ag-biotech offers important challenges to supply chain management. The success of the affected supply chains in developing solutions will significantly affect the pace at which ag-biotech will evolve in market economies. Public debate and opinion research has made it clear that consumers have preferences over numerous aspects or characteristics of the technology of origin (Weaver *et al.*, 1992). Because these characteristics are often difficult to quantify, they are uncertain for the consumer, forcing choices to be made on the basis of subjective beliefs and assessments of reputation of particular agents as well as of aggregates such as industries, sectors and science itself. Science has long made it possible to alter the private and quasi-public good aspects of private goods to change their consumption characteristics in response to consumer preferences (Weaver, 1993). In the food chain, science has a long record of altering what is available for human consumption including flavour and cosmetics, but also shelf-life, reduction in undesirable food borne content (e.g. calories or components such as fat or sugar) or desirable characteristics (e.g. non-digestible fibre) (see Weaver *et al.*, 1992; Weaver, 1995). These possibilities offer important opportunities for industry to redefine technologies and products to enhance social welfare. However, they also introduce complications into traditional markets that are organized to procure, process and deliver goods derived from homogeneous commodities.

This chapter considers economic performance of a supply chain that incorporates supply from biotechnological processes that alter quality characteristics of products. The chapter establishes suboptimal performance results occur when either consumers or producers differentiate across products by technology of origin. Institutional solutions for improving performance of the supply chain through grading, labelling or contracting are considered. Two interests are considered that may differentiate ag-biotech products: (i) consumer preferences; and (ii) technology suppliers. In either case, product differentiation will define opportunities for supply chain management. While biotech can produce positively valued characteristics, the technology itself or characteristics it produces may be negatively valued by consumers. For example, ag-biotech products may be viewed as being of lower quality, in fact as being polluted or contaminated. The presence of residues or contaminants in any one supplier's marketings may result in consumer rejection of the product or discounting of its value. For technology suppliers, ag-biotech may offer potential for productivity and quality enhancement. However, where consumers hold preferences for the technology of origin of products, biotech products may not find demand and willingness to pay at all levels of the supply chain. This generates an economic externality across all agents in the supply chain and creates incentives for self-regulation that alters supply chain performance. Supply chain management actively feeds back to the supply chain's performance depending on whether bilateral or multilateral contracts are used and depending on the transparency of those contracts. Recent experience in international commodity markets for genetically modified organism (GMO) or ag-biotech products provides an important case for analysis of the current state of opportunity for improved supply chain management.

This chapter will assess the extent of current opportunity for enhanced food supply chain performance through a spectrum of alternatives for enhanced coordination to establish value for products that are differentiated by technology of origin.

Managing Biotechnology in the Supply Chain

Product differentiation through ag-biotechnology

To consider the implications of biotechnology, and to retain a manageable scope for this chapter, we suppose that the supply chain consists of producers of primary inputs such as crop or livestock products and processors that transform those primary inputs to produce consumer products.

Suppose the ith producer produces a private good output y_i that has a continuously variable quality characteristic, q_i. This characteristic may be given a wide range of interpretations relevant to biotechnological issues. At one extreme, it could measure the quantity of a physical characteristic such as an undesirable residue or the presence of genetically modified (GM) genes. At an alternative extreme, it could measure the quantity of a desirable component (stereo-isomer fats or sugars that are not digestible and so have no nutritional effects).

A further feature of products that involves quality is the *technology of origin.* This might be either positively or negatively valued by the consumer; for example indicating use of growth hormones, use of specific types of labour (child, disadvantaged) or the use of a particular technology for which consumers hold specific preferences (e.g. laboratory based genetic modification or cell culture propagation). To allow for this, we suppose the existence of a *technology of origin label* binary indicator. We further suppose the existence of a sufficient variety of technologies of origin that allows definition of a scalar continuous indicator interpretable as a quality characteristic that reflects the production technology's characteristic that is valued in the supply chain or by consumers.

Hereafter, we drop the subscript i for convenience when it is not necessary. Primary production occurs through application of a vector of variable inputs, x, as well as a vector of quality related inputs, z. These would include inputs that effect changes in the quality, q. Suppose production involves two separate processes: $y(x, z|\theta, e)$ that results in private good output quantity (y), and one that results in a private good quality, defined by $q(x, z|\theta, e)$. Here, we define θ as a continuous variable indicating a producer type and suppose that the productivity of both quantity and quality are conditional on this type; e is defined as a stochastic event.

To allow for consideration of standards, we suppose the producer does not face a perfectly elastic demand curve allowing sale of all production at a constant price. Instead, we allow for some production to be excluded from the market or dumped. We define a dumping function as $D(x, z, \bar{q} \mid e)$ (Chong, 1996; Carpentier and Weaver, 1997). Dumping is conditional on a quality standard, $\bar{q}$, defined either by regulators or implicitly defined by processor or consumer demand as a minimum quality attribute that must be achieved as a condition of marketing. The disposal of product is not necessarily free (i.e. the technology may exhibit weak disposability in y), however, to proceed we assume it to be so. Generalization is straightforward. We suppose this function satisfies $D(x, z, \bar{q} = 0|e) = 0$, $D(x = 0, z, \bar{q} |e) = 0$, $\partial D / \partial \bar{q} > 0$, $\partial D / \partial z \leq 0$, $\partial^2 D / \partial \bar{q} \partial z < 0$, $\partial^2 D / \partial z^2 < 0$. Based on this notation, it is clear that the technology can result in product differentiation by q. Whether this occurs depends on processor or consumer valuation of q.

We suppose producers attempt to maximize profits based on a market price, P, as well as a possible incentive or differential for quality, λ. We define the most general form of profits earned from production activities as:

$$\pi = P\tilde{y} + \lambda q$$
$$- C(q(x, z \mid \theta, e), y(x, z \mid \theta, e), D(x, z, \bar{q} \mid e))$$

where

$$\tilde{y} = y(x, z \mid \theta, e)\,(1 - D(x, z, \bar{q} \mid e))$$

Together, the market price and the quality incentive define a producer differentiated 'settlement' price. In the broiler industry, the quality incentive, λ, is most often replaced by one that reflects the producer's average cost of production compared to a reference, or peer group.

We next consider the processor. Suppose that the processor earns profits by producing a single product Y_p using a supply of the raw product, Y, collected from producers, as well as variable inputs, x_p. We summarize the processor production process with a cost function. Aggregate supply is defined as $Y = \Sigma_{i=1}^{n} \tilde{y}_i$, the sum of raw product collected across a pool of n producers by prior contract or on an open market. We first view $\tilde{y}_i$ as exogeneous to the processor, consistent with the producer's technology being non-instantaneous; we suppose the processor operates under marketing agreements with producers that require the processor to buy their available supply. Here, we write processor profits as:

$$\pi_p = P_p Y_p - \lambda Q - PY - C_p(Y_p, Y, Q)$$

where Q is defined as an indicator of pool-wide quality, and λ is introduced as a supplier pool incentive paid by the processor for quality.

As in the producer case, a wide scope of specifications can be considered for Q and λ. For example, Q could be defined as the average quality-discounted quantity of product purchased, that is $Q = \Sigma_{i=1}^{n} q_i \tilde{y}_i / Y$. In general, we assume that processor cost is affected by the pool's quality index. This specification could be generalized. We specify the processing cost function as satisfying:

$$\partial^2 C_p / \partial Y_p^2 > 0, \partial^2 C_p / \partial Q^2 > 0, \partial^2 C_p / \partial Y^2 > 0$$

and cross-derivatives as non-positive.

Consumer preferences and demand

Suppose preferences define utility functions $U_j = U_j(y_j, q_j)$ where y_j is a vector of private good quantities consumed, and q_j is a vector of associated quality characteristics. We initially assume that each element of y_j is distinguished by its quality, q_{jy}, both of which are observable by the consumer. Define a consumer income constraint as $P' y_j '' I_j$.

When the consumer is fully informed with respect to associated product characteristics, the product is differentiated in the market, allowing the consumer to simply choose among distinct products, as in the textbook case. Interaction of supply and demand determines a menu of competitive prices that reflect differentials in production cost and preferences across the differentiated products. Inverse aggregate demand functions can be defined for each product h defined by a unique pair (q,θ):

$$P_h = P_h(y_h, I)$$

In truth, the markets for these differentiated products may be independent depending on their substitutability defined by technology and preferences. To proceed, we suppose that the menu of market prices is continuous in the quality, q, allowing the inverse demand to be respecified as conditional on quality:

$$P_h = P_h(y_h, I, q)$$

Or equivalently,

$$P_h = P + \lambda_h(q)$$

It follows that, in this case, unique prices for quality will be established in decentralized markets.

Market failure implications of product differentiation by biotechnology

Within this notation, we can now see how markets may fail in the presence of quality characteristics such as those imparted by biotechnology. Quality management by the producer affects the processor in several ways. First, because producer quality management affects the amount of substandard product, the supply available in the pool, Y, becomes uncertain for the processor that procures from a fixed pool of processors (e.g. through exclusive processing agreements). Alternatively, procurement cost is uncertain when additional non-pool supplies are needed to fill capacity (e.g. n is endogenous). Third, when variation in quality above the standard is relevant to the processor, the processor has a direct incentive to attempt to manage the quality delivered by each producer. Further discussion of the theoretical implications of this theory is available from the authors in related papers.

Where information is complete, the menu of prices P_h will signal the supply chain to produce quality characteristics demanded at any node of the supply chain. One caveat is that the quality characteristics must be bound to the private good's quantitative consumption of the good, making the quality characteristics exhaustible and exclusively consumable. Where information is incomplete, markets will be incomplete and the range of quality will not be priced. Suppose some markets are simply non-existent. For example, consider the case of a biotech-related quality characteristic q for which consumers hold preferences. In the absence of a market, it is clear the supply chain will achieve neither optimal profits for the supply chain nor quality supply that optimizes social welfare. This situation would clearly arrive when quality, q, is not observable. In this case, no market could exist for the quality characteristic, and it would not be priced by competitive markets.

In either case, product differentiation will define opportunities for supply chain management. For consumers, ag-biotech products may be viewed as being of lower quality, in fact as being polluted or contaminated, or as adding a margin of positive value due to changes in their characteristics.

Labelling

Within this context, the potential role of labelling must be addressed. From the perspective of the above notation, it is clear that where the search and information costs for identification of quality characteristics are high, labelling will generate improved supply chain performance by establishing a basis for product differentiation and market pricing of product characteristics. Further, labelling in itself provides an important basis for reduction of search and information costs when labelling is backed up with product performance liability or warranties. Even in the case where consumers hold preferences that may have no merit based on the product's direct consumption performance – as has arisen with respect to technology characteristics – labelling may be supportable on the basis of private economics. That is, the supply will emerge to satisfy demand at a price that covers average costs. Mandating of labelling is an issue that goes beyond the scope of this chapter; however, where products are quality differentiated and the search and information costs are high, private economics would be expected to generate demand for labelling that could finance its supply. Products that are not labelled would be shunned by consumers valuing quality characteristics that could be found labelled elsewhere in the market.

A role for standards?

The possibility that standards could be imposed in product markets affected by biotech supply is important to consider given expressions of consumer preferences and concerns already on record. The economics of the use of standards follows directly the arguments concerning standards found in the environmental economics and ag marketing literature (see, e.g. Halloway, 1998). In the latter case, grades and standards have been argued to facilitate the liquidity of markets in products with a high degree of quality variation.

More recently, while admitting this role of grades and standards, Weaver and Kim (1999) demonstrated that grades and standards leave in place substantial incentives for supply chain management of quality. While grades or standards define one limit of the range of possible quality, they do not resolve the level of quality. Where quality is valued, both producers and processors are left with incentives to find better means of managing quality demanded by consumers. One solution for the supply chain is to vertically integrate.

Contracting as a Solution

To consider alternatives for supply chain coordination when products are differentiated by quality related to biotechnology, we evaluate the profits attainable by producers, processors and for the supply chain in aggregate under a series of scenarios in which the above theoretical model is parametrized (details available from the author). Consideration of grades and standards is nearly trivial within this notation and is highlighted only in passing. Instead, focus is placed on use of contracts between producers and processors.

To do so, we use standard neoclassical shapes for the underlying functions and ensure that the Spence–Mirlees conditions are satisfied. In each case, we use computational methods to numerically solve the optimization problem given different levels of the quality incentive. For the contract design cases, we numerically solve for the optimal incentive when producers are considered homogeneous and for the nonlinear menu of incentives when producers are heterogeneous. Results are reported graphically in Fig. 26.1 and discussed below.

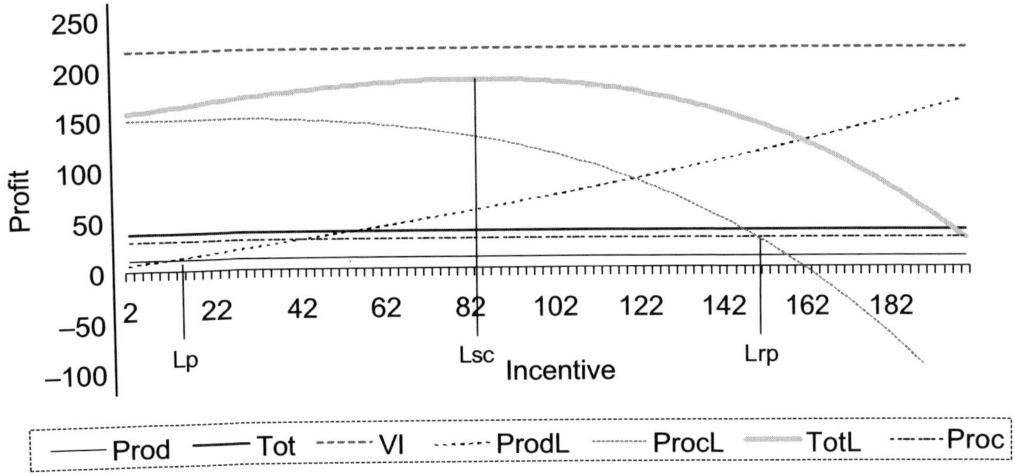

Fig. 26.1. Enhancing supply chain profits through coordination simulated results: heterogeneous incentives contracts with homogeneous producers.

Case 1: Independent decision making

Weaver and Kim (1999) present a series of propositions concerning independent decision making, noting that in the absence of incentives, quality will be produced by producers and purchased by processors if it influences productivity and costs. This follows from producer and processor first-order conditions. For the producer, introduction of an incentive greater than a reservation level, λ^r, is required to ensure participation, that is $(\pi \mid \lambda > \lambda^r) > (\pi \mid \lambda = 0)$. In Fig. 26.1, profits for the producer are graphed and labelled as Prod (when $\lambda = 0$) and ProdL (when $\lambda > 0$). We see that profits are increasing and convex in λ. From the producer's perspective, optimal λ is infinite. The exact curvature of the producer profit function depends on the quality production function specified. Here, we suppose that diminishing returns to inputs that yield quality occur slowly.

For the processor, we also find a reservation incentive at which quality incentives generate profits that exceed those attainable in the absence of incentives for quality, $(\neq^p \mid \lambda < \lambda^{pr}) > (\neq^p \mid \lambda = 0)$. In Fig. 26.1, we label the profit function as follows Proc (when $\pi_p \mid \lambda = 0$), ProcL (when $\pi_p \mid \lambda > 0$). For this case, profits are decreasing and concave in the quality incentive. At some level, quality incentives fail to generate increases in marginal revenue and profits fall to levels that make incentives unattractive. The reservation incentive for the processor is noted Lrp in Fig. 26.1.

Before closing, consider the supply chain profits attainable under independent decisions, $\pi_{ISC} = \pi + \pi_p$. In Fig. 26.1, we note $\pi_{ISC} \mid \lambda = 0$ as Tot. It is constant. $\pi_{ISC} \mid \lambda > 0$ is noted TotL and is decreasing and concave in λ, allowing a definition of a reservation quality incentive such that for $\lambda > \lambda^r_{ISC}$, $\pi_{ISC}(\lambda > 0) > \pi_{ISC}(\lambda = 0)$.

Case 2: Joint decision making: vertical integration

Vertical integration provides one means for achieving joint profit maximization (i.e. $\bar{\pi}^*$ Max $\bar{\pi} = \pi^p + \pi$), establishing benefits of internalization of the externalities generated by quality that cannot be controlled under independent action. In Fig. 26.1, we note profits from vertical integration as VI. The first-order conditions for this problem indicate that when no standards or incentives exist, that is $\lambda = 0$ and $D = 0$, independent decisions by producers and processors fail to reproduce the necessary conditions for joint profit maximization. This proposition is intuitive, though powerful. The difference $\bar{\pi}^* - \bar{\pi}^p$ defines the incentive for supply chain coordination that is feasibly attained through vertical integration or, in some cases, contracting. As is apparent from the figure, this incentive can be substantial. In the absence of contracting mechanisms, vertical integration can be expected to be pursued aggressively by firms in the supply chain.

In Fig. 26.1, we can now analyse the incentives for contracting or vertical integration. We see that supply chain profits attainable under vertical integration $\bar{\pi}^*$ (VI) cannot be attained by independent units. In the absence of incentive, the supply chain will generate $\pi_{ISC}(\lambda = 0)$ (Tot), leaving in place an incentive for vertical integration VI*= $\bar{\pi}^*$ (VI) – $\pi_{ISC}(\lambda = 0)$. Even when incentives are introduced, $\pi_{ISC}(\lambda > 0)$ (TotL) may approach $\bar{\pi}^*$ (VI), although it will not attain that level of profits. We define the incentive for vertical integration when the alternative is use of incentives as VIL* = $\bar{\pi}^*$ (VI) – π_{ISC} $(\lambda > 0)$. Importantly, Fig. 26.1 clarifies that, even when contract-based incentives are introduced, an incentive for vertical integration remains in place, though our results indicate that in general VIL* < VI*.

Implications of incentives for quality

A standard result in externality management literature is that there exists an incentive that can induce independent agents to replicate the joint profit maximum. We find the same type of result here. This highlights that while such an incentive, say λ^g, exists, λ^g will not typically maximize social welfare in our supply chain case since its determination would have ignored consumer interest in the level of quality. This result echoes a similar result associated with externality taxes or subsidies. A further problem with such incentives is that their uniformity renders them inefficient; for example paying a rent to producers who would produce quality even in the absence of incentives or given smaller incentives.

The idea of a contract for performance

From the specification above, we see that a natural incentive exists for producers and processors to communicate their interests and coordinate them across the supply chain. In a market setting, in the absence of grades and standards, where quality is observable at low cost, products will be differentiated by quality and quality will be priced creating an incentive to better manage quality. Where quality is unobservable at reasonable cost, further incentives exist for the processor to determine the quality level from the producer. In each case, these opportunities can be exploited through contracting. The potential for contracting is next illustrated by examples. To design a contract that is feasible, it must offer increased profits to both the agency and the agents. Incentives under the contract must induce the agents to participate. Intuitively, this will require that under the contract incentives, the agent earns as much profit as might be available in the absence of the contract incentives. We define the reservation profit for producers as $\pi^r_i = \pi_i(\lambda = 0)$, and for processors as, $\pi^r_p = \pi_p(\lambda = 0)$.

Processor as agency under full information (symmetry): heterogeneous incentives contracts across homogeneous producers

The contract design problem can be written as:

$$\max_{\lambda,\, Y_p} \pi_p = PY_p - \lambda Q - PY - C_p(Y_p, Y, Q)$$

subject to $\pi^c \geq \pi^r$, where $\pi^c = \text{Max } \pi = P(1 - D)y + \lambda q - C(q, y)$.

Not surprisingly, given that the quality incentive is selected to maximize processor profits, it will not typically maximize social welfare. To consider the results from this case, first note that from the processor perspective, the optimal incentive in the absence of contracting would be $\lambda = \lambda^{p^*}$ that solves: Max $\pi_p = PY_p - \lambda Q - PY - C_p(Y_p, Y, Q)$. In Fig. 26.1, this is close to zero though in general it might be any positive number depending on the curvature of ProcL. The difference $\pi^p(\lambda > 0) - \pi^p(\lambda = 0)$ = ProcL – Proc defines a strong incentive for the processor to contract, if vertical integration is not feasible.

As an agency, the processor is not free to set $\lambda = \lambda^{p^*}$. Instead, in the absence of market power over the producers, the agency must set $\lambda = \lambda_p^c$ such that $\pi^c \geq \pi^r$. That is, such that profits for the producer under the contract's λ exceed or equal the producer's reservation profits. Based on this constraint, we define the reservation quality incentive for the producer λ^r (noted as Lp in Fig 26.1.) where λ^r solves $\pi(\lambda^r) = \pi^r(\lambda = 0)$. Similarly, we define the reservation incentive for the processor, λ^{pr} (noted Lrp in Fig. 26.1), as the solution of $\pi^p(\lambda^{pr}) = \pi^p(\lambda = 0)$. Given the parametrization illustrated in Fig. 26.1, the optimal contract incentive when the processor is the agency will be: λ^c_p (Lp) = λ^r. This results in producer profits equal to those attainable without incentives and processor profits of about 150, see Table 26.1.

To consider incentives for contracting for quality in the supply chain, consider first the case where

Table 26.1. Summary of profit outcomes.

			Result			
Case	Agency	Agent	$[\lambda^*]$	$[\pi$ producer]	$[\pi_p$ processor]	$\pi_C = (\pi_A + \pi_B)$
1	Processor	Producer	10.238	10.178	149.976	160.154
2	Producer	Processor	151.063	114.531	26.285	140.816
3	Processor	Producer pool (homogeneous)	10.238	10.178	149.928[a]	160.106
4	Processor	Producer pool (heterogeneous)	28.3662	1946.5048[b]	149.9882	
5	Producer pool (homogeneous)	Processor	396.683	481.820[b]	26.285	508.105
6	Producer pool (heterogeneous)	Processor	69.9599	4779.7718[b]	139.9163	
7		Independent	0.000	10.178	26.285	36.463
8		Independent (sum)	84.000	57.386	129.080	184.467
9		Vertical integration	0.0000	n/a	n/a	216.643

[a] Increased processing volume generates increased profits.
[b] Summation of all producers' profit.
n/a, not available.

no incentive exists. In Fig. 26.1, $\pi(\lambda = 0)$ = Prod < $\pi(\lambda > \lambda^r)$ = ProdL; $\pi_p(\lambda = 0)$ = Proc < $\pi_p(\lambda < \lambda^{pr})$ = ProcL; and π_{ISC} $(\lambda = 0)$ = Tot < $\pi_{sc}(\lambda^r < \lambda < \lambda^{pr})$ = TotL < $\bar{\pi}^*$ = VI. Thus, a strong incentive for contracting by both producers and processors will normally exist depending on the technologies faced. To optimize supply chain profits (TOTL), $\lambda^* = L_{SC}$.

Producer agency: homogeneous producers

The importance of bargaining power can be explored within this framework. Suppose the agents form a pool of size n and establish agency over the processor. In this case, the contract design problem under full information would be:

Choose λ, q, y to solve $\pi^{pool} = \text{Max}\ \Sigma\ \pi_i = \Sigma[P(1 - D_i)y_i + \lambda_i q_i - C(q_i, y_i)]$.

To simplify for this exposition, we suppose all producers are identical allowing summations to be replaced by a scalar n. To emphasize the conditionality of contracting incentives on the particular parametrization of the problem, note we could parametrize the problem such that no incentives exist for producer pool agency. Instead, we explore the case where incentives exist. In this case, the condition for the contract to be of interest to both the producer pool and the processor must be designed to maximize π^{pool} subject to:

$$\pi^{pool}(\lambda^{pool}) \geq n\pi(\lambda = 0) \text{ and } \pi_p \geq \pi^r_p$$

Expressed in terms of reservation quality incentives, this amounts to $\lambda^{pool} > \lambda^r$ and $\lambda^{pool} = < \lambda^r_p$, or in terms of Fig. 26.1, Lr < Lpool < Lrp.

Given the technologies assumed in the simulation behind Fig. 26.1 we find the optimal contract incentive $\lambda^{pool} = \lambda^r_p$. At this incentive level, the processor reservation profit is met, rendering profit just equal to that available in the absence of contracting. For supply chain profits, $\pi_{sc}^{pool} = \pi^{pool}(\lambda^r_p) + \pi_p(\lambda^r_p)$.

Processor as agency: heterogeneous incentives contracts across heterogeneous producers

In this case, the optimization problem is the same as that in case where the processor has agency and uses heterogeneous incentives contracts across homogeneous producers. However, in this case, producers are not homogeneous. It follows that the processor finds the optimal contract specifications to vary across producers' varying reservation profit levels. That is, each producer has the same profit (see Fig. 26.2) although each is presented with different incentives. Figure 26.3 presents the schedule of optimal incentive for each producer type. When producers hold agency, incentives do not vary with type (flat line). When processors hold agency, incentives are decreasing with type. Further, incentives are identical under information asymmetry and symmetry

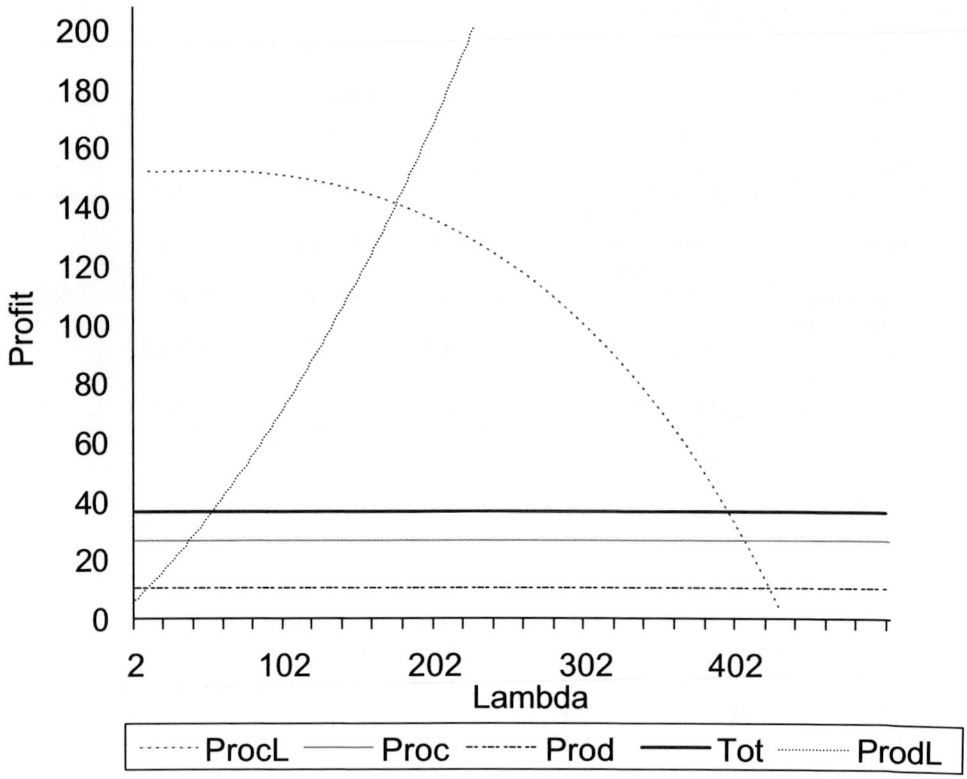

Fig. 26.2. Producer agency: homogeneous producers pool or heterogeneous producers.

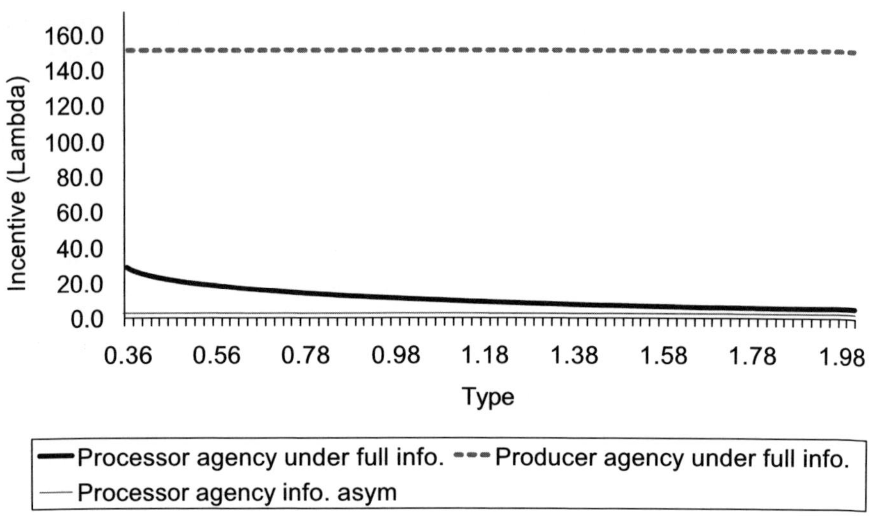

Fig. 26.3. Incentives vs. type.

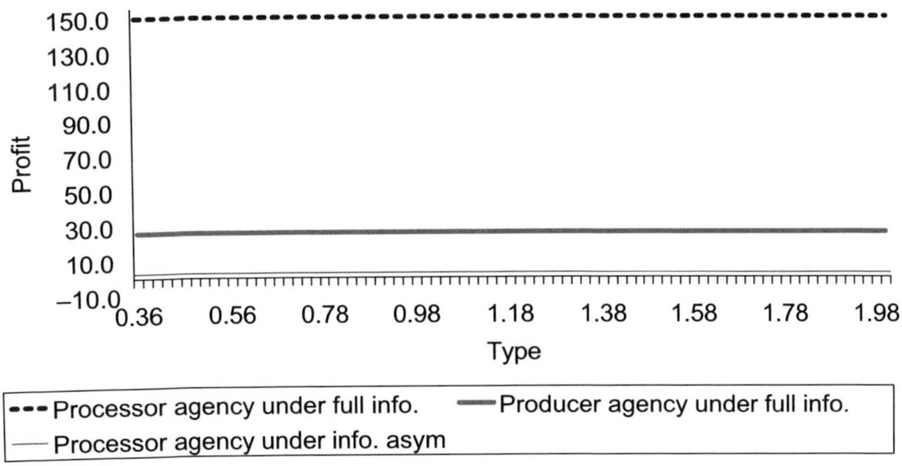

Fig. 26.4. Processor's profits.

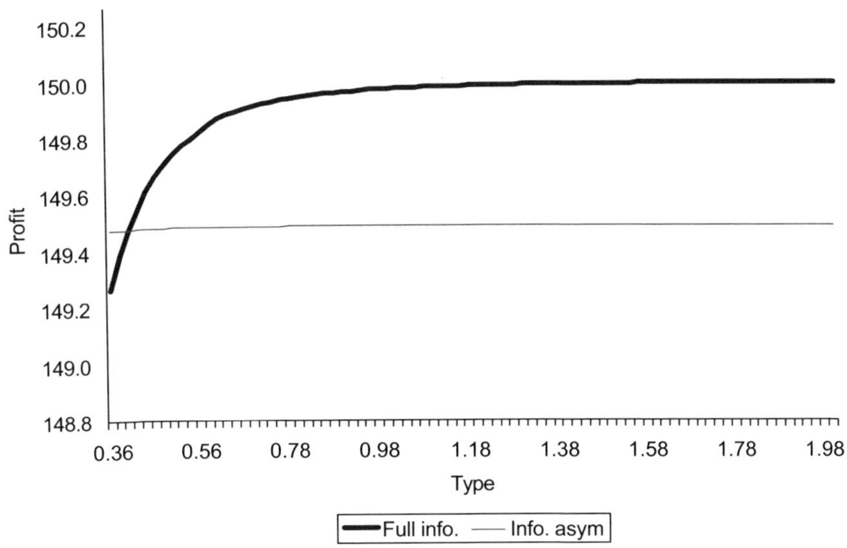

Fig. 26.5. Processor's profits under processor agency.

(Fig. 26.4). In these results, we assume bilateral contracting between each producer and the processor. Processor profits under processor agency under both full and asymmetric information are slightly increasing with producer type (Fig. 26.5). Under producer agency, processor profits are constant. These results mean that if the processor can work with producers with high quality ability, greater profit can be earned.

Producer pool as agency: heterogeneous incentives contracts across heterogeneous producers

In this case, it is intuitive that the processor's profit and the optimal incentive be fixed to the processor's reservation profit and the associated incentive. Each producer attains different profit depending on type (see Fig. 26.6). Producer profits are shown to be increasing in producer type.

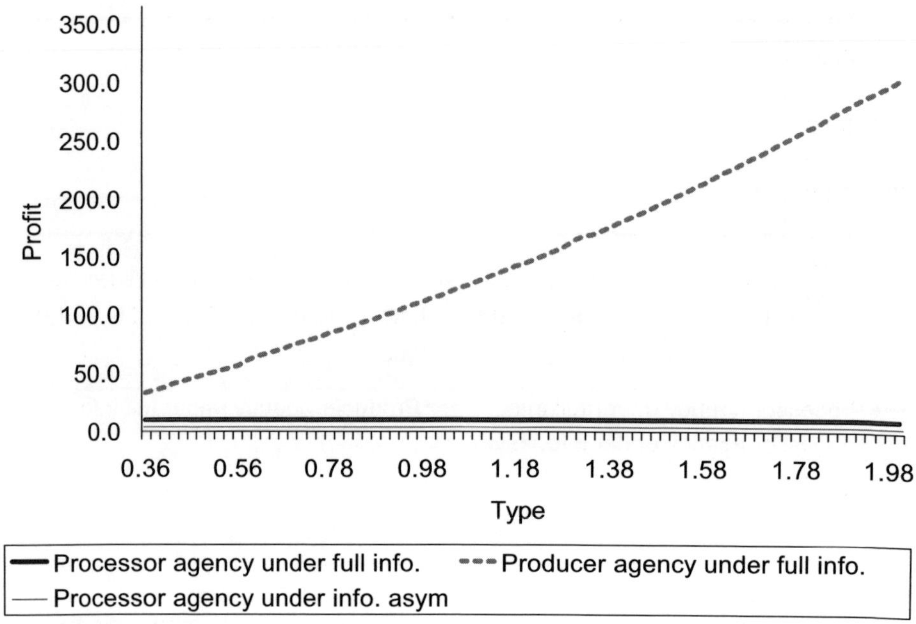

Fig. 26.6. Producer's profits.

Processor as agency: homogeneous incentive contracts across heterogeneous producers

Here, the processor has only one contract to be enforced for all types of producers. Since producer's profit is increasing in producer type, an incentive that guarantees the reservation profit for the lowest type of producer becomes the optimal incentive and each producer attains different profit levels based on this optimal incentive. In contrast to the heterogeneous incentives contract case, Fig. 26.7 shows results for a homogeneous incentive offered across heterogeneous producers. Producer's profit is also increasing in producer type.

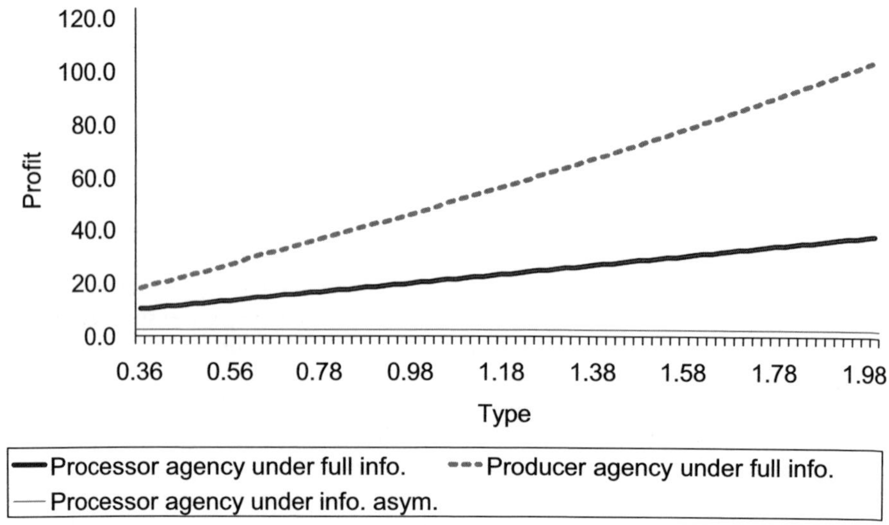

Fig. 26.7. Producer's profits in homo lambda.

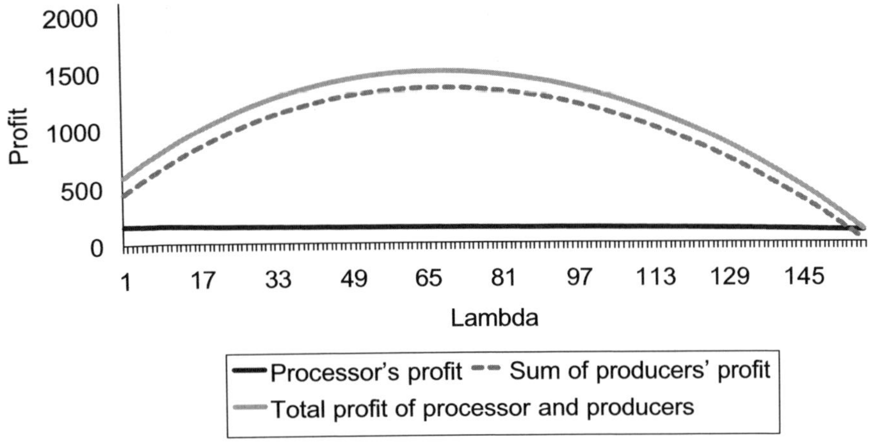

Fig. 26.8. Total profits for processor and producers.

Producer pool as agency: homogeneous incentive contracts across heterogeneous producers

In this case, the objective function for the agency is the summation of profits for producers and the constraint is that the processor profit is at least equal to the reservation profit. Figure 26.8 shows the curvature of the summation of producers' profits and processor's profit against incentive, the sum of profits of all producers increase until around 70 and decrease after 70. We see the result that when the price (incentive) is relatively small, the sum of producer profits increases as the incentive increases. When price is relatively high, the sum decreases (see Fig. 26.8). This result is intuitive from the perspective of demand–supply theory. Suppose that the incentive (price) for a specific level of quality increases. Producers will make an effort to increase their quality in order to get more profit. However, processors can try to reduce the higher quality in order to reduce the cost burden. If the price-quality elasticity of demand (processor) is greater than that of supply, then the profits of high type producers will be decreased as incentives increase. Therefore, the optimal incentive is determined at around 70, processor's profit is the profit around 70, and producer's profit is increasing in type.

Processor as agency: information asymmetry case

Under the information asymmetry case, the producer agency case has the same results as under the full information case because agency has full information. So, only the processor agency case is considered. The model is as follows:

$$\text{Max}_{\lambda,xp}\ \text{E}\pi_P = \int_{\Theta}\left[P_pY_p - \lambda Q(\theta) - PY - C_p(Y_p, Y, Q(\theta))\right] \times g(\theta)\text{d}\theta$$

subject to

$$\pi^*_i(\lambda) = P\tilde{y}(x_i^*, z_i^*) + \lambda_i q(x_i^*, z_i^* \mid \theta_i, e) - C(x_i^*, z_i^* \mid e) \geq R_A, \forall i,$$
$$Py(x_i, z_i) + \lambda_i q(x_i, z_i \mid \theta_i, e) - C(x_i, z_i \mid e) \geq Py(x_i, z_i) + \lambda_i q(x_i, z_i \mid \theta_j, e) - C(x_i, z_i \mid e),$$
$$\forall i, j \uparrow i$$

The second constraint guarantees that incentives are consistent with truth telling by the agents. That is, when each producer reveals its type truly it will have maximal profit. The objective function is the expectation of processor profit function taken over producer type.

As seen in Fig. 26.3, the optimal incentives are the same as under the full information case because the optimal incentive is determined in the constraints of producers' profit under both cases. However, the profits in both cases are different (Fig. 26.5). The slope of profits under the full information case is greater than that under the information asymmetry case because the latter is an expected value. We can regard the difference in profits between full information and asymmetry as the premium for full information. For the truth telling constraints, if a producer tells its type higher than its true type it will have less incentive than the incentive that guarantees its reservation profit. Therefore, a producer will not have an incentive to report a type value that exceeds its true type. Similarly, if a producer reports a type value that is smaller than its true type, profit will be increased above its reservation profit; however, since the processor can examine the quality of its product as a signal of type, enforcement by the processor will preclude this possibility. We proceed with the assumption that producers do not report type values smaller than the true value.

Conclusions

In conclusion, it is of interest to assess the reasons why opportunities for supply chain management of biotechnology are not fully exploited. What can be gained by contracting? In brief, we find that incentives for vertical integration are strong and offer supply chain profits that may substantially exceed those available from independent decisions in the absence of incentives for quality. Depending on the specific parametrization of the problem, strong incentives may exist for both producers and processors to contract. Whether processor or producer pools contract depends on the parametrization of the problem. We establish that whether supply chain contracting will maximize social welfare depends on its ability to replicate the vertical integration solution. This depends on whether $\pi_{ISC} \rightarrow \pi_{VI}$ for some λ.

Grades and standards as mechanisms for managing quality are inherently slow and clumsy. Worse, they fail to provide incentives for private sector led innovation to enhance quality in new directions. Within this context, grades and standards can leave in place strong incentives for firms along the supply chain to vertically integrate, as a strategy for assuring quality is supplied as needed.

While labelling is supportable on the private economics of consumer willingness to pay for related costs to find differentiated products, labelling may not be suitable for biotechnology-related quality characteristics. First, it may be difficult, if not impossible, to define labels that express characteristics of interest to consumers which at the same time are defensible. Second, quality characteristics may be too costly to quantify in markets where unlabelled products also exist.

This chapter demonstrates the potential benefits of several supply chain management strategies that use contracting to manage quality productivity of the supply chain. The framework presented has broad applicability for assessing opportunities for managing price, quantity and quality uncertainty in supply chains. Within the applied setting of managing the dairy supply chain's production of antibiotic residues, the implications of contracting solutions are illustrated. The benefits of improved quality management noted include expansion of industrial demand for commodities as well enhancing and stabilizing prices and farm producer/supplier incomes in the long term.

References

Carpentier, A. and Weaver, R.D. (1997) Damage control productivity: why econometrics matters. *American Journal of Agricultural Economics* 79, 47–61.

Chong, P.S. (1996) Modeling total quality management economics. *Total Quality Management* 7(4), 433–440.

Halloway, G.H. (1998) Contractual arrangements at the farm gate. In: Royer, J.S. and Rogers, R.T. (eds) *The Industrialization of Agriculture: Vertical Coordination in the U.S. Food System.* Ashgate Publishing, Aldershot, pp. 115–132.

Kim, K.C. and Weaver, R.D. (1995) *Biodiversity and Landscapes.* Cambridge University Press, Cambridge.

Weaver, R.D. (1993) *U.S. Agricultural Research: Strategic Challenges and Options.* Agricultural Research Institute, Bethesda, Maryland.

Weaver, R.D. (1995) Mitigation, product substitution, and consumer valuation of undesirable foodborne effects. In: Caswell, J. (ed.) *Valuing Food Safety and Nutrition.* Westview Press, Boulder, Colorado, pp. 51–68.

Weaver, R.D. and Kim, T. (1999) Supply chain contracting for quality: TQM based mechanisms. Working paper presented at *The Economics of Quality Control in Agriculture,* The University of Saskatchewan, Saskatoon, Canada, 2–4 December 1999.

Weaver, R.D., Evans, J. and Luloff, A.E. (1992) Pesticide use in tomato production: consumer concerns and willingness-to-pay. *Agribusiness* 8(2), 131–142.

Index

The following short forms are used in the index: GM, genetically modified or genetic modification; GMF, genetically modified food; GMO, genetically modified organism.
Page references in italics refer to tables or illustrations, while references to notes are in the form 183 *n*8, where 183 indicates the page number, and *n*8 the number of the relevant note.